MW01252828

Entrepreneurship in Emerging Regions Around the World

BATTEN ENTREPRENEURSHIP SERIES

Series Editor: Sankaran Venkataraman, *University of Virginia, USA*

Titles in the series include:

Entrepreneurship in Emerging Regions Around the World
Theory, Evidence and Implications
Edited by Phillip H. Phan, Sankaran Venkataraman and S. Ramakrishna Velamuri

Entrepreneurship in Emerging Regions Around the World

Theory, Evidence and Implications

Edited by

Phillip H. Phan

Rensselaer Polytechnic Institute, USA

Sankaran Venkataraman

University of Virginia, USA

and

S. Ramakrishna Velamuri

China Europe International Business School, China

BATTEN ENTREPRENEURSHIP SERIES

Edward Elgar

Cheltenham, UK • Northampton, MA, USA

Published by
Edward Elgar Publishing Limited
Glensanda House
Montpellier Parade
Cheltenham
Glos GL50 1UA
UK

Edward Elgar Publishing, Inc.
William Pratt House
9 Dewey Court
Northampton
Massachusetts 01060
USA

A catalogue record for this book
is available from the British Library

Library of Congress Cataloguing in Publication Data

Entrepreneurship in emerging regions around the world : theory, evidence, and implications / edited by Phillip H. Phan, Sankaran Venkataraman, and S. Ramakrishna Velamuri.
p. cm. — (Batten entrepreneurship series)
Papers presented at a 2006 academic conference held at the Indian School of Business in Hyderabad, India.
Includes bibliographical references and index.
1. Entrepreneurship—Congresses. 2. Entrepreneurship—Government policy—Congresses. 3. Regional economics—Congresses. 4. Venture capital—Congresses. 5. High technology industries—Government policy—Congresses. I. Phan, Phillip Hin Choi, 1963– II. Venkataraman, Sankaran. III. Velamuri, S. Ramakrishna, 1962–

HB615.E63346 2008
338′.04091724—dc22

2007045784

ISBN 978 1 84720 800 2

Printed and bound in Great Britain by MPG Books Ltd, Bodmin, Cornwall

Contents

Contributors

Thomas J. Allen is Professor of Management at the Sloan School of Management, Massachusetts Institute of Technology, and is co-director of MIT's programs on the Pharmaceutical Industry and the Lean Aircraft Initiative. He received his PhD from MIT.

Andaç T. Arıkan is Assistant Professor at the Florida Atlantic University, College of Business. He received his PhD from the Stern School of Business at New York University. His research concerns emergent entrepreneurship in a regional context and various topics on geographical clusters.

Frank Barry holds an endowed Chair in International Business and Development at Trinity College Dublin. Recent co-authored books of his include *Multinational Firms in the World Economy* (Princeton University Press, 2004) and *Understanding Ireland's Economic Growth* (Macmillan, 1999).

Dan Breznitz is Assistant Professor at the Sam Nunn School of International Affairs and at the School of Public Policy, Georgia Institute of Technology. During 2006 Breznitz has been a visiting scholar at Stanford University Project on Regions of Innovation and Entrepreneurship. He obtained his PhD from MIT.

Michael Carney is Professor of Management at the John Molson School of Business. He received his undergraduate degree in economics at Keele University and an MBA and PhD in Organization Theory at the University of Bradford. His research focuses primarily upon corporate restructuring processes during periods of institutional change.

José Cezar Castanhar teaches entrepreneurship and finance at EBAPE Business School, Brazil. He is also a consultant for SMEs in Brazil and researches entrepreneurship, firm performance and regional development. His publications include articles in *Revista de Administração Pública* (Brazil), *Revista de Administración* (Mexico) and *Revista d' Afers Internacionals* (Spain).

Young Keun Choi is a senior venture capitalist in POSCO Research Institute in Seoul and doctoral candidate in management at Yonsei University. He has been a venture capitalist for the last ten years. His research interest is on entrepreneurial development in Korean venture industry.

Seungwha (Andy) Chung is Professor of Management and Vice Dean of the Graduate School of Business, Yonsei University. He has published in the *Strategic Management Journal*, *Research Policy*, *Social Networks*, *Korean Management Review*, *Korean Venture Management Review* and *Korean Small Business Review*. He received his PhD in management from the University of Pennsylvania.

João Ferreira Dias is Associate Professor at ISCTE Business School (Portugal) and invited professor of Fundação Getúlio Vargas (Brazil), where he teaches strategic management. He is author of several articles and papers. His research areas are sustainable development and business simulation.

Marleen Dieleman is a visiting fellow at NUS Business School in Singapore. She holds a Master's degree in business administration from Rotterdam School of Management and a PhD from Leiden University, both in the Netherlands. Her research interests include business groups, Asian family business and strategy in emerging markets.

José Paulo Esperança teaches corporate finance and international business at ISCTE Business School, Portugal. He is the head of AUDAX, a center on entrepreneurship and family business. His publications include articles in the *Portuguese Review of Financial Markets*, *Journal of Multinational Financial Management*, *Management Research* and *Journal of Applied Financial Economics*.

Eileen Fischer is Professor of Marketing and the Anne and Max Tanenbaum Chair in Entrepreneurship and Family Enterprise in the Schulich School of Business at York University. She has published extensively in both the fields of entrepreneurship and consumer research.

Denise Fletcher is Director of Research at the Centre for Regional Economic and Enterprise Development at the University of Sheffield's Management School. Her work focuses on the contribution that narrative, ethnographic and social constructionist ideas have for deepening understandings of how and why entrepreneurial practices occur.

Robert Huggins is Director of Regional Development at the Centre for Regional Economic and Enterprise Development at the University of Sheffield's Management School and is also co-director of the Centre for International Competitiveness. His key areas of research include the study of competitiveness, business networks, knowledge based economic development and entrepreneurship.

S.C. Lenny Koh is Director of Logistics and Supply Chain Management Research Group at the University of Sheffield's Management School. Her research interests include enterprise resource planning, uncertainty management, modern operations management, logistics and supply chain management, e-business, e-organizations, knowledge management and sustainable business.

Jiman Lee is Associate Professor of Management at Yonsei University School of Business. His research interests include strategic human resource management and industrial relations. He has published research papers in the *International Journal of Human Resource Management, Mutinational Business Review, Personnel Review* and *Korean Journal of Management.*

Rory P. O'Shea is a college lecturer at the Smurfit Graduate School of Business, University College Dublin. Rory completed his post-doctoral studies at the Sloan School of Management, MIT. Rory received his Bachelor, Masters and PhD degrees from UCD.

Yigang Pan is Professor of Marketing and International Business at the Schulich School of Business of York University. He does research in market entry strategies and branding strategies in an international context. He has had eight publications in the *Journal of International Business Studies*, among others.

Sunju Park received her BS and MS in Computer Engineering from Seoul National University, and a PhD in Computer Science and Engineering from the University of Michigan. She was an assistant professor at Rutgers University for five and a half years, and currently Assistant Professor of Management Science at the School of Business of Yonsei University.

Phillip H. Phan is Warren H. Bruggeman '46 and Pauline Urban Bruggeman Distinguished Professor of Management at the Lally School of Management and Technology at Rensselaer Polytechnic Institute. He has published in such journals as the *Academy of Management Journal*, *Journal of International Business Studies*, *Journal of Management* and *Research*

Policy, among others. He is Associate Editor for the *Journal of Business Venturing*.

Kavil Ramachandran is Thomas Schmidheiny Fellow of Family Business and Wealth Management and Associate Dean for Academic Programmes at the Indian School of Business in Hyderabad, India. Prior to this, he was on the faculty at the Indian Institute of Management, Ahmedabad for over 15 years. His current research area covers the identification of entrepreneurial opportunities, industrial development policies, venture capital, resource building and competitive strategies of firms and family businesses.

Sougata Ray is a Professor of Strategic Management at the Indian Institute of Management Calcutta. He has research interests in strategy, international management and entrepreneurship and has contributed over 75 research papers and case studies to journals, edited books and conferences. He did his Doctorate at the Indian Institute of Management, Ahmedabad.

Frank Roche is Deputy Principal of UCD College of Business and Law, which encompasses the UCD Michael Smurfit Graduate Business School. He completed his PhD at Michigan State University, and holds the Berber Family Professorship of Entrepreneurship at UCD. Professor Roche has previously worked with the Department of Enterprise, Trade and Employment and has extensive experience of working in the areas of entrepreneurship and industrial policy.

Edward Rubesch's experience combines a mixture of practical and academic entrepreneurial experience. He has founded four companies in Thailand, and is also a member of the faculty of Thammasat University in Bangkok, where he developed the Global Entrepreneurship MBA. His research interests include entrepreneurship and distribution in developing countries, and social entrepreneurship.

Wladimir Sachs is Associate Dean for Research and Professor of Technology and Management at ESC Rennes School of Business in France. He holds a PhD in management from the Wharton School and an advanced post-masters degree in mathematics from the University of Paris at Orsay.

Hyun-Han Shin is Associate Professor of Finance at the School of Business of Yonsei University. He received his BA in Business from Yonsei University, and PhD in Finance from Ohio State University. He was an assistant professor at the State University of New York at Buffalo for four years.

Atipol Bhanich Supapol is Associate Professor of Economics in the Schulich School of Business at York University. He has published articles in a variety of fields including managerial economics, technology transfer, trade and finance. He was previously a director at PwC Securities (New York) and country manager of NatWest (Thailand).

Beata Topa is Deputy Director of the Department for Financial Market Development at the Polish Ministry of Finance. She holds a Master of Economics degree from the Warsaw School of Economics and an M.Econ.Sc in European Economic and Public Affairs from University College, Dublin.

S. Ramakrishna Velamuri is Associate Professor at CEIBS, where he teaches entrepreneurship and negotiation. Prior to joining CEIBS, he worked for four years on the faculty of IESE Business School, where he was the Academic Director of the Global Executive MBA and Inside India programs. His research has been published in the *Journal of Business Venturing*, *Journal of Business Ethics*, and the *Journal of Entrepreneurial Finance*, among others.

Sankaran Venkataraman is the MasterCard Professor of Business Administration at the Darden Graduate School of Business Administration, University of Virginia. He is the editor of the *Journal of Business Venturing* and advisor to the Entrepreneurial Forum, a program of the International Trade Administration of the US Department of Commerce, aimed at promoting trade through entrepreneurship around the world.

Introduction*

Phillip H. Phan, Sankaran Venkataraman and S. Ramakrishna Velamuri

In this book we look at the phenomenon of entrepreneurship in emerging regions in India, China, Ireland, Eastern Europe, North and South America, and North and South east Asia. The ten chapters in this book were presented in a 2006 academic conference held at the Indian School of Business in Hyderabad, India, a fast emerging entrepreneurial region. The chapters were double blind peer reviewed and completed three to four rounds of revisions before they were accepted for publication in this volume.

The book is organized into four parts to take the reader from a general framework for understanding the relationship between economic development and entrepreneurship in emerging regions to more specific examples of how entrepreneurs and their firms respond to the opportunity and threats that are dynamically evolving in such places. There are two ways to read the chapters in this book. The first is to simply read them as a series of interesting case studies, grounded in extant theories of entrepreneurship and regional economic development. This would be to short change the potential of the book. The second way is to read them for theoretical insights into why entrepreneurship is so robust even in regions that appear not to have the ingredients (such as risk capital) for such activity.

It is not surprising that self-employment naturally arises where the opportunities for meaningful employment are few, such as in rural economies. This kind of self-employment has been referred to as subsistence entrepreneurship. However we are particularly taken by the fact that value-creating entrepreneurship can occur in these regions and note that rural value-creating entrepreneurship is a distinct form of economic activity. For example opportunities are identified almost always as the fulfillment of human needs with technology (broadly understood as knowledge embedded

* The editors would like to acknowledge the Darden School of Business at the University of Virginia and the Wadhwani Center for Entrepreneurship Development at the Indian School of Business for financially supporting and hosting the conference that led to this book. The opinions of the authors and editors are theirs alone and not of the sponsoring organizations.

in products and services), whereas traditional notions of technological entrepreneurship allow for technology to create its own markets. This may be due to the fact that consumers in rural regions do not have the resources for discretionary consumption. A second explanation could be that the rural entrepreneurs have gateways to larger domestic and international markets, as is reflected in the case of the Brazilian furniture cluster (Chapter 2). One implication may be that the explanatory power of individual differences for the level of entrepreneurial activity may be higher in rural regions than in urban areas, where institutions may matter more.

In addition to the possibility that the chapters in this volume offer interesting extensions to theories in use, they consistently cite the role of government intervention in emerging economies. Without expositing on the political economy of such regions, we surmise that the reason governments figure so prominently is because emerging regions often lack the critical mass of inputs (capital, human talent and technology) required to ignite entrepreneurial action. Hence, in models of entrepreneurship in emerging regions, government bridges the supply side causes of market failure in entrepreneurial activity. Hence a common theme in these chapters is the notion that the occurrence of entrepreneurship in emerging regions is not accidental. The chapters consistently show that the development of entrepreneurial activity can be traced to a confluence of initial conditions comprising resource endowments, institutions and markets. This is exemplified by the first two chapters of the book in which the authors discuss the formation of what have become easily identifiable entrepreneurial regions in such diverse places as New York State, Ireland, Brazil, the Balkans and India. This pre-eminent role of governments in the emergence and sustenance of entrepreneurial regions is consistent with and confirms the findings of a large body of research in this area (for example, Alesina and Giavazzi (2002); Bresnahan et al. (2002); Saxenian (1994)).

Part 1 focuses on explicating parts of the dynamic model articulated by Venkataraman (2004). Venkataraman (2004) tries to explain how the institutional and resource endowment conditions in a defined geographic area create the market opportunities that form the basis for new firm formation. As a consequence, in contrast to traditional foreign direct investment theories of economic development, the starting conditions also determine the prospects for sustainable economic growth in a region. This 'virtuous cycle' is the framework for understanding Part 1. The chapters in this part also suggest a level of dynamism in the phenomenon that has largely been ignored in the literature. Studies that rely on standard linear models typically cannot deal with notions of path dependence, recursive interactions between the factors that drive entrepreneurial activity, and agent decisions that are conditioned on institutional constraints (Phan, 2003). Models of entrepreneurially driven regional development often begin with discussions of initial conditions, as many of the chapters in this volume, but do not always allow for those conditions to evolve with the introduction of agent actions. The chapters in this part try to address this issue by offering a view of the entrepreneurial development process that is both recursive as well as evolutionary.

Part 2 takes off from the previous one. Here the authors are concerned about the role of a particular stakeholder in the transformation of a region. In most parts of the world, including some emerging regions in the United States, governments at all levels figure prominently in the allocation of resources toward entrepreneurial development. Early attempts by government to influence economic development, a legacy of Keynesian economic thought, were manifest in policies on public expenditure including military spending, encouraging foreign direct investment, non-tariff incentives for technology and knowledge transfers, and monetary policy to encourage exports. In recent years, with the successful transformation of regions like Ireland and Scandinavia through the technological entrepreneurship, government policy has become more nuanced, targeting the traditional macroeconomic factors like the money supply but also focusing on microeconomic factors like labor supply (through education policy), risk capital (through bankruptcy law reform and publicly funded and privately managed venture funds for specific areas, such as information and communications technologies (ICT) and biotechnology), personal incentives (through tax policy) and innovation capacity (through intellectual property rights protection). Much of the research on the policy implications of technology entrepreneurship have to do with the identification of successful formulae. Recently, the research has become more fundamental, concentrating on characterizing the dynamics of growth, explaining new firm formation, and offering limited models that test conjectures from evolutionary economics, knowledge spillovers theory and absorptive capacity theory, with less concern for building 'all-weather models' for policy purposes. This does not mean that the research is less useful to policymakers. It does imply that we are more careful in deriving prescriptive implications. This is a good thing for, as the chapters in this part demonstrate, it can be problematic when government expends public resources to manage a phenomenon that is at best partially understood, as we shall see in Part 3.

In Part 3 the chapters more narrowly focus on the role of government policy in encouraging or impeding the development of a major input into the entrepreneurial production function – venture capital. Venture capital has been a prime focus for researchers in regional economic development because it has been closely associated historically with the rejuvenation and sustenance of high technology regions such as Silicon Valley, USA and more recently with their emergence (such as the ICT cluster specializing in security related applications in the Washington DC area). The two are nowadays closely related because risk capital, of which venture capital forms a major part, is seen as a key input for innovation projects with difficult to forecast cash flows, and therefore subject to the market failure of traditional forms of financing such as bank debt and corporate equity. This function was earlier performed by governments. For example in the 1940s and 1950s the Office of Scientific Research and Development (OSRD), a US federal agency, awarded US$ 330 million in research contracts and played a significant role in the emergence of what we now know as Route 128 (Rosengrant and Lampe, 1992). Thus an implication for the development of technological entrepreneurship in emerging regions is the necessity for either government funding with a sufficiently long time horizon or the rise of a robust venture capital industry with the appropriate risk-reward profile.

In one example, Ireland, the impact of government is assessed to be positive, generally creating the conditions that encouraged the creation of a risk capital market, whereas in the other example, South Korea, the influence of government is less positive. In the latter case the authors believe that the government's impact has been at best neutral, and more likely to be negative given the subsequent lack of development of a risk capital free market in the post-Internet boom era. In reading Part 3 one must also take into consideration the way that economic production has traditionally been organized in the two countries, and by extension in other jurisdictions. In Ireland the concentration of market power in the hands of a few large companies is relatively low, which provides the basis for a dynamic economy characterized by high levels of firm foundings that can absorb the supply of venture capital.

In South Korea, in particular, the occupation of the *chaebol* or family network form of organization through all niches of the economy and the heavy influence of the family enterprise dominant logic with its focus on loyalty and reciprocity probably precludes the development of an independent risk capital market. Risk capital markets depend on the existence of information on opportunities to the reorganization of production and the ability for production to be reorganized into more efficient forms. However, in South Korea, the inability for firms to restructure because of social mores and government policy causes a market failure on the demand side for risk capital. The population of all conceivable market niches by the *chaebols* crowds outs the supply side driver of new firm formation, which implies lower demand for risk capital. Finally, the internal financing of new production by large corporations reduces the availability of capital to domestic institutional risk investors, which in turn makes the capital market for foreign investors who have one less avenue for exit (that is, the IPO market). Hence, in spite of the government's efforts to encourage venture capital, the structure of the country's economy tends to attenuate the incentives offered.

Part 4 turns our attention to the entrepreneur and venture level of analysis. Here we look at the how entrepreneurs and their firms respond to the opportunities and threats inherent in the dynamism of emerging regions. The authors have picked a wide range of examples to illustrate different aspects of the entrepreneurial mindset (flexibility, persistence, commitment to a vision and adaptability), while also demonstrating the evolutionary dynamics common to all the chapters in this book, which provide us with valuable insights on how the entrepreneur influences her environment. The chapter on initial conditions by Supapol, Fischer and Pan is particularly notable in this regard because it explicitly links the starting conditions given by the institutional environment of an emerging region with the entrepreneur and firm level responses to opportunities as they are impacted by these conditions. The chapter also demonstrates how starting conditions are not static and therefore a key entrepreneurial challenge, and a potential focus for future research is the adaptation of the entrepreneurial dominant logic to these changes. The chapter by Rubesch provides a counter-intuitive finding on the ability of border traders to compete based on value added services with authorized distributors, in spite of being burdened by a price disadvantage.

In conclusion we believe this book takes a significant step forward in the research on entrepreneurship in emerging regions. The most important, we believe, is the theoretical orientation of the studies. In this regard the chapters are exemplars of the type of research that will enhance our ability to disconfirm hypotheses and hence advance the field. The choice of subject areas and geographic locations additionally demonstrate the generalizability of the extant theories in the field, which moves us closer to acceptable models that will eventually form a paradigm for the domain. Finally, because this book demonstrates how unique aspects of the entrepreneurial phenomenon in specific locations, such as Indonesia, the Balkans, Korea and Silicon Alley, can be highlighted for further study by first explaining the institutional environment and its relationship to the general theory on new firm formation.

REFERENCES

Alesina, A. and F. Giavazzi (2002), 'Europe's defense investment gap, Project Syndicate', April, www.project-syndicate.org/commentary/842/1.

Bresnahan, T., A. Gambardella and A. Saxenian (2002), ' "Old economy" inputs for "new economy" outcomes: cluster formation in the new silicon valleys', *Industrial and Corporate Change*, **10** (4), 835–60.
Phan, P.H. (2003), 'Entrepreneurship theory: possibilities and future directions', *Journal of Business Venturing*, **18**, 617–20.
Rosengrant, S. and D.R. Lampe (1992), *Route 128: Lessons from Boston's High Tech Community*, New York: HarperCollins, Basic Books.
Saxenian, A.L. (1994), *Regional Advantage: Culture and Competition in Silicon Valley and Route 128*, Cambridge, MA: Harvard University Press.
Venkataraman, S. (2004), 'Regional transformation through technological entrepreneurship', *Journal of Venturing*, **19**, 153–67.

PART 1

Institutional determinants of entrepreneurship in emerging regions

1. The dynamics of an emerging entrepreneurial region in Ireland

Frank Roche, Rory O'Shea, Thomas J. Allen and Dan Breznitz

INTRODUCTION

> I don't think it's coincidence that Ireland and Dell share the same character and connection. Every success we have achieved around the world has been due to the old Irish recipe of big dreams, hard work and strong relationships. At the heart of both the Irish and Dell character is big dreams, a passion for building and re-building and the tenacity to adapt to challenging circumstances.
>
> Michael Dell, Chairman of Dell, University of Limerick, 2002

> Central to Ireland's economic development is its ability to grow its own enterprises businesses that are grounded in value-added products and services that can be exported.
>
> Enterprise Ireland Annual Report, 2003

There is a growing recognition among policy makers of the need to place more emphasis on knowledge creation and knowledge exploitation, and specifically on technology-based entrepreneurship, which converts new scientific discoveries into new opportunities (Phan and Foo, 2004). An important feature of these clusters is that knowledge-intensive production and urban places provide the central focal point for entrepreneurship activity to emerge (Feldman, 2005). Economic development is increasingly linked to a nation's ability to acquire and apply technical and socio-economic knowledge and the process of globalization is accelerating this trend.

Since the early 1980s a large amount of research has been devoted to industrial clusters and their dynamics. In recent years a subset of individual creativity and entrepreneurial-based knowledge-intensive clusters generated a great deal of attention in the literature (Phan, 2004; Acs and Varga, 2005; Florida, 2005). Nonetheless, while we know quite a lot about the specific dynamics and the different kinds of clusters, to date we do not have an accepted theory as to how clusters are formed in the first place, and hence, what policy makers can do to encourage cluster creation. Nowhere is this truer then in high-technology entrepreneurship-based clusters (Phan, 2004; Bresnahan and Gambardella, 2004).

This chapter seeks to understand the emergence cluster of a high growth dynamic transitioning region such as Dublin given the role of the indigenous sector in explaining the high-technology story in Ireland. According to the Irish Software Association, some 600 indigenous Irish software companies exported over US$ 2 billion worth of sales in 2005. This represents double the rate of start-ups since 1999. Furthermore, in the last five years, the contribution of indigenous start-ups now represents one-tenth of the $15 billion worth of software shipped out of Ireland. Furthermore the 2005 Global Entrepreneurship Monitor report (Bygrave and Hay, 2005) showed that Ireland had become one of the leading countries in Europe in terms of entrepreneurship and is fast approaching the levels of early stage entrepreneurial activity prevalent in the United States. Overall the research found that almost one in ten of the adult population living in Ireland is actively planning or has recently set up a new business. This conclusion was supported by an earlier report by *Forbes Magazine* (2001), which ranked Ireland as the fourth best place in the world to start a business.

Ireland has transformed itself over the last 15 years due to large-scale Foreign Direct Investment (FDI) and more recently high-technology entrepreneurship. Dublin is the capital hub of Ireland's internationally renowned high-technology cluster. Much of the research has to date attempted to understand why Ireland has been successful at attracting FDI (Begley et al., 2005; Grimes and White, 2005; O'Riain, 2004; Sands, 2005). This chapter takes an alternative perspective and builds on Venkataraman's virtual cycle of the regional innovation model (2004) to explore in a systematic way the factors behind the emergence of the successful indigenous software cluster in Dublin.

THEORETICAL DEVELOPMENT

The literature on clusters is vast and heated debates exist between various research traditions. Nevertheless, in our view, most of this disagreement occurs because of the specific focus of the different researchers, and not because of inherent theoretical contradictions. As our main aim in this chapter is to analyse the emergence of the software cluster in Dublin, we argue that a constructive way forward would be to group the different theories into theories about the needed infrastructural factors for a successful high-technology cluster, and theories about the specific institutional dynamics that make such a cluster more entrepreneurial and innovative.

Doing so we quickly realize that there is wide agreement on what factors and infrastructures we must have if they are to succeed. Heading most of these lists are access to information, customers, education, communication, transport and specialized services, including specialized financial investors, such as venture capital (Feldman and Martin, 2005; Avnimelech and Teubal, 2004; Koepp, 2002). However most researchers agree that such conditions are necessary, but not sufficient. What truly makes industrial clusters successful are their unique institutional systems that allow companies as well as individuals within them to constantly share and recombine knowledge and resources, lowering the inherent market failures associated with industrial R&D and collective action (Antonelli, 2000; Capello, 1999; Florida, 1995; Green et al., 2001; Keeble et al., 1999; Lester and Piore, 2004; Morgan, 1997; Piore and Sabel, 1984; Saxenian, 1994). For these reasons companies in clusters not only have lower transaction costs, but also access to resources and capabilities that are not available outside the cluster. This enables the cluster's actors to innovate and offer productive services in new and more efficient ways (Braczyk et al., 1998; Cooke and Morgan, 1998; Porter, 1990). In order to do so most researchers agree that clusters need to create and institutionalize a culture that allows individuals and companies to both fiercely compete in the market place, while at the same time constantly cooperate with each other. Indeed some have argued that we should recognize the commonality of our argument, and term such institutions as institutions that generate collaborative public spaces (Breznitz, 2005).

Public policy, it is often argued, can have a large role in the creation of clusters. It can aid in the supply of the necessary resources, from an educated workforce to venture capital, but even more importantly assist in the creation and maintenance of the institutional dynamics of clusters by stimulating and supporting collaboration, helping to solve the collective action dilemma and tackling the R&D market failure (Braczyk et al., 1998; Cooke and Morgan, 1998; Keeble et al., 1999; Kenney, 2000; Morgan, 1997).

In this chapter we are building on the new approach for cluster analysis offered by Venkatamaran (2004) to investigate the determinants of regional transformation through the unique lens of technology-based entrepreneurship. He argues for the need for national policy makers to think more about the important role of 'intangible' factors in promoting Schumpeterian (1976) type entrepreneurship within emerging regions. Agreeing with most of the cluster theorists, Venkataraman argues that critical determinants for the creation of successful high-technology regions, such as Silicon Valley, are 'intangible' institutional variables. Going a step further he points out that these normative behaviors and values take time to develop and require continual encouragement in order for a 'virtuous circle of innovation' to emerge within the regions. The author encapsulates these findings into a coherent framework that identifies seven conditions important for technology entrepreneurship to flourish. These include: (1) a focal point for creating novel ideas and to provide the stimulus for technological ideas to emerge; (2) development of the right role models to allow people to look at a challenge and see it as a feasible reality; (3) informal networking forums for entrepreneurship to promote a less formal face to face exchange; (4) the need for region specific ideas to be created; (5) the need to create safety nets; (6) the need to have access to large markets; and (7) the need for top-down leadership to execute the policy plans. A central tenet of this stream of research is that regional entrepreneurial activity is a reflection of both institutional behavior, and the social relationships and institutions in which individuals are embedded.

Throughout the rest of this chapter we incorporate Venkataraman's (2004) model, together with a detailed analysis of the specific infrastructural context of the Irish case and it relationship with the global IT industry, to provide a theoretical underpinning of the notion that a fast growing cluster lies in the positive feedback loop that drives the formation and development of Schumpeterian-like entrepreneurial ventures. This enabled us to develop a bottom-up micro-level perspective and deal with this question from the perspective of the actors in the region. The framework provided is illustrated in Figure 1.1.

Using this framework of analysis, we trace the development of entrepreneurship and the development of the Dublin technology cluster over a 25 to 30 year period, from the 1970s to 2005. There were major changes over that period, both in terms of the characteristics of the cluster, and in terms of the global industries of which Dublin technology companies form a part. The global industry has grown considerably, and has been through a series of major technological, industry and market changes. These factors have been reflected in the Dublin cluster, and in entrepreneurship within the cluster. The time dimension has been addressed by looking at the cluster, and at the wider industry context, through four time windows: a formative phase from the early 1970s to the early 1980s; an early development phase from the early 1980s to 1990; a growth and internationalization phase from 1990 to 1997, culminating in the cluster's first NASDAQ flotation; and a rapid growth phase from 1995 to early 2005, while overcoming the effects of the global technology market slowdown in this period.

THE IRISH SOFTWARE INDUSTRY CONTEXT

In 1989 the Irish economy was in severe crisis, with excessive budget deficits and minimal growth. Furthermore GNP per capita amounted to only 65

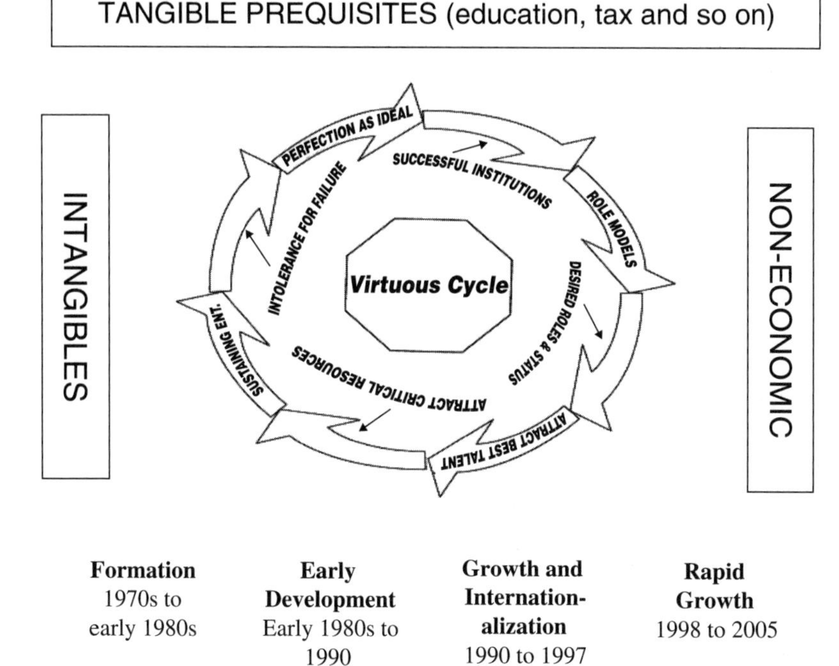

Figure 1.1 Framework of analysis

per cent of the EU level and Irish unemployment stood at 17 per cent, even with 1.1 per cent of the total population leaving the country in the same year. *The Economist* declared Ireland as the 'poorest of the rich', due to its under-performing economy, high taxation, mass emigration and high unemployment. Yet a mere decade later *The Economist* (1997) argued that Ireland is 'Europe's Shining Light' and The Economist Intelligence Unit went further to claim that Ireland is 'the best place to live in the World', ranking Ireland first in 2005 because it successfully combined the most desirable elements of the economics factors (the fourth highest GDP per head in the world in 2005, low unemployment and political liberties) with desirable social elements, such as a stable family and community life (EIU, 2005) (see Table 1.1).

The Irish economy outperformed all other European economies in the 1990s, recording a growth rate throughout that period of three times the EU average with GDP per capita climbing from fourth lowest in the EU to

Table 1.1 Best place to live in the world

Rank	Country	Score
1	Ireland	8.33
2	Switzerland	8.07
3	Norway	8.05
4	Luxembourg	8.02
5	Sweden	7.94
12	USA	7.62

Source: Economist Intelligence Unit, www.economist.com/media/pdf/QUALITY_OF_LIFE.pdf.

the third richest in the world within a period of 15 years (Breathnach, 1998; Sands, 2005). The *IMD World Competitiveness Yearbook* (2004) ranked Ireland third for GDP per capita at Purchasing Power Parity (PPP), ahead of the USA (fifth) and Switzerland (eleventh). Ireland was ranked fifth by the *IMD World Competitiveness Yearbook* (2004) in terms of exports of goods as a percentage of GDP. Ireland's unemployment rate of 4.1 per cent is the fifth lowest within the EU 25 and compares to a Eurozone average of 8.8 per cent.

Over this period Ireland became the largest exporter of software in the world (second place when price transferring from MNCs is taken into account). Eight of the top ten software suppliers in the world have operations in Ireland and it continues to be a leading European location for technology companies. Furthermore the latest investment are by MNCs, such as Apple, Amazon, Cisco, eBay, Google, IBM, Microsoft, Siebel, SAP and Yahoo. These newer investments tend to be more strategic in nature and in some cases have European or global mandates for product development, high level pan-European technical support, revenue accounting and financial shared services (Arora et al., 2004; Breznitz, 2007). Ireland is an important base for software development, semiconductor manufacture and assembly of electronic products by North American and European companies. It is the main base from which North American MNC software companies address the European market, with software localization, technical support and other market specific functions (Arora and Gambardella, 2005; Sterne, 2004). Growth and development in these industries, along with parallel growth in the manufacture of pharmaceuticals and medical devices, appears to have been the primary driver behind very rapid economic growth. GDP growth averaged 9.9 per cent per annum between 1995 and 2004 (CSO, 2004). Over 300 overseas companies in the ICT sector have a presence in Ireland directly employing approximately 61 000 people with

Table 1.2 Economic performance of Ireland

Rank	1993	2003
Unemployment	15.7	4.7
Government Debt as % GDP	93	34
Corporation Tax %	10/40	12.5/25
Personal Tax (lower and higher)	27/48	20/42
Irish GDP per capita as a % of E	69 125	

Source: Department of Finance, Ireland (2004).

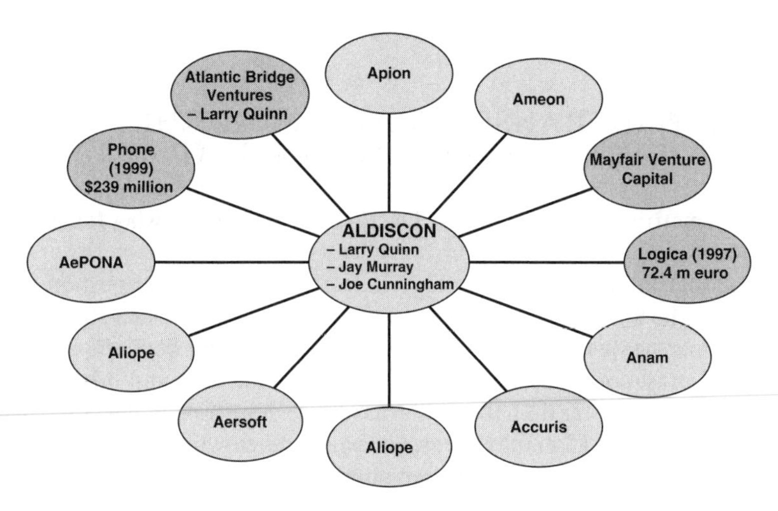

Figure 1.2 Larry Quinn, Jay Murray – Aldiscon and second order effects

seven of the world's top ten leading ICT companies having a substantial base in Ireland (IDA Ireland, 2004). (See Table 1.2).

Ireland's indigenous software sector is unique within the Irish high-technology industry in that Irish owned companies, and not only MNCs, play a very prominent role. There have been 11 Irish founded companies listed on the Nasdaq or the London Stock Exchange (see Appendix), and another few in the closely related area of integrated circuit design. In addition, many other software companies have undergone trade sales (see Appendix) and made their founders multi-millionaires. Eighty-three per cent of software sector employment is located in Dublin or adjacent counties. A recent review of Enterprise Ireland Supported High Potential Start-ups (HPSUs) 1989–2004 found the average number of start-ups per year has also increased significantly between 1989 and 2004. The level of start-up activity has risen from an average of 17 new HPSUs per year in the 1989–96 period to an average of 58 HPSUs per year in the period 2001–3. During 2004, 65 new HPSUs were started with a projected employment in these businesses of 1900 within three years, the majority of jobs in which are high-skilled positions. In 2005 Enterprise Ireland reported a total of 75 new start-up companies – the highest number in any year to date. Total financial assistance approved to these projects was in excess of 17 million euro which it is anticipated will result in the creation of over 1450 new jobs with an estimated 183 million euro in additional exports over the next two years. The economic importance of HPSUs has also been documented. In 2003 the 357 companies still trading as Enterprise Ireland clients had a combined direct employment of 7458 people – the majority of positions in which were highly skilled jobs.

REGIONAL TRANSFORMATION AND THE EMERGENCE OF A HIGH-TECHNOLOGY CLUSTER

The Formative Years (1970–80) – The Emergence of a Growing Industry

The global information technology industry growth has important early effects to the understanding and growth of the Irish IT cluster. In the USA employment growth averaged 11.2 per cent per annum between 1972 and 2000. In most countries employment growth in software and other IT services was much faster than that in the hardware industries, and was the main net source of new IT employment. Globally, the growth of the IT industry had concentrated disproportionately in places with a good supply of technically competent people, where English is used as a business language.

The first great transformation of the global industry to affect Ireland was the advent of the minicomputer revolution, which disrupted the established mainframe computer industry order in the 1970s. In that period companies, such as Digital Equipment Corporation located in Ireland, launched lines of minicomputers whose power was comparable to that of some mainframes, but whose price was very much lower. At the time, however, operating systems were proprietary without a de facto standard, such as Windows, and thus software had to be written for each specific type of machine. Manufacturers of minicomputers, hence, faced a competitive disadvantage by not having many of the most popular software packages that existed for the mainframe. In order to bridge this gap many of them opted for a new strategy by developing external networks of software developers – initially through providing bespoke development services, and later through marketing software products.

It was this transformation that first allowed Irish software companies to grow and reach markets outside Ireland. Timing and the movement of the MNCs into the republic were important catalysts in this development. In the 1970s, the MNCs led the progress in Ireland away from the mainframe platform into minicomputers. As the opening of the new and smaller Irish MNC subsidiaries coincided with the rise of minicomputer technology, relative to its population Ireland became one of the world's largest users of minicomputers. Minicomputers were: (1) a different platform, and hence, did not use the same software employed at the MNCs' foreign headquarters and (2) a new platform, and hence did not have an established base of software. As a result, many local companies managed to get development projects that could easily be packaged (a new concept at the time) into software products. The more successful of these packages quickly secured a few more sales in Ireland and the UK. However a second and much more important transformation of the local software industry happened when the computer manufacturers, such as Digital, IBM and ICL, invited some Irish companies to market their products together as a bundled solution. In the 1970s and early 1980s, the supply of technical and business graduates far exceeded domestic demand over the period. There was no shortage of attractively priced labor at industry entry level, but the domestic supply of very experienced technical people was limited. Experienced technical people were mainly confined to the sales and marketing organizations of multinationals, as well as the computer departments of a small number of companies (particularly financial institutions and manufacturing multinationals). However, given that English was the traditional language of the computer industry, this provided Irish technology companies with an advantage in serving international markets. Ireland's traditional close economic relationship with the UK has given Irish technology companies an advantage in exporting to the UK, and in cooperating with UK companies. Since Ireland's accession in 1973, membership of the EU has facilitated access to continental European markets.

The number of technology entrepreneurs during this period was low. However, in a country as culturally unfriendly towards entrepreneurship as Ireland, this was a critical transformation. What appears to have made the difference, and allow software entrepreneurs to emerge in Ireland at all, was an exceptionally strong supply of people with the business understanding of leading multinational companies, which was necessary to develop competitive applications software and to market them to multinationals. This was a direct result of the Irish FDI-based industrial development strategy that brought many of the leading MNCs to open operations in Ireland, coupled with the large number of Irish graduates who emigrated in search for jobs and subsequently decided to return to Ireland. Working with these

MNCs provided the initial focal point for technical ideas to emerge to form start-up companies.

The IT divisions of financial institutions in Ireland were a source of many of the early software entrepreneurs. In establishing their firms they utilized their positions in the heart of the financial center of Ireland to recognize promising opportunities. Not surprisingly, business applications software was the main product market area addressed by the industry in that period. The combination of multinational experience and experience in a small entrepreneurial company gave entrepreneurs to be the base of knowledge that they required to service customers, along with the exposure to an entrepreneurial environment. Therefore the typical technology entrepreneur was a person with experience in a major multinational company (in Ireland or overseas) or a financial institution, followed by experience in an indigenous technology company. In most cases, while entrepreneurs were technically aware, and some had qualifications in engineering, the skills and experience that they brought to the enterprise were more business related than technical.

Accordingly, toward the early 1980s, Dublin software companies had become more intimately connected to multinational companies and Irish financial institutions. These related industries were both a key source of entrepreneurs and key customers. The Irish sales and marketing operations of multinational minicomputer companies, such as Digital and ICL, were also central in enabling Irish software companies to move into overseas markets. Irish software companies, which already had strong relationships with the local sales and marketing organizations of the main computer companies, were invited to participate in trade fairs and in sales pitches by the international marketing organizations of these companies. It became apparent that much of their software, developed in the first instance for the Irish market, was competitive in international markets. Especially important in that regard was the British ICL. The result was that a number of companies had significant success in the late 1970s and through the 1980s in penetrating markets in the UK, continental Europe and in British Commonwealth countries through piggybacking on the sales activities of computer vendor partners.

As this formative stage progressed, the beginnings of a virtuous circle driving innovation appeared. As entrepreneurial technology companies developed, they became a training ground for more entrepreneurs, and for the middle ranking and senior staff that would be required in new enterprises in future years. One positive consequence of this, and of the fact that many of those working in entrepreneurial technology companies had studied together in college, was that an industry community started to develop, based more on social contacts than on systematic attempts to network. The existence of this community gave entrepreneurs and their staff access to an extensive informal network of contacts, reminiscent of the informal networks that Saxenian (1994) describes in Silicon Valley, although on a much smaller scale. Access to this network appears to have been an important source of learning and innovation for companies, at a time when there were few formal learning resources available. The Irish Software Association, an industry grouping of software companies, was founded in 1978 under the name ICSA (Irish Computer Services Association). ICSA provided a more formal forum for contacts between senior people in software ventures.

However it is important to note that while there were notable early stage successes in high-technology entrepreneurship, they were the exception and not the rule. A critical issue remained the low technological-entrepreneurial spirit and the tendency toward risk averseness. Ireland was caught in what is referred to a 'vicious cycle' (Venkataraman, 2004) whereby great ideas and bold bets within the region did not emerge and pursuing a career as an entrepreneur was seen as something trivial or unattractive. Instead, the majority of Irish aspired toward the traditional established professions, such as medicine, civil service, teaching and law. Many people in Ireland shunned away from risk taking and failure was seen as a shame for life, with a widespread cynicism toward risk taking. In addition banks and other financial institutions were reluctant to provide funding for high risk new technology ventures. In short both industry and its financiers were far from Schumpeterian in their behavior and the culture did not support making bold bets at the time.

Early Development of the Dublin Technology Cluster

The global context of the 1980s was again a time of rapid technology industry change. The most important of which was the decision by IBM to launch the PC in 1981. In addition the 1980s were the decade of the high performance workstations, especially for graphic-intensive design work. The biggest change for the Irish software industry was the standardization of operating systems that these changes brought toward Unix on one side and Microsoft DOS on the other. This process of standardization made it commercially feasible to develop a software application to run on computers from different manufacturers, vastly increasing the market for software products.

Globally, these changes drove continued rapid industry growth, although the rate of growth moderated toward the end of the decade. Demand for independently developed software continued to grow rapidly, driven by the new opportunities to address market needs created by technological developments, and by the fact that standardization was shifting market power from hardware to software enterprises. The 1980s was the first time in Irish history that a few technological entrepreneurs had gone through the whole new business creation cycle: starting, building and financial 'exiting' technology companies. This period represented a critical juncture in the development of the Irish software cluster, as people became engaged in the process of ensuring that good ideas were being developed, and acquired experience about how to start the companies, made prototypes, gained first customers, developed the products and placed them into a competitive product market situation (Venkataraman, 2004). These experienced entrepreneurs have made a central strategic contribution to the sector's development by bringing their accumulated knowledge and earnings, and starting, advising and funding new ventures. Toward the end of the 1980s an increasing number of entrepreneurial technology companies sought to export. However, in what seems to strengthen Venkataraman's emphasis of the intangible and the importance of the dynamics of entrepreneurship, during our interviews many entrepreneurs highlight the less tangible factors, including Ireland's image overseas, the compatibility of Irish culture with the cultures of key markets and the affinity that many in the USA feel for Ireland. They credit these factors with 'getting them in the door' to make sales pitches, and with facilitating them in developing mutually beneficial relationships with customers.

During this period the main contribution of universities and other third level colleges continued to be in supplying graduates. Our universities played an increasingly important role in terms of regional economic development. University research was showing signs of relevance. NASDAQ quoted Baltimore Technologies (IPO, 1999) was founded by a college lecturer in the 1980s to exploit research, and to employ some graduates in high level technology work. NASDAQ quoted Iona Technologies (IPO, 1997) was founded in 1990 to exploit research that had been undertaken by its principals during the 1980s. At this time, since no venture capital was available, and as the Irish state was still focusing most of its attention on the MNCs, what was key to the success of the university related companies was their ability to raise funding through the European Communities research projects (Breznitz, 2007; Grimes and Collins, 2003).

The virtual lack of venture financing in Ireland meant that most ventures continued to grow organically. To the extent that there was a venture capital industry, it focused almost exclusively on non-technology ventures providing development type capital and funding management buyouts. However the more successful companies needed some external finance to develop, and most of the entrepreneurs from that period interviewed indicated that they had received financial backing at critical times from one, almost visionary, individual financier – Dermot Desmond (see Appendix), He appears to have almost single handedly accelerated the entrepreneurial technology sector's growth over a period of five to ten years by offsetting the worst effects of the lack of a financing infrastructure.

An important policy change was the Finance Act of 1984, which extended the favorable tax advantages available to manufacturing companies to data processing services and software development services. This reduced the corporation tax rate to 10 per cent, improving the economics of software enterprises. During the latter half of the 1980s the Irish industrial promotion agencies, which had earlier focused on developing manufacturing industry, broadened their strategy to promote 'internationally traded services' industries. While software was not initially the main focus of the strategy, significant development funds (grants, loans and latterly equity) were being disbursed to entrepreneurial software companies by the end of the 1980s (Breznitz, 2007; Grimes and White, 2005). The funds were important in themselves, but also had a multiplier effect through presenting potential financial backers with what they could interpret as an expert 'seal of approval' (O'Riain, 2000; O'Riain, 2004).

The Irish financial institutions also continued to have an important influence through acting as high quality early customers for entrepreneurial financial technology companies. For example one of the country's largest banks was an early customer for the credit card verification technology developed by Trintech, now one of Ireland's most successful companies and NASDAQ quoted. Throughout this period most of the software companies in Dublin continued to be founded by people with a background in business, and less so by technologists. The most common background for entrepreneurial success continued to combine experience in a major technology company (in Ireland and/or overseas) or a financial institution with experience in an entrepreneurial Irish technology company. However, as the decade progressed, the number of important companies founded by people whose primary skills were in technology increased. This shift toward much more technology-based software companies was critical to the later rapid growth of the industry.

The result of our study show a very small number of highly successful ventures were of central importance in forming the landscape on which later entrepreneurial technology ventures moved beyond the business applications software product markets that supported most of the entrepreneurial technology sector at the time. Probably one of the most important in that respect was Glockenspiel (Sterne, 2004). Glockenspiel, developing and marketing the first widely accepted C++ compiler (a software development tool), was the first Dublin-based company to break into global markets with a technology focused product. While the company was eventually taken over, it is credited by key industry people consulted with establishing a pattern that was followed by other Dublin companies, often employing Glockenspiel alumni, and with advice from ex-Glockenspiel managers. John Carolan, the founding entrepreneur, was a software engineer who recognized early that an implementation of AT&T's C++ specification might significantly improve software development productivity and quality in Dublin software companies to which he provided consultancy services. While the company's origins were in the existing sector, its strategy and the circumstances of its formation were distinctly different. Where most other companies were founded by people whose primary skills were in business, Glockenspiel was founded by a technologist. Where other companies addressed markets that were fragmented geographically, and focused mostly on the UK and continental European fragments, Glockenspiel addressed a global market in which a strong US market presence was a condition for success. It left a legacy of marketing, customer support and technology know-how that many other entrepreneurial technology ventures exploited during the next growth and internationalization stage of the sector's development.

Another company established in that period that helped turn around the Irish software industry is Silicon & Software Systems (S3). S3 is a Philips owned, Irish managed company established in 1986 to design integrated circuits and develop software. The company was established by Professor Maurice Whelan of Trinity College Dublin. Whelan is a professor of electronic engineering, who had previously worked with Philips in a senior technological role. Whelan's rationale was that many of the best technology graduates were emigrating to get good quality technology work, and that it would be better if they could be employed in Dublin. The core of S3's strategy was to provide development services for application specific integrated circuits to Philips and other major electronics companies. By the end of the 1980s S3 had established itself as a credible service provider, and had established Dublin as a location with a supply of credible integrated circuit design engineers, facilitating the establishment of more integrated circuit design ventures, which were able to draw on S3 for their core employees.

However probably the most successful company established in that era, and the first ever Irish software company to list on NASDAQ was CBT Systems. CBT, now part of the NASDAQ quoted company Smartforce, was founded by Patrick (Pat) McDonagh, formerly a school teacher, in 1983 to develop and market computer-based training materials. CBT's first financier, customer and the source of the domain knowledge for its first product, a training kit for money market dealers, was Dermot Desmond. The company addressed a number of different market opportunities during the 1980s, but ended the decade concentrating on training software for professional IT skills. As with Glockenspiel, this was a market in which a strong US presence was required for success. CBT and Patrick McDonagh are also responsible for most of the Irish e-learning industry, including Riverdeep, another company who floated on NASDAQ in 2000, but has since been made private.

The entrepreneurial origins of the Irish involvement in telecommunications software also lie in the 1980s. The presence in Ireland of a number of telecommunications software development operations of major telecommunications technology companies appears to have facilitated the formation of telecommunications software ventures in Dublin through developing a supply of people with knowledge of telecommunications technology and the telecommunications industry. It also provided entrepreneurial ventures with opportunities to undertake contract work. Under these circumstances it is not surprising that many of the entrepreneurs were from engineering backgrounds. However some were also from business backgrounds, notably the founders of Aldiscon – probably the biggest success of the period, and now a major division within UK-based Logica. Aldiscon grew out of an Irish cabling venture. After entering the US cabling market it identified Short Messaging Service (SMS) software for mobile telephony and moved to address this market by developing software in Dublin (see Appendix).

As this development stage progressed, the virtuous circle driving innovation became stronger, powered ultimately by market growth. Employment opportunities drew more graduates into technology ventures, in time producing greater numbers of experienced technologists and managers and more potential entrepreneurs. Entrepreneurs gained experience, and grew their ventures. Social and business contacts made this fast-growing pool of know-how available to entrepreneurs and their ventures, and to prospective entrepreneurs. The pool of entrepreneurs who had exited their original ventures grew gradually, and this both accelerated the sharing of know-how and increased access to capital.

This decade represented the beginning for a small number of like-minded individuals with entrepreneurial tendencies who overcame the fear of failure, and were rewarded for trying and being successful. These people became for many successful role models who knew the nuts and bolts for the new process of new venture development and success, which had important implications for the next generation of start-ups to emerge in the 1990s.

Growth and Internationalization of the Dublin Technology Cluster – Early 1990s

The first five years of the 1990s were difficult for the global technology industry. Overall industry growth slowed below the long-term trend, and many market segments shrank, as economic growth slowed and as the end of the Cold War led to lower defense technology spending. Many minicomputer vendors that had been very successful during the 1980s suffered as their existing products were displaced by lower cost solutions using standard operating systems running on networked servers and PC-based hardware. While the early 1990s was a difficult time for the global technology sector, it was a time of rapid growth and rapid internationalization for indigenous Irish software companies, most of them based in Dublin.

During the first half of the 1990s third level colleges were producing graduates in computing and electronic engineering with skills relevant to industry needs, and in numbers that exceeded demand up to the end of the period. Irish technology companies and the Irish development operations of technology multinationals were mostly using recently educated technologists to apply the latest technologies, rather than using people with dated skills to continue and maintain old product lines. As a result the skills mix in the labor force was more modern and better suited to innovation than that of most other European locations. Furthermore, after two decades of development, the sector had a base of experienced technological entrepreneurs.

An important virtuous cycle process of role model imitation, mentorship and angel finance began for new entrepreneurs. In some cases entrepreneurs found that they were able to get access to good guidance from Irish-Americans in areas such as Boston. One major gap in skills was that the number of people with a track record of growing technology companies rapidly was very limited. There was a particular shortage of senior sales and marketing staff with experience of building and servicing US sales. Sources of finance were beginning to take shape. As the Irish industrial promotion agencies finally focused specifically on software, increasing amounts of funding were put into grants, loans and equity participation for companies thought to have good prospects. In addition the state actions have created a new technological focused Irish venture capital industry at the end of the decade.

In 1991 the National Software Directorate (NSD) was founded within the main industrial development agency to promote the development of the industry, headed by Barry Murphy, a successful industry figure. For the first time the software sector had a direct voice in policy formation. The NSD produced a strategic plan for the sector in 1992, focused mainly on upgrading the sector's factor conditions. During this period leading Irish companies moved beyond domestic sources of funds, attracting funding from international venture capital funds, and moving towards NASDAQ flotation. The NASDAQ flotations of CBT Systems (1995) and Iona

Technologies (1997) were landmarks in bringing Irish ventures into the US funding mainstream.

At the beginning of this five year period, external funding was still a major problem for entrepreneurial technology ventures needing to grow quickly. By the end of the period, while the funding system was not yet well developed, it was still superior to that of most other European locations. Research activity in colleges was gathering pace, still mainly funded under EU research programs. It was during this period that college research first had a high-profile impact on technology entrepreneurship in Dublin. During the latter half of the 1980s a number of staff at Trinity College Dublin, worked on an EU funded project in the course of which they contributed to the development of the CORBA standard for sharing 'objects' (program blocks produced by object oriented programming) across networked computer systems. The team founded Iona Technologies to produce middleware software based on the CORBA standard, and succeeded in developing and marketing the first Object Request Broker adopted widely by major corporations. This led to an investment from Sun Microsystems, and with Sun following its policy of selling all the shares it owns in public companies, Iona's IPO became the fifth largest software offering at the time on NASDAQ.

Technology ventures became more professional in their approach and increasingly focused on overseas markets from the start. A market need for well engineered and configurable software products meant that software products were increasingly likely to be built from scratch, rather than adapted from bespoke software developed for a local client. While high quality local demand continued to contribute to competitiveness in some instances, entrepreneurs increasingly used other sources of insight into market needs (such as direct contact with international customers with whom Dublin software companies already had a relationship) as a basis for competitive advantage.

The profile of entrepreneurs continued to evolve in this period. If before many of the successful entrepreneurs had experience within a multinational, in this period more and more entrepreneurs came directly from within the Irish owned industry. Where the key skills possessed by entrepreneurs had typically been in business, the proportion whose key skills were technological increased. This appears to have reflected both increased opportunities to gain high quality technological experience within Ireland, and a shift in the type of product markets addressed by new ventures toward areas where technological capabilities were more competitively important. Entrepreneurial technology companies were still predominantly concentrated in business applications, but activity in other areas, such as telecommunications software, middleware, e-learning and integrated circuit design was increasing.

The distinctive feature of strategy in this period was that a relatively large number of entrepreneurial technology companies rapidly grew, with the growth being driven primarily by exports. Increasingly, where companies had often specialized in particular geographic niches, the tendency was now to move into global niches, or at least into the North American market which represented about half of the global information technology market. Companies faced two major constraints that shaped their growth strategies and their strategic execution. One was that there was a very limited supply of people with experience of managing rapid growth available in Ireland. A second was that the limited availability of external finance meant that the execution of the strategy had to be very cost-effective.

During this period the entrepreneurial technology sector continued to fragment into related industries. The inflow of overseas technology companies increased. Electronics companies with Irish manufacturing operations increasingly added or grew software development operations. Ireland was increasingly seen as the base for US software companies seeking to address the European market. An infrastructure of printing, packaging, kitting and logistics companies made it possible to outsource physical operations. A supply of software graduates, along with a number of support ventures (in areas such as translation or contract localization), and a base of experienced localization professionals made Dublin a very attractive localization center.

Sharing Dublin with major localization operations was a mixed benefit to entrepreneurial software companies. The positive side was that a proportion of technologists recruited by localization companies got very good experience that was transferable. The downside was that many of those recruited got fairly narrow and limited experience, making them unattractive as employees to the entrepreneurial sector. Multinational companies also raised the expectations of software development staff in terms of pay and options. While this raised costs in entrepreneurial ventures, it also gave a positive view of the broad technology sector to potential new entrants (Arora et al., 2004; O'Riain, 2000; Sterne, 2004).

During this growth and internationalization phase the virtuous circle driving the industry forward continued to strengthen. As before, there was a progressive accumulation of human capital, with numbers of experienced technologists, managers and entrepreneurs compounding over time. Similarly to the 1980s, the driving force was strong growth in market demand. The industry grew very rapidly over this period, and the overall growth was complemented by a high rate of new venture formation. There appear to have been three main reasons for the rapid growth of the Irish industry at the same time that the global industry slowed. First, market conditions were favorable to well designed business applications software products. Second, the typical capabilities of Dublin's entrepreneurial technology companies were collectively passing a threshold that allowed them to successfully enter international markets. Third, developments in the 1980s, which had established the conditions for companies in the sector to move beyond business applications software, came to fruition across a number of market areas, facilitating the emergence of a number of companies that were among the leaders in some global niches.

The growth of tight social networks continued to make the sector's pool of know-how available to new and old entrepreneurs. The conspicuous success of a number of companies during this period had a major influence on the ambitions of other companies. By 1995 the software sector became the first ever sector in Ireland in which technological entrepreneurs became rich, leading to ever larger waves of entrepreneurs who hoped to follow in their footsteps. Furthermore these companies, by path blazing specific business models and modes of operation in the USA, supplied established routes to follow.

Rapid Growth of the Dublin Technology Cluster with Cyclical Bumps (1995–2005)

The period 1995 to mid 2000 was a period of very rapid technology industry growth globally. However a major slowdown in global technology spending started around November 2000. An understanding of some of the factors driving this global growth is necessary to understand the specific response of Dublin's entrepreneurial technology companies. One key driver of global growth was that business process change supported by information technology was starting to produce significant returns to investment, when measured at both the firm and national levels. The Internet presented new opportunities for business process change, addressing the processes connecting companies as well as those within companies. New opportunities for business process change brought a continuing stream of new innovation, productivity improvement and cost cutting. In some cases a need emerged for software to encapsulate or support entirely new business processes.

The sector appears to have weathered the technology market downturn well. Relatively few of Dublin's entrepreneurial technology companies depend on the markets that have been hit worst – those that rely ultimately on obtaining a commercial return from consumer use of the Internet, and those that rely directly on capital spending by telecommunications providers. Instead they are concentrated in areas that ultimately facilitate business process change, and in developing technologies that are required for the telecommunications infrastructure of the future. However it does seem as if the Dublin cluster is slow to go back to a high growth trajectory, especially with regards to growing firms big enough to seek listing on foreign exchanges. Since 2000 there have been no new listings of Irish companies on the NASDAQ or LSE (London Stock Exchange), while four public companies were either bought, delisted or went out of business, including Baltimore Technology, Riverdeep and Parthus. However during this period Dublin still experienced labor shortages, driven mainly by a high rate of employment growth both in indigenous and overseas owned software companies. This coincided with the emergence of high skill labor shortages in other developed countries. The Irish government and third level colleges (which are mostly state funded) responded with a series of programs to increase the output of third level programs in computing, starting in 1996. All in all these efforts coupled with conversion courses and new graduate degrees in electronic engineering more then trebled the output of software programmers. While these measures have not been sufficient to avoid software and integrated circuit design staff shortages, they have allowed both indigenous and overseas owned parts of the technology sector to grow more rapidly than would otherwise have been possible. They have probably allowed many more start-ups to successfully recruit critical staff and start operations.

Dublin's attractiveness as a place to live and work changed. During the early 1990s Dublin increasingly gained a reputation as an attractive, lively, culturally interesting place to live. This reputation peaked around 1995 and 1996, but by 1997 the problems of growth were becoming a major issue for Dublin's residents and prospective residents. During this period the base of venture capital companies in Dublin matured quickly as the city gained an international reputation as a center for technology ventures. Existing indigenous venture capital companies became active in technology, and a number of new companies and branches of overseas companies were established. In 1999 there were 15 Irish venture capital companies which provided 66 million euro to 76 technology ventures (Matheson Ormsby Prentice, 2000). By the end of 2000 Ireland was well supplied with venture capital firms capable of supplying first and second round funding, although it might still have been necessary to look to the USA for substantial third and fourth rounds prior to a possible IPO. Access to equity markets improved greatly over the period. After CBT Systems and Iona Technologies floated in 1995 and 1997, respectively, there were another five NASDAQ or Neuer Market flotations in 1999 and 2000. Local stockbrokers invested significantly in developing capabilities in technology equities, and developed relationships with the main international investment banks.

The Irish industrial promotion agencies continued to support technology ventures, channeling research and development funds, and marketing support funding and equity funding to companies showing good potential. They established incubator style facilities in key markets, making space available to Irish technology companies establishing sales offices. They targeted prospective entrepreneurs with grants to investigate the feasibility of their idea, and with a High Potential Start-up Program.

Research activity in third level colleges formed the basis for an increased number of technology start-ups in the period from 1995 to 2005. Many of the universities have industry or innovation centers in operation to assist with technology transfer from the departments and research groups. Our analysis shows the effect of how one good success story can act as an encouragement to other start-ups; those colleges that have produced leaders in the past tend to show a stream of follow-on development companies. For example Trinity College Dublin (TCD) – several members of the Trinity College Distributed Systems Group founded IONA Technologies as a campus company in 1991. Other companies whose founders were staff at TCD include Baltimore Technologies, Havok, New World Commerce, Trintech and Wilde Technologies. Current Trinity campus companies include Broadcom (a joint venture between Ericsson, Eircom and TCD), Simtherg, Tolsys and XCommunications. In University College Dublin (UCD) WBT Systems was the first big success. Since then the college has helped with the creation of software development companies, including ChangingWorlds, Massana and Sephira.

During this period the proportion of new entrepreneurs with experience only within the entrepreneurial Irish owned sector continued to increase, although the proportion of these had experience with technology companies overseas, or with overseas-based companies in Ireland. With limited domestic opportunities in electronic commerce, Dublin had relatively few successful start-ups in the area. Thus, unlike many other centers that grew quickly at the time, Dublin did not have a large influx of entrepreneurs from the non-technology sector backgrounds that were often suited to building electronic commerce-based ventures.

Entrepreneurial technology companies established after 1995 were more likely than those of earlier times to be positioned in very high growth niches in which competition was global, and in which the technology embodied in the product was a critical determinant of competitive success. The most successful appear to have been in areas where there was already an established capability in relevant technologies and markets in Ireland, and which were also recognized internationally as high growth areas by start-ups and established companies. Thus the key companies established since 1995 are mostly in telecommunications (including mobile telephony and mobile applications), middleware (including XML tools) and customer relationship management, with some still appearing in more general business applications

Table 1.3 Academic spinoffs – Iona Technologies and second order effects

SteelTrace	Shinka Technologies
Prediction Dynamics	Bind Systems
Rococo Software	Brand Aid
Technology Sales Leads	Ebeon
Wilde Technologies	Cyrona
Zenark	Wolfe Group
LeCayla Software	Fields Point

software. Where specific vertical markets have been targeted, they tend to be in the Irish sector's traditional areas of financial firms and telecommunications companies, and also among fellow technology companies.

Much of the activity in middleware can be traced to the influence of Iona Technologies, which has spun off at least 15 'second order' spin-offs (most, but not all in Ireland; most, but not all in software – see Table 1.3). The specialization in customer relationship management appears to have emerged from a combination of the traditional specialization in business applications software with at least one spin-off from a systems integrator. The specialization in telecommunications comes out of both the established base of indigenous and overseas owned telecommunications companies, and the integrated circuit design sector. The design of integrated circuits for mobile telephony is one of the main constraints on the performance of mobile devices. Mobile telephony companies often outsource the development work. Increasingly, since the mid 1990s, they have relied on licensing of independently developed intellectual property to achieve competitive product designs. Irish integrated circuit design companies, including the NASDAQ quoted Parthus (IPO, 2000), now a part of Ceva, have been active both in contract development and in licensing of intellectual property. While entrepreneurial technology companies founded since 1995 have had greater access to equity finance than their predecessors, they have also faced a stronger imperative to use it. Time to market is a critical determinant of success in most of the markets they have targeted, and resources for development and marketing are a critical determinant of time to market. A significant proportion of the most prominent new firms founded since 1995 have been founded by people who have previously built and exited successful technology companies.

The industry grew very rapidly over this period, increasing the numbers employed much faster than in the previous five year period. The overall growth was again complemented by a high rate of new venture formation. It appears that the sector's exceptionally high rate of growth was rooted in

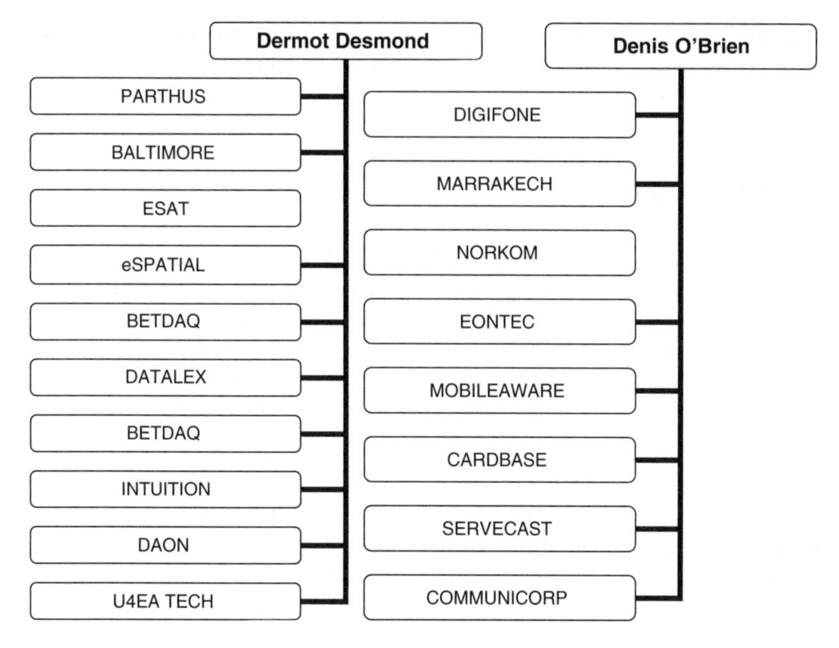

Figure 1.3

the continued development of its capabilities, complemented by very favorable market conditions. The product markets targeted by new ventures tended to be in high growth niches, in which competition is global and in which the technology is a significant determinant of competitiveness. Companies of this sort first appeared in Ireland in the mid 1980s, since which time they have gradually overtaken business applications software as the main focus of new technology company formation in Dublin. Over this period the number of people in the sector grew, increasing the numbers involved in informal social and business networks. New groups such as the Irish Internet Association and Wireless Wednesday emerged to complement the work of the Irish Software Association in arranging more formal networking events and contributing to the development of the networked state. However the number of people from the mainstream of entrepreneurial technology ventures who have participated regularly in formal events is small relative to the numbers active in the sector. Most entrepreneurs interviewed also rely on established relationships for networking rather than seeking to participate in wider sectoral or industry level events. Conspicuous successes by significant numbers of entrepreneurial technology ventures during this period increasingly validated the idea of starting and growing a technology company. Where technology entrepreneurship was unusual, and almost the preserve of a subculture up to the mid 1990s, in the latter half of the 1990s it became a part of Dublin's social, cultural and economic mainstream.

DISCUSSION

This chapter builds on Venkataraman's work of the entrepreneur as an input completer and the Austrian approach which attributes a central role to the entrepreneur in explaining economic development. We aim to build on a new approach for cluster analysis and to build further upon this theoretical perspective which argues that local social conditions ('intangibles') play an important role in the genesis and assimilation of technology-based entrepreneurship in a region.

Using the case of Dublin, Ireland as our focal point, the results of the study indicate that the Irish owned technology sector in Dublin did not appear overnight – it took 30 years to accomplish. In contrast to previous studies on this topic we demonstrate in fact that Ireland's transformation was largely a 'bottom-up' story of successful early enterprising individuals, who played a central role acting as role models for budding entrepreneurs looking to start promising new ventures. These early stage ventures provided a feasibility proof ('If that person can do it, I can do it, too') for other peers to seek entrepreneurial success. The implications of our research suggest that regions with cultures that support commercialization activity with a tacit approval of entrepreneurs is a key factor in explaining the rise of regional entrepreneurship in Dublin. This finding highlights the role of social norms, expectations and local group norms in cluster formation.

Second, the early entrepreneurs who founded the technology sector and created the dynamic were not technologists, but business people. Typically, they had sales and marketing backgrounds, experience of working in technology companies in Ireland and overseas and they had a deep appreciation of the technology. Thus they came from the population that was most likely to produce the entrepreneurs who could spot the opportunities in the relevant niches. As Irish owned entrepreneurial companies became established an increasing number of entrepreneurs came from these technology companies. The pure technology enterprises that emerged in the late 1980s and 1990s were initially set up by people leaving multinational companies in Ireland and later by people working on EU funded projects in universities. Some of the most prominent new firms set up since 1995 were founded by entrepreneurs who had previously built and exited successful technology companies. The evidence suggests that most entrepreneurs were 'pushed' into entrepreneurship. The evidence also shows that the opportunities for entrepreneurship changed over time and that the people who responded also changed over time.

Third, the role of informal personal networking is a key feature of the technology cluster in Dublin, but there is little by way of formal networking. However these informal meeting points, like bars and restaurants, appeared to play an important role for Irish entrepreneurs to exchange ideas, meet role models and acquire the wish to create and finance ventures. As argued by many authors, including Saxenian (2000), personal networking plays an important role in accessing resources, such as finance, people, specific skills and space. Personal networks were also used extensively to gain access to potential customers in Ireland, the USA and other European countries, and in many cases existing networks were used to leverage other networks and thus extend their influence. The access to role models mostly occurs in informal forums.

Fourth, the role of government in the development of the software and technology cluster in Dublin was that it initially attracted multinational companies to locate in Ireland, including Dublin, but without any particular thought for the development of a software sector. As the sector emerged, and at critical points in its development, the government responded with specific policies and activities to support what GEM (Bygrave and Hay, 2005) describe as the entrepreneurial infrastructure or framework conditions. A key elements of government policy included introducing policies in the late 1970s to support start-ups in attractive market segments and in good areas of technology. This is now known as the High Potential Start-up Program. In 1984 software was included among other internationally traded service activities which qualified for low rates of taxation on corporate profits (10 per cent) similar to those applied to manufacturing companies. This was followed in the early 1990s with financial and other specially targeted support to help internationalization. In the post-1995 period of skill shortages the government significantly expanded the output of the education sector to meet the increase in demand for graduates. Our findings also support the work from Bresnahan et al. (2001) that suggest accommodative government policies that enable conditions like the creation of suitable demand and markets can be an important part of cluster development.

Fifth, the vast majority of software firms in Dublin did not use venture capital to develop their businesses. To a large extent this reflects the nature of the businesses established, particularly in the earlier periods which were dominated by bespoke software followed by the gradual development and marketing of products. These businesses did not, typically, require significant up front funding. The funding required generally came from the entrepreneurs themselves or their families and friends, and in some instances from an outside investor. Once started, ongoing development was funded from revenue. This changed in the 1990s as the opportunities to exploit global niches with high-technology-based products arose. The development of these products required significant upfront investment and time to market was critical to success. These businesses required significant venture capital to develop and in the period between 1995 to the present a number of venture capital firms were set up in Ireland to respond to the opportunities. These firms are now complemented by a number of UK and US venture funds that are active in Ireland. The technology type software firms would not have developed in Dublin without venture capital.

Finally, our chapter highlights the central role of research led universities in stimulating Schumpeterian type high growth and knowledge intensive companies from university research. Much has been documented about how formal research activities of universities, such as MIT in Cambridge, contributed to emerging industries and clusters, such as Kendall Square in Boston (O'Shea et al., 2005, 2007). It is noteworthy from a Dublin context that most successful firms did have their origin in university grounded research and were an important component in the stimulation of high-technology entrepreneurship in the region. The government has allocated over 4.7 billion euro over the period to 2010 to develop Ireland as a center for world class research excellence in strategic niches of biotechnology and information and communications technologies.

The Dublin technology cluster exhibits all the characteristics of a dynamic innovative cluster. Driven by rapid growth in market demand, a number of technology driven entrepreneurs set up new firms and many achieved conspicuous success as a result of major trade sales or IPOs. The compounding effect of this success created a self-reinforcing dynamic or virtuous circle where the accumulation of entrepreneurs, human capital and networking all contributed to collective learning and innovation, thus fueling growth in the Dublin region. The study finds that multinational companies can play a very positive role in the development of a cluster. In the Dublin technology cluster the multinational companies contributed significantly as early customers. The impact of FDI (Foreign Direct Investment) in the transformation of Ireland's entrepreneurial economy cannot also be underestimated. It has been a major source of knowledge transfer in technology and management know-how, international market trends, global investment trends and financial expertise. An increasing number of entrepreneurs responded by setting up or help scale new firms. They were the core training ground for the early entrepreneurs and are still an important source. In addition, the multinational companies formed a part of the collective system of learning along with the entrepreneurial companies as people moved back and forth among various firms. This helped kick-start a self-reinforcing dynamic of entrepreneurship and innovation, and positioned Dublin as a very important center for software and technology internationally.

REFERENCES

Acs, Z.J. and A. Varga (2005), 'Entrepreneurship, agglomeration and technological change', *Small Business Economics*, **24**, 323–4.

Antonelli, C. (2000), 'Collective knowledge communication and innovation: the evidence of technological districts', *Regional Studies*, **34**, 535–47.

Arora, A. and A. Gambardella (2005), *From Underdogs to Tigers: The Rise and Growth of the Software Industry in Some Emerging Economies*, Oxford, UK: Oxford University Press.

Arora, A., A. Gambardella and S. Torrisi (2004), 'In the footsteps of the Silicon Valley? Indian and Irish software in the international division of labor', in T. Bresnahan and A. Gambardella (eds), *Building High-Tech Clusters: Silicon Valley and Beyond*, Cambridge, UK: Cambridge University Press, pp. 78–121.

Avnimelech, G. and M. Teubal (2004), 'Venture capital start-up co-evolution and the emergence and development of Israel's new high technology cluster', *Economics of Innovation and New Technology*, **13**, 33–60.

Begley, T.M., E. Delany and C. O'Gorman (2005), 'Ireland at a crossroads: still a magnet for corporate investment?', *Organic Dynamics*, **34**, 202–17.

Braczyk, H.-J, .P. Cooke and M. Heidenreich (1998), *Regional Innovation Systems: The Role of Governances in a Globalized World*, London: UCL Press.

Breathnach, P. (1998), 'Exploring the "Celtic Tiger" phenomenon: causes and consequences of Ireland's economic miracle', *European Urban and Regional Studies*, **5**, 305–16.

Bresnahan, T. and A. Gambardella (2004), *Building High-Tech Clusters: Silicon Valley and Beyond*, Cambridge, UK: Cambridge University Press.

Bresnahan, T., A. Gambardella and A. Saxenian (2001), ' "Old economy" inputs for "New economy" outcomes: cluster formation in the new Silicon Valleys', *Industrial and Corporate Change*, **10** (4), 835–60.

Breznitz, D. (2005), 'Collaborative public space in a national innovation system: a case study of the Israeli military's impact on the software industry', *Industry and Innovation*, **12**, 31–64.

Breznitz, D. (2007), *Innovation and The State: Political Choice and Strategies for Growth in Israel, Taiwan, and Ireland*, New Haven, NJ: Yale University Press.

Bygrave, W. and M. Hay (2005), *Global Entrepreneurship Monitor: 2005 Summary Report*, Boston, MA and London: Babson College and London Business School.

Capello, R. (1999), 'Spatial transfer of knowledge in high technology milieux: learning versus collective learning processes', *Regional Studies*, **33** (4), 353–65.

Central Statistics Office (CSO) (2003), *Measuring Ireland's Progress. Vol. 1, Indicators Report*, Dublin, Ireland.

Cooke, P. and K. Morgan (1998), *The Associational Economy*, New York: Oxford University Press.

Enterprise Ireland Annual Report (2003), Government Publication, Ireland.

Economist (1997), Edition 15 May 1997.

Economist Intelligence Unit (EIU) (2005), 'World quality of life index rankings', www.economist.com/media/pdf/QUALITY_OF_LIFE.pdf.

Feldman, M. (2005), 'The entrepreneurial event revisited: firm formation in a regional context', in S. Breschi and F. Malerbo (eds), *Clusters, Networks, and Innovation*, Cambridge, MA: Oxford University Press

Feldman, M. and R. Martin (2005), 'Constructing jurisdictional advantage', *Research Policy*, **34**, 1235–49.

Florida, R.L. (1995), 'Toward the learning region', *Futures*, **27**, 527–36.

Florida, R. (2005), *Cities and the Creative Class*, Oxford, UK: Routledge.

Forbes Magazine, 28 May 2001, J. Anderson, J. Doebele, S.H. Hanke and R. Heller, 'Clusters of excellence', 70–2, www.forbes.com/legacy/golbal/2001/0528/070tab1_talde.shtml.

Green, R., J. Cunningham, I. Duggan, M. Giblin, M. Moroney and L. Smyth (2001), 'The boundaryless cluster: information and communications technology in Irelend', paper in *Innovative Clusters: Drivers of National Innovation Systems* Paris: OECD, pp. 47–64, http://fp.tm.tue.nl/ecis/papers/iii_2_1.pdf.

Grimes, S. and P. Collins (2003), 'Building a knowledge economy in Ireland through European research networks', *European Planning Studies*, **11**, 395–413.

Grimes, S. and C.M. White (2005), 'The transition to interrational traded services and Ireland's emergence as a "Successful" European Region', *Environment and Planning A*, **37**, 2169–88.

Industrial Development Authority (IDA) Ireland (2004), www.iclaireland.com.

IMD World Competitiveness Yearbook (2004), Switzerland: IMD Business School.

Keeble, D., C. Lawson, B. Moore and F. Wilkinson (1999), 'Collective learning processes, networking and institutional thickness in the Cambridge Region', *Regional Studies*, **33**, 319–32.

Kenney, M. (2000), *Understanding Silicon Valley: The Anatomy of an Entrepreneurial Region*, Stanford, CA: Stanford University Press.

Koepp, R. (2002), *Clusters of Creativity: Enduring Lessons on Innovation and Entrepreneurship from Silicon Valley and Europe's Silicon Fen*, Chichester, UK: Wiley.

Lester, R.K. and M.J. Piore (2004), *Innovation – The Missing Dimension*, Cambridge, MA: Harvard University Press.

Matheson Ormsby Prentice (2000), 'The Irish venture capital report on venture capital activity in Ireland', report, Dublin.

Morgan, K. (1997), 'The learning region: institutions, innovation and regional renewal', *Regional Studies*, **31**, 491–503.

National Software Directorate, *The Software Industry in Ireland*, Dublin, Ireland: National Software Directorate, www.nsd.ie/htm/home/home.php 3.

O'Riain, S. (2000), 'The flexible development state: globalization, information technology, and the "Celtic Tiger" ', *Politics and Society*, **28**, 157–93.

O'Riain, S. (2004), *The Politics of High Tech Growth: Developmental Network States in the Global Economy*, Cambridge, UK: Cambridge University Press.

O'Shea, R.P., T.J. Allen, A. Chevalier and F. Roche (2005), 'Entrepreneurial orientation, technology transfer and spinoff performance of US universities', *Research Policy*, **35** (7), 994–1009.

O'Shea, R.P., T.J. Allen, C. O'Gorman and F. Roche (2007), 'Delineating the anatomy of an entrepreneurial university: the MIT experience', *R&D Management*, **37** (1), 1–16.

Phan, P. and M.D. Foo (2004), 'Technological entrepreneurship in emerging regions', *Journal of Business Venturing*, **19** (1), 1–5, special edition on techno-entrepreneurship.
Piore, M.J. and C.F. Sabel (1984), *The Second Industrial Divide: Possibilities for Prosperity*, New York: Basic Books.
Porter, M. (1990), *The Competitive Advantage of Nations*, New York: Free Press.
Sands, A. (2005), 'Eye of the Tiger: evolution of the Irish software industry', in A. Arora and A. Gambardella (eds), *The Rise and Growth of the Software Industry in Some Emerging Economies*, Oxford, UK: Oxford University, Press.
Saxenian, A. (1994), *Regional Advantage: Culture and Competition in Silicon Valley and Route 128*, Cambridge, MA: Harvard University Press.
Saxenian, A. (2000), 'Networks of immigrant entrepreneurs', in C. Lee, W.F. Miller, M.G. Hancock and H.S. Rowen, *The Silicon Valley Edge: A Habitat for Innovation and Entrepreneurship*, Stanford, CA: Stanford University Press, pp. 248–75.
Schumpeter, J.A. (1976), *Capitalism, Socialism and Democracy*, New York: Harper and Row.
Sterne, J. (2004), *Adventures in Code: The Story of the Irish Software Industry*, Dublin, Ireland: The Liffey Press.
Venkataraman, S. (2004), 'Regional transformation through technological entrepreneurship', *Journal of Business Venturing*, **19**, 153–67.

APPENDIX

The following sample of 30 companies provides an overview of the range of venture backed technology start-ups that are located in Dublin. To obtain the entire database on ICT software companies in Ireland, please go to the following website www.nsd.ie/htm/home/home.php3.

Company	Website address	Company description
CapeClear	www.capeclear.com	Cape Clear Software provides a proven way to implement a service oriented architecture
Qumas	www.qumas.com	Recognized as a world leader in enterprise compliance management
Silicon & Software Systems	www.s3group.com	Silicon & Software Systems Ltd (S3) is an electronics design company that develops pioneering integrated circuits and embedded software solutions
AEP Networks	www.aepnetworks.com	AEP Networks offers a comprehensive range of innovative, secure networking and application access products to meet the most demanding requirements of commercial and public sector customers
Corvil Networks	www.corvil.com	Corvil Networks is developing next generation solutions for network monitoring, analysis and control that enable the assured delivery of new applications and services over the Internet
Innovada	www.innovada.com	Innovada has unique technology and expertise targeted at the fast growing market for devices that communicate information,

Company	Website address	Company description
		including among others applications such as wireless sensor networks (WSN), automated meter reading (AMR), point-of-sale terminals (POS) and set-top boxes (STB)
Straatum	www.straatum.com	Straatum, based in Dublin and Santa Clara (California), is a leading provider of real-time fault detection and classification (FDC) solutions to the semiconductor sector worldwide
Frontier Silicon	www.frontier-silicon.com	Frontier Silicon is a pioneer and leading provider of complete semiconductor and modular solutions for digital multimedia products
Prime Carrier	www.primecarrier.com	Prime Carrier provides wide margin optimization software for telecommunications companies
PXIT	www.pxit.com	PXIT manufacture high performance optical and electrical PXI modules used to test a wide range of photonic components in the telecommunications, data communications and lighting markets
Xiam	www.xiam.com	Xiam is the industry expert in information routing technology
Alphyra	www.alphyra.com	Alphyra is the leading supplier of mobile payment solutions
Openet	www.openet.com	Openet, a world leader in real-time charging,

Company	Website address	Company description
		advanced mediation and network-edge rating
Valista	www.valista.com	Valista is now the leading provider of electronic and mobile payments software
Voxpilot	www.voxpilot.com	Voxpilot is a provider of IMS compliant Voice XML and video in Voice XML platforms
Havok	www.havok.com	Havok creates and licenses physics software to computer games developers, publishers and to motion picture projects
Changing Worlds	www.changingworlds.com	Changing Worlds is the market leading provider of intelligent personalization and portal solutions for the wireless telecommunications industry
Anam Mobile	www.anam.com	Anam Mobile provides a complete messaging services solution to mobile operators worldwide, covering core network SMS and MMS delivery platforms and additional service control products
Shenick	www.shenick.com	Shenick is an emerging leader in the delivery of combined network services plus applications emulation and performance test systems
Xsil	www.xsil.com	Xsil is a world leading developer of innovative laser micro-machining systems
eSpatial	www.eSpatial.com	eSpatial is a world leader in enterprise strength spatial information management technology

Company	Website address	Company description
Mapflow	www.mapflow.com	Mapflow provides technology to enable real-time location-based services for Internet enabled devices
Betdaq	www.betdaq.com	Online betting Internet company
Aepona	www.aepona.com	Aepona is a market leader in the delivery of software solutions to telecommunications operators
Soft-ex	www.soft-ex.net	Soft-ex is a leading provider of communications management solutions
Mobile Cohesion	www.mobilecohesion.com	Mobile Cohesion is a provider of VASP and premium VAS management infrastructure
MobileAware	www.mobileaware.com	MobileAware enables operators, ASPs, portals and corporate enterprises to exploit the full potential of mobility services through its flagship product suite, Everix™
Marrakech	www.marrakech.com	Marrakech provide enterprise e-purchasing and e-payment solutions as a single service
Copperfasten	www.copperfasten.com	Copperfasten is a technology leader in the next generation intrusion detection market
Automsoft International	www.automsoft.com	Automsoft International is an innovative provider of plant information systems to the pharmaceutical and process-based industries
Aran Technologies	www.arantech.com	Aran Technologies is a telecommunications company that specializes in the development of software for the emerging mobile systems

Company	Website address	Company description
IPOs		
Norkom	www.norkom.com	Norkom is an industry leader in business intelligence software and fraud detection spyware applications
Iona Technologies	www.iona.com	Iona Technologies provides leading edge integration and infrastructure platforms
Baltimore Technologies (Out of Business)	n/a	Baltimore Technologies developed security products and services to enable companies to develop trusted, secure systems
Horizon Technologies	www.horizon.ie	Horizon Technologies specialize in delivering application solutions to its clients
Riverdeep Interactive Learning	www.riverdeep.com	Developer of educational software, including software for mathematics instruction
Trintech	www.trintech.com	Trintech is a leading provider of transaction management, payment infrastructure, reconciliation software and risk management solutions to financial institutions
Smartforce (Skillsoft)	www.skillsoft.com	Skillsoft is a leading provider of enterprise e-learning, with learning resources targeted to business and IT professionals
Parthus (takeover by CEVA)	www.parthus.com	ParthusCeva is the leading licensor of DSP cores, multimedia and storage platforms to the semiconductor industry
PaddyPower	www.paddypower.com	Online betting Internet company
Conduit (Investcorp Tech.Ventures)	www.conduit.ie	Conduit is an industry leader in providing call center solutions and is one of the

Company	Website address	Company description
		fastest growing information service providers in Europe
Datalex	www.datalex.com	Supplier of reservation software and online booking engines for tour operators and travel agencies
Recent Trade Sales		
Aldiscon (Logica)	www.aldiscon.interweb.ie	Logica Aldiscon is a leading developer of innovative infrastructure elements for digital cellular, PCS and satellite network operators worldwide
Apion	www.phone.com	Apion has particular expertise in GSM intelligent networks, wireless data, and WAP technology
Voicevault (Biometric Security)	www.voicevault.com	Voicevault is a world leader in voice verification. Voicevault's proven solutions radically enhance security, deliver substantial cost savings and improve levels of customer service
Performix Technologies (Nice Systems)	www.nice.com	Performix Technologies is a provider of employee performance management solutions for the contact center industry
Alphamosaic	www.broadcom.com	Alphamosaic Ltd, a leading developer of advanced mobile imaging, multimedia and 3-D graphics technology optimized for use in cellphones and other mobile devices
Amphion (Conexant Systems, Inc)	www.conexant.com	Amphion Semiconductor Ltd (Amphion) is a company specializing in developing video compression technology

Company	Website address	Company description
Massana (Agere Systems)	www.agere.com	Massana Ltd, a privately held developer of gigabit ethernet-over-copper physical layer device (PHY) technologies
WBT Systems (Horizon Technology)	www.wbtsystems.com	WBT Systems provide proven e-learning solutions to rapidly create, deploy and manage online learning content across the extended enterprise
Stockbyte (Getty Images)	www.gettyimages.com	Web-based company called Pixel Images Holdings Ltd
Needahotel (Cendant)	www.NeedaHotel.com	International online reservation company
Similarity Systems (Informatica Corporation)	www.similaritysystems.com	Similarity Systems is a leading provider of enterprise data quality management software
New start-ups (Dublin region) – 2005		
Gas Sensor Solutions	www.gss.ie	GSS is a manufacturer of low cost, non-invasive, oxygen sensors and sensing equipment
Accuris Ltd	www.accuris.ie	Innovative wireless cellular convergence products, offering seamless mobility with a single number across all fixed and wireless technologies
Acquis Ltd	www.acquisinc.com	Enterprise location software enhances applications with unique real-time, mapping and location information data capture, analysis and dissemination
Advanced Field Solutions Ltd	www.advancedfieldsolutions.com	Web-based solution that enables optimal productivity for field service engineers

Company	Website address	Company description
Blueface	www.blueface.ie	Development of a White Label Voice over IP platform
Broadcast Learning Ltd	www.broadcast.ie	Application software development, which offers a comprehensive suite of e-assessment and e-learning solutions
Cicero Networks Ltd	www.ciceronetworks.com	VoIP solutions which enable telecommunications operators and service providers to deliver all calls (fixed, cellular and VoIP) on mobile phones and PDAs
Dualtron	www.dualtron.ie	Smart card payment processing and cash handling technology systems
Lecayla	www.lecayla.com	Engaged in the development and sale of 'pay as you go' billing for packaged software
PaceMetrics Ltd	www. pacemetrics.com	Business activity monitoring solutions to the global Tier I investment banks
RedMere Technology Ltd	www.redmeretechnology.com	Fabless semiconductor company supplying high speed communications chips to the consumer electronics industry for use in high definition TVs and DVD recorders
Silansys Technologies	www.silansys.com	Embedded software chips for mobile digital TV, digital radio and mobile phones
Advanced Modular Solutions Ltd	www.ams-ireland.com	Design and sell energy efficient power technology for the telecommunications, automotive and industrial markets
Bianamed	www.biancamed.com	Software for the analysis of electronic signals for the sleep breathing disorder market

Company	Website address	Company description
Nualight Ltd	www.nualight.com	Advanced lighting solutions for the retail display case market using Nualight proprietary technology
Digisoft TV	www.digisoft.tv	Digisoft TV designs, develops, implements and supports a suite of interactive TV application software programs
PixAlert	www.PixAlert.com	PixAlert is a leading provider of innovative image monitoring software and e-security solutions to clients in the banking, financial services and pharmaceutical sectors

2. The entrepreneurial drivers of regional economic transformation in Brazil

José Cezar Castanhar, João Ferreira Dias and José Paulo Esperança

INTRODUCTION

The study of entrepreneurship has usually been carried out in a somehow paradoxical approach by two different streams of literature: the economics literature and the management literature. The economics literature has been emphasizing the contribution of entrepreneurship for economic development through quantitative, econometric studies in which the entrepreneur is an aggregate variable. In this type of study entrepreneurship is measured by a proxy variable, usually the creation of new firms, and the aim is to evaluate the impact of different firm creation rates on some measures of economic development, usually job creation. Also a regular feature in this type of study is the use of some regional space unit (different countries or different regions within a country) as the level of analysis. In a different path the studies in management literature emphasize the individual entrepreneur, and/or the firms created by entrepreneurs. In this type of study the aim is usually to evaluate the impact of entrepreneurs' attributes, firm's characteristics and strategies as well as environmental conditions on the firms' performance. The economic impact of the entrepreneur at local or regional level is taken for granted.

These two different approaches for studying the entrepreneurship phenomenon can be seen to contain a curious paradox: on the one hand, the economics literature is more concerned in studying the aggregate effects of entrepreneurship, disregarding the dynamics of the entrepreneurial process itself; on the other hand, the management literature is dedicated to study the dynamics and process of entrepreneurship, with little or no concern at all to the macro-economic impacts of the entrepreneurial activity. There are a few exceptions in the management literature, represented by studies that examine the conditions that favored the emergence of high-technology firms agglomerations in specific regions (Ding and Abetti, 2001; Hitt et al.,

2001; O'Gorman and Kautonen, 2001; Saxenian, 1994). It is worth also mentioning some studies that examined the effect of a benign environment on the emergence and growth of industries in specific regions (Eisenhardt and Schoonhoven, 1990). But even in these studies the emphasis was either on government policies or environmental and regional conditions that worked as stimuli to the creation of firms and the emergence of agglomerations of firms, usually in the high-technology industry in particular regions. The analysis of the entrepreneurs, the processes of firm creation, their strategies and evolution, as well as their particular contribution to the regional economic development and change were seldom considered.

One important exception in the prevailing approach in the extant literature is the contribution of Venkataraman (2004), who explicitly suggests that technological entrepreneurship is an effective lever to rescue regions trapped in a 'virtuous cycle' that tend to perpetuate traditional economic activities, with an inefficient use of strategic resources (especially skilled human resources) and, as a consequence, make it more difficult for the region to break its present constraints and develop. Venkataraman goes one step further and defines a set of 'intangible' factors, as opposed to 'tangible' factors (usually related to infrastructure and capital), that should be assembled in order to ensure that the resources available in the region would be channeled to enhance technological entrepreneurship, and in the process foster regional transformation and development.

One pertinent question that emerges from Venkataraman's approach is, 'would non-technological entrepreneurship, that is, entrepreneurship in a mature sector, with medium levels of technology, also be able to make a major contribution in the process of transforming and developing former traditional regions?' A second and related question is 'if entrepreneurship in a mature sector is to play a strategic role in the process of transforming traditional regions, are the "intangible" factors, defined by Venkataraman (2004), also strategic tools to ensure that the entrepreneurial initiatives result in regional transformation, as in the case of technological entrepreneurship?'

These are the main questions that we will try to discuss with the present case study. It is important to mention in advance that the case study will not be used to test a hypothesis on Venkataraman's model. The case study will be used as an illustration of the operationalizational possibilities of Venkataraman's model concepts and categories, as well as the possibility of using the model in non-technological entrepreneurship situations. Note that even in Venkataraman's approach there remains an apparent gap, which is the connection between the strategies and trajectories of individual entrepreneurs and their firms, and the emergence of collective actions and institutions that, ultimately, will result in regional transformation. With the purpose of making a contribution to fill this gap we shall examine simultaneously the case of the transformation of a Brazilian region from a traditional agricultural economic base to an industrialized economic base, and the trajectory of one specific entrepreneur of the region and how he related to the overall regional process.

We shall do so by studying one particular case of a furniture manufacturing cluster located in a medium size city in the countryside of Brazil, and the pioneer company of the cluster, named MOVAL, that we call the 'anchor' entrepreneur. Based on the retrospective historical accounts of the key actors involved, as well as secondary documentation, we shall track the trajectory and key events for the emergence and development of the cluster and the pioneer company, as well as the connections between the two trajectories. A third and subsidiary research question in this study is 'to what extent do the four principles of the Effectuation Theory, proposed by Sarasvathy (2001), help to illustrate and explain the strategy and decisions of the entrepreneur along his trajectory?'

There were three reasons for the choice of this particular case. First, the tremendous impact that the emergence and development of the cluster had on the regional development in a rather short time span (less than three decades). Second, the emergence and development of this cluster resulted unequivocally from a drive of local entrepreneurs, especially from a particular entrepreneur, whose firm trajectory and strategies shall be highlighted in the study, with very little support from governmental policies. Third, this cluster refers to a mature and traditional (in contrast to a high-technology) industry. This last feature is important. First, because most of the studies on the impact of entrepreneurship on regional economic development emphasize high-technology industry in the context of broad governmental technological supporting policies. Second, because we believe that it would be easier to replicate in other regions a successful experience in a traditional and mature industry rather than one in a sophisticated technological industry.

In the next section we review the economics literature on entrepreneurship and regional growth and the management literature on entrepreneurship, especially that related to venture creation, entrepreneurial strategies and performance. We then present the method of analysis used and describe how data were collected, followed by a presentation and discussion of the results. The chapter concludes with a summary and interpretion of the results where we state their implications.

LITERATURE REVIEW

As mentioned earlier, there is a very clear distinction between the two branches of literature on entrepreneurship. One approaches the field from an economic perspective and emphasizes the aggregate impact of entrepreneurship on economic development at a national or regional level of analysis. The management literature emphasizes strategies and outcomes, adopting individual entrepreneurs and their firms as the unit of analysis. In this section these two different approaches are reviewed separately.

The Economics Literature

The very first discussion on the role of the entrepreneur on the economy came from economic theory related to explanations of economic growth. It is worth mentioning the classical work of Schumpeter (1934) who saw the entrepreneur as responsible for the introduction of radical innovations in the economic system previously in balance. These radical innovations produced the 'creative destruction' of old industries for new ones, triggering new cycles of economic growth. The so-called Austrian School (Hayek, 1984; Kirzner, 1973) attributes to the entrepreneur the role of an arbitrageur, who identifies and exploits opportunities associated with under allocated economic resources. In an attempt to develop models of endogenous growth, neo-classic economists, such as Solow (1956), Lucas (1988) and Romer (1994), emphasized the role of technical knowledge and human capital which, through the mechanism of spillover, could work as an effective engine to foster economic growth, although no explicit mention of the entrepreneur is made in such models. Implicit is an assumption, resembling the Austrian School approach, that market mechanisms will come up with an instrument to transform knowledge in economic or business opportunities.

A recent stream of theoretical and empirical research in economics explicitly emphasizes the role of the entrepreneur in this process. Acs et al. (2005a, 2005b) suggest that the entrepreneur is the missing link in classical economic models. According to the authors, entrepreneurship is an effective channel to transform technical opportunities created by technical knowledge into commercial (business) opportunities. Thus, to promote endogenous economic development, governmental policies should support entrepreneurship as much as technical knowledge. Audretsch and Fritsch (2002) and Fritsch and Muller (2004) suggest that regions should be characterized by a different growth regime, resembling the theory of a technological regime (Dosi, 1988). The growth regime of each region should be defined in terms of its entrepreneurial intensity. Audretsch and Keilbach (2004) proposed the concept of entrepreneurship capital, in the same fashion as human capital and social capital, as a relevant production factor that should be developed in each region in order to improve economic development.

Acknowledging the growing importance of entrepreneurship as a strategic factor to promote economic growth, Audretsch and Thurik (2004) suggested that the economic and institutional changes of the 1970s and 1980s, such as the deregulation trend, globalization of markets and economies and the development of information technology, among others, led to a radical change in the very nature of the economy, which shifted from a managed economy model, based on production of scale and scope and regularities, to an entrepreneurial economy model, based on flexible production, innovation, market specialization and frequent turbulence. Following this renewed attention to the 'entrepreneurship factor' a perusal of empirical studies was carried out to measure the impact of entrepreneurship on economic growth. Most of these studies are econometric and can be classified into three categories. The first group of studies, using the Global Entrepreneurship Monitor[1] as the database, use the level of entrepreneurship as the independent variable (along with other independent and control variables) and the countries' GDP as the dependent variable (Van Stel et al., 2004, 2005a, 2005b; Wennekers et al., 2005). The second group of empirical research uses a proxy for entrepreneurship and relates it to the countries' GDP or employment growth (as the dependent variable). The proxy most commonly used in the empirical studies of this kind is the (gross or net) number of firm creation, and the unit of analysis is the country. Most of the studies utilize data from OECD countries and are of a cross-sectional nature (Audretsch et al., 2001; Carree et al., 2000; Acs et al., 2005a). The third group of empirical studies also uses a proxy for entrepreneurship and relates it to GDP or employment growth. The difference to the previous group is that the studies refer to a single country and the units of analysis are sub-national regions (planning regions, districts, local market areas, among others). The studies of Audretsch and Fritsch (2002), Fritsch and Muller (2004), Baptista et al. (2005), Van Stel and Storey (2004), Acs and Armington (2004) are representative of this kind of study.

Most of the empirical studies mentioned found a positive and significant relationship between entrepreneurship (measured by firm creation) and regional economic development (measure of GDP growth or employment growth). As a result, a common policy recommendation in the studies is the support for the entrepreneurship activity, whether by creating a proper environment for the existing business, or by stimulating the continuous creation of new firms. These studies make no mention, however, of the challenge and effort needed to create a new venture, and most important, the strategies, management needs and challenges that would lead this new venture to a successful path and, consequently, make an effective contribution to growth and job generation. The implicit assumption is that a 'normal' number of businesses will succeed and the most important task is to keep the flow of entrepreneurs opening new businesses. Paradoxically, then, although these new economic approaches rescue and emphasize the strategic role of entrepreneurship, for socio-economic systems and for economic development they do not consider important or relevant, or maybe even possible, the need to focus on the entrepreneur.

The Management Literature

The evolution of theoretical and empirical studies relevant to the management literature experienced some important shifts in the last 30 years or so. The first wave of studies focused on the individual entrepreneur. The basic research questions were: (1) who is the entrepreneur? and (2) in which ways does the entrepreneur have a different behavior from non-entrepreneurs? This approach produced a vast literature that discussed such aspects as the identification of characteristics and personality traits of the entrepreneur (McClelland, 1967; Hornaday and Aboud, 1971; Carland et al., 1984; Timmons, 1989); taxonomy of entrepreneurs (Vesper, 1979; Schollhammer, 1982; Cooper and Dunkelberg, 1986; Miner, 1996); and social, cultural and economic conditions favorable to the development of potential entrepreneurs (McClelland, 1967; Kunkel, 1971; Grilo and Thurik, 2005).

The early study of Miller (1983) on entrepreneurship proposed a radical shift in the approach to this theme that influenced most of the literature and research afterwards. He proposed that instead of focusing on traits and abilities of the 'Entrepreneur', usually a dominant organizational personality or the owner-manager who makes the strategic decisions for the firm, the emphasis would shift to the entrepreneurial activity of the firm. That is, a change that should shift the focus from the individual entrepreneur to the process of entrepreneurship itself and the organizational factors which foster and impede it. He then suggested that an entrepreneurial behavior is one that 'engages in product market innovation, undertakes somewhat risky ventures, and is *first* to come up with "proactive" innovations, beating competitors to the punch'[2] (Miller, 1983, p. 771). From Miller's (1983) definition of entrepreneurial behavior, three salient dimensions emerge: innovativeness, proactiveness and risk taking. Covin and Slevin (1989) referred to it as strategic posture but also defined it in terms of innovation, proactiveness and risk taking. Zahra and Covin (1995, p. 44) used the concept of corporate entrepreneurship and suggested that 'it provides a potential means for revitalizing established companies. This is accomplished through risk taking, innovation, and proactive competitive behaviors'.

Lumpkin and Dess (1996) made an impressive effort to consolidate the most acknowledged theoretical and empirical contributions to that time and presented the challenge of linking this entrepreneurial behavior with firm performance. Lumpkin and Dess started by proposing a more comprehensive definition of the entrepreneurial processes within organizations, adding to the original dimensions of innovativeness, proactiveness and risk taking proposed by Miller (1983), two others: autonomy and competitive aggressiveness. The organizational behavior characterized by these five dimensions was called entrepreneurial orientation (EO) by Lumpkin and Dess (1996), a definition that was subsequently adopted in most of the studies in the entrepreneurship field (Aloulou and Fayolle, 2005; Covin et al., 2006; Wiklund and Shepherd, 2005).

A second distinctive shift in the management literature on entrepreneurship was introduced by Shane and Venkataraman (2000), when they proposed that the domain of entrepreneurship is the nexus of two phenomena: the presence of lucrative opportunities and the presence of enterprising individuals. Accordingly, the fundamental role of entrepreneurs is the discovery and exploitation of entrepreneurial opportunities. A significant stream of theoretical and empirical studies followed Shane and Venkataraman's (2000) study. These ranged from developing and testing operational measures for the opportunity recognition construct (Brown et al., 2001), identifying the factors that affect the ability of entrepreneurs to recognize and exploit opportunities (Shepherd and DeTienne, 2005) and developing taxonomies for the opportunity recognition process (Alsos and Kaikkonen, 2004). Also, some authors questioned an apparent dichotomy between the identification and exploitation phases of the opportunity recognition process and proposed the notion of opportunity development, as a continuous and proactive process (Ardichivili et al., 2003). Using the same line of reasoning, some studies identified another dichotomy, between the notion that opportunities are objective (and should then be discovered) and the notion that they are subjective (and must then be enacted or created by the entrepreneur). According to these studies, the organizational learning capability could provide a way to reconcile these different views of the opportunity recognition process (Dutta and Crossan, 2005; Lumpkin and Lichtenstein, 2005).

A more recent strand of literature challenged the traditional approaches of the entrepreneurship theoretical framework. According to these studies, most of the extant literature on entrepreneurship takes a theoretical and methodological stance based on the functional-positivist paradigm, which leads to 'numerous dichotomies in entrepreneurial research and irreconcilable differences in the nature of entrepreneurship: independence vs. dependence, process vs. personal attributes, revolution vs. evolution, vision vs. action, and social vs. business orientations' (Chiasson and Saunders, 2005, p. 749). Alternative theoretical and methodological frameworks, based on different paradigms, especially the interpretive paradigm, are suggested to deal and reconcile these differences. A distinctive contribution was made by Sarasvathy (2001), who proposed a theoretical shift from economic inevitability to entrepreneurial contingency, contrasting the constructs of causation and effectuation. The latter is an alternative process to the formation and development of the entrepreneurial firm, based on the notions that pre-existent goals should be challenged in the decision making process (March, 1982), the evidence against planning and prediction in favor of synthesis and action suggested by Mintzberg (1994), and enactment and retrospective sense-making (Weick, 1979).

In all these theoretical and empirical contributions, it is clear that the other side of the paradox existing in the entrepreneurship literature initially stated that the focus is on the individual entrepreneur, on the firm performance, and the internal and external factors that explain firm creation and performance. The aggregate impact of the entrepreneurial action on local and regional development is not a considered issue. Implicit is the notion that this impact is obvious, taken for granted. As mentioned earlier, a few exceptions are the studies on the formation of cluster technology, where the emphasis is on identifying the external (environmental) conditions that enable the development of these company agglomerations, instead of examining the impact of the cluster emergence on the regional conditions (Eisenhardt and Schoonhoven, 1990). Other studies examine the strategies of firms in developing countries, especially in high-technology industries, and how they succeed in local and global markets, as well as the role of governmental policies to support their strategies (Park and Bae, 2004).

As mentioned in the introductory section, a more explicit contribution in linking regional transformation with a purposive entrepreneurship strategy is offered by Venkataraman (2004), who discusses the threat for a region of becoming trapped between a virtuous and vicious cycle pendulum movement. According to Venkataraman (2004, p. 158):

> A region exists in a state of 'virtuous equilibrium' when it has been conducting economic and cultural activities for long periods and has settled into a predictable and comfortable position. Such a state exists when patterns of activity have formed and evolved through historical and local contingencies and through ceaseless competition. . . . Indeed, in these regions, equilibrium is defined as much by product-market competition as it is by social and political competition.

This competition process yields a limited number of very successful institutions (firms, government organizations or social organizations in the region). These institutions, which are the survivors of competitive selection, practice repeatedly certain cultural and economic activities in the same way, contributing to the establishment over time of certain normative behaviors and values. It is through the continual maintenance of these norms that the 'virtuous circle' becomes established. In this 'virtuous circle', the leaders of these institutions become role models, leading the people from the region who aspire to be like them. According to Venkataraman (2004), such role models may be bureaucrats, politicians or leaders and managers in certain admired companies. Talented people are attracted to these admired positions, and the critical resources in the region's economy flow toward the talent, feeding this type of sustaining entrepreneurship (Christensen, 1997) or the 'weak entrepreneurial force' (Venkataraman, 1997). That is, the entrepreneurial force that exists in the region is directed to activities that support the existing and successful firms in the region. In this environment experimentation should be considered extraordinary, and if it fails, the failure is associated to a deviation from the established model. Intolerance to failure and a deal of perfection then emerges, closing the 'virtuous circle'.

According to Venkataraman, attempts to break with this equilibrium and foster technological entrepreneurship as a way to transform and develop the region are limited by a culture that does not support bold bets, which is implied in technological entrepreneurship. This is when a region becomes trapped in a vicious cycle: new ideas and bold bets cannot and do not emerge. Along this circle, the fear of failure will lead to a prevalence of 'push' rather than 'pull' entrepreneurship, resulting in the emergence of low quality firms and low status for entrepreneurship, negatively reinforcing the lack of a culture of entrepreneurship. In such an environment a poor deal flow is produced, which fails to attract risk capital than close the circle.

Governmental policies directed at breaking this vicious circle, observed by Venkataraman, rely on single solutions, and usually emphasize the injection of risk capital in the region's economy. There is a great probability that these funds will be directed to low quality entrepreneurs, reinforcing instead of breaking the vicious circle. According to Venkataraman (2004), the ineffectiveness of these policies results from the single solution approach adopted, which relies in what he calls the 'necessary' condition for technical entrepreneurship, that is, the 'risk capital' tangible factor. Venkataraman (2004) suggests that an effective approach to break with this vicious circle should consider multiple and simultaneous solutions and rely on 'intangible' or 'sufficient' factors.

THE CASE OF THE FURNITURE MANUFACTURING CLUSTER OF ARAPONGAS AND THE DISTINCTIVE ROLE OF AN 'ANCHOR' ENTREPRENEUR

In the following sections we shall present the events and processes that lead to the emergence of an important cluster of furniture manufacturing in a medium sized city in the countryside of Brazil. The process, as shall be argued, resulted in intense and purposeful entrepreneurship, led by an 'anchor' entrepreneur and his pioneering company, whose history shall also be described and analysed. With this analysis we hope to make a useful contribution in linking the actions and decisions of private and individual entrepreneurs to the development and transformation of a significant region. The case study shall be used to illustrate the extent to which entrepreneurship in mature sectors can also play an important role to foster regional transformation, expanding the notion proposed by Venkataraman (2004) of associated regional transformation to technological entrepreneurship. Furthermore the case study shall be used to illustrate the extent to which the intangible factors defined by Venkataraman (2004) are a strategic tool to promote regional transformation through entrepreneurship in mature (or medium technology) sectors.

As a subsidiary or complementary research issue, we shall use the case study of the 'anchor' entrepreneur to illustrate the extent to which the strategy and decisions of the entrepreneur along his trajectory find support in the four principles of the 'effectuation theory' proposed by Sarasvathy (2001).

The furniture manufacturing cluster which is the subject of the present case study is located in Arapongas, a medium sized city in the state of Paraná in the southern region of Brazil. The city currently has a population of approximately 90 000 inhabitants. The city is considered to be in the countryside since it is significantly distant from the main economic centers of the country. The city is approximately 450 km from the state capital (Curitiba), 600 km from the city of São Paulo, the most important Brazilian city, and more than 1000 km from Rio de Janeiro, the second biggest Brazilian city.

The furniture manufacturing cluster comprises approximately 150 companies, all family firms, in an area of 1.5 million square meters. The companies in the cluster have annual revenues of more than US$ 800 million, of which approximately US$ 40 million come from exports to more than 40 countries in Latin America, the Caribbean, North America, Europe and the Middle East. None of them existed 35 years ago. In this time span, a small number of local micro- and small companies, involved in the production of very primitive furniture pieces for the local market, managed to become the second largest furniture manufacturing cluster in Brazil, radically transforming the region by changing the economic base from the agricultural sector, based on coffee plantations, to a very densely industrialized region. In this process the role of local entrepreneurship was central and determining in particular, as we shall see, the actions and decisions of one entrepreneur, the owner and executive of one of the pioneers and presently the biggest company in the cluster, the MOVAL company.

The MOVAL company was, less than 40 years ago, literally, a backyard furniture maker, employing less than ten people, producing 600 pieces of solid wood furniture per month, using artisan processes and grossing around US$ 100 000 per year. In less than 40 years the company became the second largest furniture maker in Brazil, with more than 1000 employees, producing 50 000 pieces of furniture per month, with a gross revenue of around US$ 100 000 per year and exporting 15 per cent of its production to more than 30 countries. The emphasis on the role and actions of a particular entrepreneur is consistent with Sarasvathy's (2004) recommendation that the entrepreneurship research should refocus attention to the agency role of the entrepreneur: 'The first thing that leaps out at us when we examine the phrase "Making it happen" is the necessity of agency – the idea that "it", whatever it might be, might not "happen" if it were not for someone making it happen' (Sarasvathy, 2004, p. 520).

In the rest of this chapter we shall identify and analyse the trajectory and growth of this particular entrepreneur and his company, and how its role was a determinant in the emergence, consolidation and success of the cluster, and how the emergence of the cluster promoted the economic transformation and development of the region.

RESEARCH METHOD

According to the taxonomy proposed by Miller and Friesen (1982), the present study falls in their type 1 category: 'Longitudinal, broadly focused, and non-quantitative studies of single organizations'. They suggested that 'perhaps the most common longitudinal studies are those which have been performed upon individual organizations'. Through such studies important organization information, such as changes in size, management, product market strategy, competition and markets are documented, and often many details are obtained about the rationales for key decisions. Miller and Friesen (1982) argue that a prime strength of all type 1 studies is that they provide a basis for real insights into how organizations make decisions, adapt to their environments, enact new environments and restructure themselves. This kind of study provides the researcher with a wealth of detail on the sequences of decisions and events which affords much knowledge on the time priority of change in variables of strategy, structure, environment, decision making methods and leadership style and personality.

They also remind us about the weaknesses in these studies. First, they do not allow generalizations since they are built upon a sample of one. Second, it is difficult to overcome this problem by comparing different studies, since the authors usually have dealt with different kinds of processes, attributes and variables. So, according to Miller and Friesen (1982), the net result of combining or comparing many studies is a better intuitive feel for how organizations operate, which cannot be easily translated into a testable or firmly grounded set of hypotheses. Finally, the purely historical studies tend not to be very systematic or analytical and do not generate or test hypotheses. An alternative to minimize these limitations and to ensure the general qualities of the findings, according to Pettigrew (1973), is to give the study more theoretical content, not only using pre-existing theory to explain the research setting, but also to attempt to develop a general theoretical approach that could be applied in other studies of social process.

The case choice in the present study was not random. It was a judgment sample, a type of purposive sampling used in exploratory research in which the researcher selects a sample to meet specific criteria (Dess et al., 1997). The selection of the specific cluster for study in this chapter aimed to meet the following criteria: (1) the economic importance for the specific region and for the country as a whole; (2) the emergence and development of the cluster in a relatively short time span, which could allow for a more reliable historical account, preferably with actors that were involved in all phases of the processes; (3) the cluster should preferably be from a mature industry (as opposed to an emergent industry), the products of which were not of high-technology content (the purpose of this condition is to allow for a more comprehensive replication of the analysis for other cases in Brazil or abroad); and (4) the possibility of full and continued access to relevant actors (entrepreneurs, local political officials and other stakeholders). The selection of the MOVAL owner as an individual entrepreneur to be researched followed the same criteria, and moreover, the undisputable fact that he was considered the inspiring influence and the most influential entrepreneur in the emergence and development of the furniture cluster (we refer to him and his company as the cluster's 'anchor' entrepreneur).

The field research was conducted over two years, between 2004 and 2006, and consisted of several rounds of detailed interviews with the owner and founder of MOVAL and all the most ranked managers. Some interviews were unstructured and others followed a check list of answers connected to the variables we were investigating (environmental conditions, strategy, strategy making process, structure at different points in time). Considering that we had no intention of conducting a quantitative analysis, we preferred this alternative to the formal application of questionnaires. For comprehensive understanding and information on the development process of the furniture cluster, local personalities, such as politicians, public officials and five other business owners were also interviewed, since we were interested in the role of different stakeholders. For the same reason representatives of some of the most important company's suppliers and costumers were also interviewed. The data gathering were complemented with company's documents, such as financial reports and annual reports, as well as history books written about the city and the local industrial sector, and the legislation of the city council concerned to policies of support to the cluster.

ANALYTICAL FRAMEWORK

For the analysis of the present case study we defined two different, but interrelated levels of analysis. The first, and more specific, is the individual entrepreneur level. In our case the entrepreneur is the owner and founder of MOVAL, referred to here as the 'anchor' entrepreneur. The second level of analysis is the furniture cluster. In this case we shall study the factors, or institutional drivers, associated to its trajectory, because of the emergence to the consolidation, as well as to its decisive role in positively transforming economic and entrepreneurship culture in the region.

For a historical account of the emergence and development of the 'anchor' company, MOVAL, as well as the furniture cluster as a whole, we shall use an event driven approach, based on process theory (Van de Ven et al., 2004). We shall identify key events in the history of the 'anchor' company and entrepreneur, and that of the cluster, and discuss the outcomes associated with these events. Van de Ven et al. remind us that the event driven process approach implies the following:

- Entities participate in events and may change over time
- Time ordering of independent events is critical
- Explanations are layered and incorporate both immediate and distal causation
- Entities, attributes and events may change in meaning over time
- Generality depends on versatility across cases.

Analytical Framework for the 'Anchor' Entrepreneur Trajectory and Role

For the emergence process of the 'anchor' company, MOVAL, we shall use the effectuation theory proposed by Sarasvathy (2001) as a framework of analysis. As mentioned earlier, the case study will not be used to test hypotheses on the effectuation theory, but to illustrate to what extent the concepts and categories that comprise the theory, especially the four principles that constitutes the baseline of the theory, are operational and could explain the processes and decisions of the entrepreneur's trajectory.

We have so far referred to the emergence of the 'anchor' company and of the cluster, and not to the creation of it. This is not a semantic choice. We take the stance that the whole process of initiating and developing a firm, or a collective institution, like a manufacturing cluster of companies, is one of emergence in which the actions of the entrepreneur (or agents in general) are in a constant relationship with resources and structures (environments) and co-evolve in this process. So, even if there is a formal event of the legal creation of a company or of an association of business people, the consolidation of the phenomenon (a firm or an institution) will be done gradually, through an emergence process (Lichtenstein et al., 2006; Sarason et al., 2006). As in the effectuation theory framework, this process is essentially one of exploiting contingencies instead of exploiting knowledge (Sarasvathy, 2001).

We shall confront the key events in the emergence process and development of the 'anchor' company, MOVAL, to the four principles that form the core of effectuation theory, as proposed by Sarasvathy (2001):

1. Affordable loss rather than expected returns: as decision making criteria: effectuation predetermines how much loss is affordable and focuses on experimenting with as many strategies as possible with the given limited means. The effectuator prefers options that create more options in the future over those that maximize returns in the present.
2. Strategic alliances rather than competitive analysis: causation models emphasize detailed competitive analyses. Effectuation emphasizes strategic alliances and precommitments from stakeholders as a way to reduce and/or eliminate uncertainty and to erect entry barriers.
3. Exploitation of contingencies rather than exploitation of pre-existing knowledge: when a pre-existing knowledge, such as expertise in a particular new technology, forms the source of competitive advantage, causation models might be preferable. Effectuation, however, would be better for exploiting contingencies that arose unexpectedly over time.
4. Controlling an unpredictable future rather than predicting an uncertain one: causation processes focus on the predictable aspects of an uncertain future. The logic for using causation processes is that to the extent we can predict the future, we can control it. Effectuation, however, focuses on the controllable aspects of an unpredictable future. The logic for using effectuation processes is that to the extent we can control the future, we do not need to predict it.

Analytical Framework for the Cluster Trajectory and Role

In analysing the emergence and development of the furniture manufacturing cluster and its impact on the local and regional economy, we shall consider three different levels of analysis: the economic environment conditions, the institutions and the intangible factors proposed by Venkataraman (2004).

The Economic Environment Conditions

The environment conditions prevailing in the country has, of course, an impact on the outcome of entrepreneurial ventures and can foster or impede it. So, in order to get a more precise evaluation of the impact of the other factors on these ventures, it is important to control (or at least identify) the conditions prevailing in the environment in which the entrepreneur is inserted.

We shall measure the prevailing environment conditions using the definition adopted by Covin and Slevin (1989). They defined two polar situations for the environment: the benign environment and the hostile environment. The benign environment was defined by the prevalence of the following conditions: very safe, little threat to the survival and wellbeing of the firm and rich in investment and marketing opportunities. The hostile environment is characterized, according to Covin and Slevin (1989), as very risky, stressful and exacting, in which it is very difficult for the firm to keep afloat and a false step can mean the firm's demise.

Based on the retrospective evaluation of the entrepreneurs and other stakeholders interviewed, and the evaluation of the authors in considering the macro-economic indicators prevailing at the time of the events defined along the cluster trajectory, the economic environment condition was qualitatively classified from intensively hostile (− − −) to intensively benign (+ + +).

Local Institutions

Venkataraman (2004) points out the importance of local institutions as mechanisms that can either foster or impede entrepreneurship. In the present case study we identified five institutions that played different roles in the process of the regional transformation at different points in time.

Cultural, social and economic models

Venkataraman (2004) observed that 'a region exists in a state of "virtuous equilibrium" when it has been conducting economic and cultural activities for long periods and has settled into a predictable and comfortable position' (p. 158). This was, to a large extent, the case of the city of Arapongas. For this situation to change, as we shall see, other institutions had to be developed in order to replace or modernize the existing ones.

Public policies

The mayor that was in office in the 1965–70 period was known as a very energetic and visionary politician. He foresaw the city industrialization as a way to face the economic instability and weak economic dynamic associated with the agricultural economic base and as a trigger to stimulate changes in cultural and social values. His effective contribution was the approval of a law creating the industrial district of Arapongas. To foster industrialization City Hall segregated a vast portion of land and provided the basic physical infrastructure.

The entrepreneurs interested in the city would have land in the industrial district donated by City Hall and, in addition, would get tax exemption from local taxes for a period of three to ten years, depending on the amount of investment. We shall argue that this decision proved to be very effective in the end. But in the beginning it aimed at something else: to attract big companies from other cities and countries, benefiting from a dynamic environment prevailing in the country at that time, which was experiencing a fast and intense process of industrialization. It is also interesting to note that the mayor's initiative was associated with providing what Venkataraman (2004) called, the tangible (or necessary) factors, in this case physical infrastructure. As we shall see, this decision alone, without the emergence of the intangible factors, would be ineffective.

The 'anchor' entrepreneur

The role of the 'anchor' entrepreneur transcended that of a private entrepreneur aiming to making a profit with that business and assumed an institutional role in the process of transforming the region of Arapongas. Not only did he contribute decisively to the creation of other institutions that were crucial for the process, but he also became a role model and, as we shall show, the executive leadership that played a crucial role in fostering entrepreneurship and allowing the regional transformation to take place.

The Furniture Makers Association

As mentioned before, in 1978, ten years after the creation of MOVAL and six years after the decision to build new facilities in the industrial district, the 'anchor' entrepreneur played a decisive role in the creation of the Arapongas Furniture Makers Association, being their first chairman. The formal objective of the association, at his inception, was to carry out technical and political actions that could benefit the companies of the cluster collectively, such as employee training, technical courses for prospective workers, sharing information on suppliers and coordination of actions before public authorities.

The Furniture Trade Show

One of the most important achievements of the Furniture Makers Association was the establishment of the National Furniture Makers Trade Show, which has taken place in the city of Arapongas every odd year, since 1997. The trade show congregates most of the local industries and is open to customers (wholesale and retail traders) from all around the country. The trade show takes place in the Arapongas Furniture Trade Show pavilion, a building of 40 000 square meters, built by the initiative and with funds of the association, once again a pioneer idea of the 'anchor' entrepreneur. The project of building the trade show pavilion was decided in 1991 and took five years to be completed. So, in February 1997, the first Furniture Makers Trade Show took place. The project expanded and included a Trade Show for Equipment, Raw Materials and Implements for the Furniture Industry, which takes place in every even year.

These five institutional drivers were confronted with the key events of cluster trajectory and they were, in the same fashion of the economic environment conditions, qualitatively classified in terms of their negative (− − − to −) or positive (+ to + + +) impact on that trajectory.

The Intangible Factors

Venkataraman (2004), identifies a set of seven intangible factors for improving regional technological entrepreneurship. In the following section we present these seven intangible factors with a brief definition or description extracted from the discussion of each intangible factor developed by Venkataraman (2004).

1. Focal points capable of producing novel ideas. Novel ideas originate from bright and knowledgeable individuals. And these knowledgeable individuals are often in the neighborhood of their region's great institutions: this is where talent congregates and where ideas are produced.
2. The need for the right role models. If risk capital is combined with novel ideas the result will be success for a few people. They become the new role models who show their peers that entrepreneurial success is not a theory.
3. The need for informal forums of entrepreneurship. Access to role models mostly occurs in informal forums. Information, stories and celebrations about real entrepreneurship rarely happens in company offices or in routine jobs. Informal forums are necessary to discuss the trench wisdom that is required to execute these ideas.
4. The need for region-specific ideas to be created. This intangible is related to idiosyncratic value. Sustained success often comes when it is

based on some idiosyncratic or special ingredient that the regions have to offer the world. Such idiosyncrasy may be based on the region's core competence, natural resource or some other source of idiosyncratic advantage.

5. The need for safety nets. Attempts at novelty are always accompanied by failure. Unless there are mechanisms and institutions that address these failures, new trials will dry up. The need is for safety nets for entrepreneurs who may fail in their attempts to create something new. These may take the form of a job in an existing company, sometimes even better than the one the entrepreneur had before they embarked on an entrepreneurial venture. In addition, in these places, people do not attach stigma to a failed business and even if a venture was not successful for an individual, they still retain human and social capital.
6. The need for gateways to large markets. Unless easy access to larger markets are provided to aspiring and potential entrepreneurs in regions that do not have size and density advantages, both the quantity and quality of new enterprises will be affected. Not only the physical infrastructure is important and necessary to provide such access to larger markets. Even more vital for this aim is the intangible social network infrastructure. In addition to social and economic ties, the creativity with which a region provides access to product and financial markets will make a difference to its success. According to Venkataraman (2004), this brings us to the final intangible.
7. The need for executive leadership. The executive leadership is the kind that roll up their sleeves and does the hard work. This is the kind of leadership that will do the necessary work to ensure that potential entrepreneurs have (1) access to institutions that produce new knowledge; (2) access to risk capital; (3) access to the right role models; (4) the necessary informal forums for entrepreneurial education and experience; (5) the necessary safety nets and the culture of accepting failure; and (6) access to gateway cities and large markets for their products and services. It is clear from the definition proposed by Venkataraman (2004) that this is a focal and essential intangible factor. In fact it is the presence and effectiveness of the executive leadership that will ensure that the other intangibles will be developed and work. They will also play a central role in the region to solve vicious cycle problems.

In the case of the intangible factors, instead of measuring the intensity of the factors, as in the previous levels of analysis, we shall identify which of the intangibles emerged in the different key events, and the institutional drivers that allowed them to emerge. The illustration of whether these factors were present in the process of the emergence and development of the furniture cluster in Arapongas, Brazil, is particularly interesting since it is not a technological cluster. We could then suggest that Venkataraman's instrument can be used to foster regional entrepreneurship, based on even more traditional and mature industries, in what is an alternative and more accessible instrument for developing countries, in which the scarcity of strategic resources, such as risk capital, technology and skilled labor, is more severe.

THE DATA AND RESULTS

In this section we shall present the information obtained in the field research, using the analytical framework discussed in the previous section and following the different levels of analysis defined in that section.

The MOVAL Trajectory: Mapping the Key Events

The MOVAL company was founded in 1968 and is ranked today as the second largest Brazilian furniture maker. It must be said that the selection was not random. One of the authors had previous contacts and knowledge with the owners and with some of the company's managers, as an instructor of the Executive MBA Programme, which facilitated access to the company and a degree of openness and depth that was necessary for the research.

The company is family owned and its shares are equally owned by two brothers. The company was founded by a merger of two small companies inherited by the two brothers: a backyard furniture manufactory and a small home appliances and furniture retail shop. In 1968 their father decided to step down and donate the business to the sons, and the elder brother become co-owner and general manager. He was, at the time, 26 years old and the younger brother was 18 years older. At that time, the furniture manufactory was, literally, a backyard manufactory, making around 600 pieces a month of low cost furniture (basically solid wood pine pieces, such as tables, chairs, wardrobes and kitchen cabinets). The company had ten employees and the revenue was no more than US$ 100 000 per year.

The new owner-executive was determined to change the business and make it grow. Through informal research with customers he knew that they were demanding more sophisticated furniture products, produced at that time by manufacturers from bigger cities in the state of Paraná and from other states. The problem was that he didn't have the products, nor the knowledge to make them, nor a known brand, nor capital for such a chance. He solved the problem with three bold moves. First, he hired an experienced and retired furniture plant manager that worked for an important company in the state. Second, he hired a very experienced sales man who represented, at that time, one of the most important furniture makers in the country, offering him twice the fee he previously earned in the other company. Third, he used his own previous experience as deputy manager of a local branch of an important wholesale chain to broaden his base of suppliers and to extend his credit with them.

After one year the company had changed the production line to more well finished four-piece bedroom sets, with greater added valued, and had extended its market from its home town to more than 30 cities in the states of Paraná, São Paulo and Mato Grosso do Sul. The production rose from 600 pieces to 6000 pieces a month, with greater added value. The gross revenue rose to US$ 1 000 000 a year, and the number of employees to 30.

In the following sections we change to a more itemized description linking the company to specific events.

1970

The company makes two important moves: a radical change in the product technology, replacing solid wood as the basic raw material with a combination of fiberboard and veneer finishing which significantly improved the quality of the furniture and the company efficiency and productivity. At the same time the owner-executive decided to integrate some activities, buying a fleet of trucks to deliver the furniture to customers, thus becoming independent of outsourced transportation that was more expensive and scarce at that time, and eventually leading to delays in delivering the products as well as product damage. The production increased from 6000 pieces to 10 000 pieces a month, the gross revenue rose from US$ 1 000 000 to US$ 1 500 000, and the number of employees increased to 35.

1972

The company moves to new facilities in the city's industrial district, using land donated by City Hall. Another radical change was introduced in the production processes with the import of new equipment from Germany. In two years the production increased five-fold from 10 000 pieces a month to 50 000 pieces a month. The market base continues to extend to other Brazilian states.

1978

As a result of the efforts and initiative of the owner-executive of MOVAL, the Arapongas Furniture Manufacturers Association was created. The owner-executive of MOVAL was the inspiration and first president of this association. At this time the city had around 40 companies in the industry. The purpose of the association was to promote collective actions, such as employee training and technical courses for prospective workers, as well as sharing information on suppliers, coordinated actions before public authorities and the like.

1980

New radical changes in product technology and process. The firm adopts a new synthetic component, called MDF (Medium Density Fiberboard), and promotes a radical automation in the production lines. These changes allow a significant increase in production volume and efficiency. At the end of the 1980s the production is approaching 200 000 pieces per month. The firm intensively explores a low cost leadership strategy. Its market expands to the whole country.

1988

A radical threat to the firm's survival. Pressed by a turbulent, heterogeneous and hostile environment, the firm is suffocated by a financial liquidity shortcoming and decided to file for the Brazilian Bankrupt Protection Law. The financial problems were aggravated by the hyper-inflation period in the country leading the firm to a mismatch between the credit received from suppliers and the credit granted for customers. Operating in an economic environment in which the inflation reached rates as high as 80 per cent a month, in a few months liquidity was drained and the company's survival was threatened. The problem occurred and was aggravated by the ineffectiveness of some aspects of the company structure, especially the financial information systems. At that time, the company had an efficient and sophisticated manufacturing operation, a sharp entrepreneurial orientation expressed by constant process innovation, aggressive competitiveness and risk taking propensity, but failed to adjust its formal structures to a more complex mode of operation and to a more threatening environment.

In six months the company balanced its financial flows, recovered its working capital and lifted the bankrupt protection. This was a bold move, since the law ensured a protection (renegotiation of suppliers and government debts for two years under special conditions) and at that time this meant an important capital inflow. The Bankrupt Protection Law allowed all the debt to be repaid over two years with an 'official interest rate' significantly below the market rates. The view of the owner-executive was, however, that he could not profit from this situation to the detriment of his suppliers who had been, for many years, an important support of the company's growth strategy. This move restored the trust between the company and its suppliers and allowed it to operate normally and to grow again in no time. The firm initiated an effort for modernization, improvement and formalization of its basic information systems.

1992

A subsidiary was created, named IRMOL, to manufacture more popular furniture aimed at low income classes.

1994

The company began exporting, initially to Mercosur countries, and in the following years to countries of the Caribbean, Mexico, Portugal, Italy and Spain.

1996

New technology change. The veneer finishing was replaced by paper finishing and painting, improving efficiency and reducing costs.

2006

In the last decade the company has kept a more stable innovation strategy. The technology is updated every two years. In terms of product market strategy the company consolidated its cost leadership strategy and it is considering new moves into more value added markets. In this direction one strategic goal is to increase exports. Presently the company exports around 15 per cent of its production and the goal is to reach 25 per cent. The company produces 50 000 pieces a month, its gross revenue is US$ 100 000 000 (approximately) and it has around 1000 employees.

The Principles of the Effectuation Theory in the MOVAL Trajectory

In Table 2.1 we examine the extent to which the events of the MOVAL trajectory described above can be related to and eventually explained by the four principles of the effectuation theory proposed by Sarasvathy (2001). As mentioned earlier, the purpose of this part of the analysis is to illustrate, with this specific case study, how these principles may better explain the

Table 2.1 How the key events of the MOVAL trajectory correlate with the principles of effectuation theory

Period/Events	Principles of the effectuation theory underlying the entrepreneur's decisions/strategies	Information on the MOVAL trajectory that illustrates how the principle was considered
1968 Birth	P1: Affordable loss as decision criteria	As mentioned by the founder entrepreneur, the decision to create the firm was not preceded by a careful analysis of markets, or by a feasibility evaluation; they decided to invest an amount of money, and they were aware that this money could be entirely lost if the enterprise failed
	P2: Strategic alliances	The alliances with the sales representative and with suppliers were essential to make the start-up viable, since it solved two problems: access to new customers, provided by the sales representative and scarcity of funds, complemented with credit lines granted by suppliers
	P4: Controlling an unpredictable future	A consequence of P1 + P2: since the entrepreneur was not willing or capable to predict the future through elaborated planning or market research, he made decisions that would minimize the risks, enhance the chances of success and to a certain extent, craft his own future
1970 Change technology	P1 + P2 + P4	Same as above. Additionally, the decision to import new equipment resulted from alliances with international equipment suppliers and reinforced the commitment to control the future, since it would improve dramatically the company's productivity, enhancing its capacity to survive and grow in the future
1972 New facilities/ improve technology	P1 + P2 + P4 + P3: Exploiting contingencies	Same as above. Additionally, the decision to build and move to new facilities characterized an opportunity to exploit the contingency of benefiting

Table 2.1 (continued)

Period/Events	Principles of the effectuation theory underlying the entrepreneur's decisions/strategies	Information on the MOVAL trajectory that illustrates how the principle was considered
		from the support that City Hall was granting for the companies to move to the industrial district
1978 Leadership in creating the Furniture Makers Association	P2 + P3 + P4	This event is obviously related to the formation and strengthening of strategic alliances (P2), and makes an additional contribution to control the future. Furthermore it was a decision/strategy that explored two contingencies: the exhausting of a benign environment and the risk of predatory competition. The association was an important step in preventing these two contingencies to harm the companies and the cluster irremediably
1980 Change technology/ product and process	P1	The decision to invest in new technology follows the usual criteria: the money invested was an 'affordable loss'
	P2	The decision was made possible through alliances with suppliers of equipment and raw material
	P3	The growing hostile economic environment was a contingency to be faced with improving productivity and expanding markets
1988 file for bankrupt protection	P1	The decision was based on the lesser of two evils: file for bankrupt protection could harm the company's image; not file could make the company close the doors
	P3	Self-explaining: the aggravated financial problems created a contingency that needed to be faced with the instruments available
	P4	Self-explaining: either the companies file for bankrupt protection or would

Table 2.1 (continued)

Period/Events	Principles of the effectuation theory underlying the entrepreneur's decisions/strategies	Information on the MOVAL trajectory that illustrates how the principle was considered
		bankrupt effectively. The decision to file for protection was to ensure the company's future
1988 Lift bankrupt protection	P1	Lifting the bankrupt protection had a financial cost in the short run, but restored the public image of the company
	P2	The decision aimed at restoring alliances, especially with suppliers, who were stakeholders more harmed by the bankrupt protection
	P3	The protection provided by the law created conditions for the company to recover in a few months, making it possible to lift the protection
	P4	The entrepreneur considered that restoring the strategic alliances was more important to the company's future than the benefits of using the bankrupt protection extensively
1990 Improve management systems	P3	The availability of software for financial and administrative control and the rapid spread of information technology created the conditions for a radical change in the company's information systems
	P4	The aggravation of the inflationary environment make it clear to the entrepreneur the importance of improving the information systems to ensure the company's survival
1992 Create subsidiary	P1 + P2 + P3 + P4	The decision aimed at exploiting contingencies represented by the growing market for low cost furniture (P3) and to reinforce the ability of the company to control its future by expanding its marketing (P4); the decision, like many

Table 2.1 (continued)

Period/Events	Principles of the effectuation theory underlying the entrepreneur's decisions/strategies	Information on the MOVAL trajectory that illustrates how the principle was considered
		others made by the entrepreneur, was implemented through the intensive exploration of strategic alliances and the decision criteria were, as always, establishing an affordable loss associated with the project
1994 Initiate exports	P1 + P2 + P3 + P4	Same as above, with the difference that the expansion was towards external markets, instead of domestic low income markets
1996 Change technology	P1 + P2 + P3 + P4	Similar to situations described in 1972 and 1980 events
2006 Increase exports/ improve design	P1 + P2 + P3 + P4	Similar to situations described in 1992 and 1994 events

Source: Field research.

process and logic underlying the decisions and strategies adopted by the entrepreneur in different stages of his undertaking.

The information gathered in the case study, and summarized in Table 2.1, yielded, we think, some interesting results. Through the MOVAL case study, we were able to make an empirical illustration of the theory of effectuation proposed by Sarasvathy (2001), and confirmed that the principles that embodied the theory are not only operational, but in the case reported here showed an interesting explanatory power for the actions and decisions of an entrepreneur, not only in the decision of creating a new venture, but also in the very process of managing and growing his venture. In fact, as can be seen in Table 2.1, in every crucial event the decisions of the entrepreneur could always be explained in terms of strategic alliances formed and contingencies exploited.

Also, most of the times it was possible to show that investment decisions, or organizational decisions, were not guided primarily by an expected maximization of return. This was clearly the case when the entrepreneur decided to lift the bankrupt protection 18 months in advance of the time allowed by the Brazilian legislation. If the decision had followed the criteria of maximum return, the MOVAL's owner should have kept the bankrupt protection to completion, since it would have a significant financial benefit with it. Instead the entrepreneur opted to minimize the benefit of the bankrupt legislation, and recover the trust of his stakeholders, especially suppliers, rebuilding and strengthening his strategic alliances. So, he worked a trade-off between an affordable financial loss and the strengthening of his public image and future strategic alliances.

Although we acknowledge that limitations of the research methodology prevent us from making generalizations, the results presented here are, at least, encouraging in terms of the applicability and possibilities of the effectuation theory proposed by Sarasvathy (2001) and invites additional research that could explore and test these possibilities.

The Cluster Trajectory: Mapping the Key Events

The city of Arapongas, in the southern state of Paraná, Brazil, was founded in 1947. Its economic base was formerly, and until 1975, mainly dependent on the agricultural sector, especially coffee production, as was all the northern part of the state of Paraná. At the beginning of the 1970s the city's population was approximately 30 000 people, of which almost 50 per cent lived in the rural areas, working in the agricultural sector as small farmers or hired workers. After a period of rapid growth since its inception, the city was experiencing some rough times in terms of economic growth, due to the dependency of the agricultural sector and the turbulence of the economic sector. High volatility of international prices or frustration of crop production due to more severe winters kept the local economy ranging from periods of intensive growth to periods of recession. The local mayor saw, at that time, the industrialization of the city as an alternative to increase the dynamism of the local economy, and to fight the volatility arising from within the agricultural economy. Thus a municipal law was created in 1966, the Industrial Expansion Plan, to promote the development of the industrial sector in the city. The main instrument for attracting industries to the region was the donation of land to companies interested in investing in the local industrial district, as well as exemption of local taxes for a period ranging from three to ten years, depending on the amount of capital invested.

The initiative had no practical effect in the mayor's first five years of existence, since no local or foreign company decided to move to the industrial district. This began to change radically in 1970, when the MOVAL owner, who was experiencing huge growth for two years in a row as a result of his aggressive strategies, decided to invest in new facilities in the industrial district.

In the following sections we summarize the main events associated with the development of the furniture cluster in the industrial district of Arapongas.

1971

MOVAL began its operations in the new facilities in the industrial district. The initial area of MOVAL estates was 5000 square meters.

1972–75

The success of MOVAL had a lever effect on other small local companies. Either by the inspiring example of a local company being successful in outside markets or by the initiative of the MOVAL owner-executive who permanently supported other entrepreneurs to invest and improve their firms. As a result, in this period some ten new firms had grown sufficiently to be stimulated to move to the new facilities in the industrial sector. Most of these companies were in the furniture industry or similar (like mattresses, upholstered furniture and so on).

1975

The destruction of the coffee plantations in the region. This year was marked by the most severe winter of the century, which led to the destruction of nearly all coffee plantations in the region. As a result, the former local economic base was devastated, and unemployment and migration to the urban areas soared. The up-side of this unhappy event was that the cost of labor decreased, coupled with an abundant supply, and many farmers sold their farms and moved to towns with some capital looking for an opportunity. At that time, the existence of an incipient furniture sector in the city, and the fact that the furniture industry demanded a relatively low level of technology and knowledge, stimulated other entrepreneurs to initiate enterprises in the industry. In three years the number of companies in the industrial city had risen from 10 to 40.

1978

As a result of the efforts and initiative of the owner-executive of MOVAL the Arapongas Furniture Manufacturers Association was created. The owner-executive of MOVAL was the inspiration and first president of this association. At this time the city had around 40 companies in the industry. The association's purpose was to promote collective actions, such as employee training and technical courses for prospective workers, as well as share information on suppliers, coordinated actions before public authorities and the like.

1991

After little more than a decade after the creation of the Furniture Manufacturers Association, and 20 years after the first company (MOVAL) set foot in the industrial district, approximately 100 firms were operating in the cluster, manufacturing all kinds of furniture, and reaching all the 27 Brazilian federation units (26 states and the Federal District). At this time, once again following a suggestion and initiative of the owner-executive of the MOVAL company, the Furniture Association decided to implement an ambitious undertaking: the construction of a pavilion where a national furniture trade show would take place annually. The objective of the project was two-fold: (1) to provide a space and an environment where furniture manufacturers from different furniture clusters could meet, and meet customers of the whole country in one place and time; and (2) to expand and consolidate the brand of the furniture produced in the Arapongas cluster, so far the one with the least tradition in the country.

1997

The pavilion of the National Furniture Trade Show with 40 000 square meters is built and the first National Furniture Trade Show is held. Since then, in every odd year the National Furniture Trade Show is held, and in every even year the International Trade Show for Quality on Equipment, Raw Materials and Implements for the Furniture industry is held.

2000

Sponsored and coordinated by the association and, once again, following a suggestion of the owner-executive of MOVAL, the cluster's company began a project for obtaining an environmental certification (the green stamp) that enables local firms to export to European countries.

2006

The furniture manufacturing cluster comprises approximately 150 all local, family companies in an area of 1.5 million square meters. The companies in the cluster have annual revenues of more than US$ 800 million, of which approximately US$ 40 million comes from exports to more than 40 countries in Latin America, the Caribbean, North America, Europe and the Middle East.

In a time frame of little more than 30 years the city of Arapongas changed its economic base radically from a traditional agricultural region to a sophisticated industrialized city. The population more than tripled. The cluster alone generates 8000 direct jobs, and another 25 000 jobs are indirectly linked to the cluster firms (local suppliers, services and so on). The per capita value added in the industrial sector in Arapongas is almost 50 per cent higher than the most important cities of the state and the local per capita industrial value added is more than 50 per cent higher than the state average. Also, the value added in the industrial sector in the city represents almost 60 per cent of total local value added. Finally, the city has the lowest unemployment rate in the state of Paraná.

THE RELATIONSHIP BETWEEN THE CLUSTER TRAJECTORY AND THE ENTREPRENEURIAL DRIVERS

We previously analysed the cluster trajectory along three different levels: Economic Environment Conditions; Local Institutions (which comprises Cultural, Social and Economic Models, Public Policies, Anchor Entrepreneur, Furniture Makers Association and the Trade Show); and Intangible Factors. These three different levels adopted for the cluster analysis, translate into seven drives that fostered entrepreneurship in different ways, and transformed the region's economic basis. Table 2.2 is used as a template to illustrate how the key events defined and described above correlate with the seven entrepreneurial drivers. In the following analysis we consider as a first step the intangible factor as a 'dependent variable', in the sense that its emergence along the cluster trajectory is explained by the effect of other institutional drivers, especially the local institutions. In a second step we evaluate how the emergence of the intangible factors contributed to the growth and consolidation of the cluster, thus enabling the transformation in the region.

As shown in Table 2.2, the economic environment conditions were classified as significantly hostile when the law creating the Arapongas industrial district was approved. At the time the region was still highly dependent on the agricultural economic base and suffered with the volatility associated with it. Moreover the country was still recovering from a severe economic recession in 1964 and 1965 and the general economic

Table 2.2 The relationship between the entrepreneurial drivers and the cluster trajectory

Period/Event	Economic Environment Conditions	Institutions					Intangible Factors Associated with the Event
		Cultural, Social and Economic Models	Public Policies	'Anchor' Entrepreneur	Furniture Makers Association	Trade Show	
1966: Approval of legislation creating the industrial district	– –	– – –	+				I3: The emergence of informal forums of entrepreneurship I4: Support for region-specific ideas to be created
1971: The 'anchor' entrepreneur is the first to invest in building facilities in the industrial district	+ +	– –	+ +	+ +			The previous ones plus I2: The emergence of a role model to be followed I7: The emergence of an executive leadership
1972–75: Period of slow expansion	+ +	–	+ +	+ + +			The previous ones plus I1: Support for focal points capable of producing novel ideas

Table 2.2 (continued)

Period/Event	Economic Environment Conditions	Institutions					Intangible Factors Associated with the Event
		Cultural, Social and Economic Models	Public Policies	'Anchor' Entrepreneur	Furniture Makers Association	Trade Show	
1975: Devastation of the regional coffee plantations	+ +	+	+ +	+ + +			All the previous ones: I1, I2, I3, I4 and I7
1978: Creation of the Furniture Manufacturers Association	+ + +	+ +	+ +	+ + +			All the previous ones plus I5: Safety nets for displaced entrepreneurs I6: Gateways to large markets
1978–91: Period of rapid expansion	78–80: + + + 81–91: − − −	+ + +	+	+ + +	+ +		The whole set of intangible factors: I1 + I2 + I3 + I4 + I5 +I6 + I7
1992: Decision to build a pavilion for trade shows for the Arapongas furniture	− − −	+ + +	+	+ + +	+ + +		Same as above

1997: The pavilion, owned by the Furniture Manufacturers Association is opened to the public and the first National Furniture Trade Show is held	–	+ + +	+	+ + +	+ + +	+ + +	Same as above
2000: The project for environmental certification is undertaken to enable the cluster firms to export to Europe	+	+ + +	+	+ +	+ + +	+ + +	Same as above
2006: The cluster is the second most important in the country	+	+ + +	+	+ +	+ + +	+ + +	Same as above

Note: The impact of the factors (columns) on the key event considered (rows) range from – – – extremely negative to + + + extremely positive.

Source: Field research.

environment was not favorable for new business, especially in more competitive sectors. From the beginning of the effective occupation of the industrial district, marked by the decision of the MOVAL owner to build a new plant in the district in 1971 to 1978, when there were more than 40 new companies operating and the Furniture Makers Association was created, the environment turned from hostile to significantly benign.

The country grew in these ten years at an average rate of 10 per cent per year, characterizing what was called at the time 'the Brazilian economic miracle'. In such an environment the national market was intensively growing, allowing positive conditions for the start up of business, even in the regions more distant of the dynamics centers of the country, located in the southeastern region. Nevertheless there were not specific policies to support the furniture industry and, as mentioned before, the emergence and consolidation of the Arapongas furniture cluster had to rely on local initiatives at the policy and entrepreneurial level. These conditions lasted at least to 1980.

During the second period, from 1980 to date, the economic environment conditions changed from significantly benign to significantly hostile. In fact 1981 is the landmark of the end of the 'Brazilian economic miracle'. A succession of international events, such as the second oil shock, the interest rate hike in the USA in 1980 and the external debt crisis of the emerging countries at the beginning of the 1980s, led the country to a combination of external debt crisis, fiscal crisis and hyper-inflation that has harmed economic growth since then. During this whole period in only one year the country's GDP growth rate was above 5 per cent, and only in three years above 4 per cent. Moreover the macro-economic policies, faced with the set of crises mentioned above, implied in the rising interest rates and taxes a worsening environment for business activity. Nevertheless the city of Arapongas was able to make the journey of this long and negative period not only with no harm, but with steady and above average economic growth. As we shall argue, the economic transformation occurring in the region, driven by local entrepreneurial forces, was the main factor underlying this particular economic dynamism.

Only in recent years have the overall economic conditions showed some improvement, following the consolidation of the economic stability that was implemented in 1994, and the aggressive growth of Brazilian exports, which benefited the economy as a whole. We see, then, that if in the first years of existence of the cluster, the local companies benefited from a favorable environment, in the period of growth and consolidation of the cluster, the companies faced a hostile environment. As mentioned earlier, in the period from 1980 to more recent years, the country experienced low GDP growth rates, which was in part the result of a very volatile and risky international economic environment and in part the consequence of a sequence of changes in the macro-economic policies aimed at confronting external threats and stabilizing the economy, and was undergoing a period of hyper-inflation. Nevertheless the cluster grew steadily in this period, either in terms of the number of companies, which was counted at around 150 at the end of the 1990s, and in revenue growth, job generation and participation in the regional economy, representing today more than 60 per cent of the regional GDP, confirming a radical change in the economy's region profile. Therefore the growth and consolidation of the cluster could not only be attributed to favorable external conditions, but also, as we shall see, to local forces, entrepreneurship being the most important.

These local forces are represented in Table 2.2 by the five institutions that we defined in the analytical framework: cultural, social and economic models, local public policies, the 'anchor' entrepreneur, the Furniture Makers Association and the trade show. As shown in table 2.2, out of the five institutions defined, only the cultural, social and economic models of the region had a negative impact on the process of creating and developing the furniture cluster and, even so, only at the beginning of this process. As mentioned earlier, the economic base of the region was, from the founding of the city to the early 1970s, dependent on the agricultural sector, especially the coffee plantations. Following this, the role models for the city's talented people were the successful farmers and the prevailing culture stressed conservative values. Thus the most dynamic urban economic activities were connected to coffee industrialization and distribution and were led, almost exclusively, by the owners of the big plantations.

The remaining local entrepreneurship was, mainly, related to local services and commerce, such as retail shops, hotels and a very incipient financial sector. These initiatives were also to some extent indirectly linked to the agricultural economic base and many of them had the owners of coffee plantations, or their relatives, as their founders. As such, much in the way predicted by Venkataraman (2004), the prevailing cultural, social and economic models delayed more dynamic entrepreneurship. Even the creation of the industrial district that was intended to change the economic base of the region and, along with it, its cultural and social models, was not a sufficient factor to promote that change. Moreover the predominant objective of the local mayor, when he created the industrial district, was to attract industries from other regions of the country to the city. This objective reflected, to some extent, the skepticism of the mayor in the probability that the traditional leadership and entrepreneurs that dominated the region and formed its role models had the entrepreneurial drive to promote the effective industrialization and change of the economic base of the region. As suggested by Venkataraman (2004), the industrial district was a necessary but not a sufficient factor for that change to take place. For this purpose other conditions that could enable the emergence of the intangible or sufficient factors should be created.

As shown in Table 2.2, it wasn't until the 'anchor' entrepreneur came into play that the social and cultural models, following the initial changes in the local economic structure, started to change and began to play a positive role. In fact, as was clear from the historical accounts on the cluster development process, the decision of the 'anchor' entrepreneur to move to the industrial district created a different role model to inspire and to be followed. One that was based on risk taking and innovative initiatives and that was not dependent on the traditional economic activities prevailing in the region, or relying on the traditional leadership of the coffee plantation owners' families. Thus the 'anchor' entrepreneur had a positive impact in two directions. On the one hand, he was decisive for the social and cultural models to start to change; on the other hand, his decision to move to the industrial district also had a decisive influence in turning specific local public policy into a more effective instrument for promoting the region's industrialization.

The influence of the 'anchor' entrepreneur, as seen in Table 2.2, not only played a decisive role in influencing the existing institutions, but it was also fundamental for the creation of two other institutions that played a vital role in the process of growth, development and consolidation of the cluster. These were the Furniture Makers Association and the Arapongas Furniture Trade Show. It is interesting to note that these two institutions have a common characteristic. Although they were created following the initiative of the 'anchor' entrepreneur, they were both the result of a collective action of the new 'class' of local entrepreneurs, and were aimed at promoting common benefits and goals. The association, by creating the conditions for the emergence of informal forums for entrepreneurship, providing support for region-specific ideas and a safety net for newcomers, allowed a rapid and intense growth of the cluster, based on the continuous expansion of the market. The trade show, created 15 years later, played a fundamental role in strengthening the local furniture brand and helped local companies, on the one hand, to defend their positions in the domestic market in a time of a hostile economic environment and, on other hand, supported the initiative of the leading companies of the cluster to expand to foreign markets.

In a way it can be said that the collective institutions contributing to the cluster became increasingly independent from the 'anchor' entrepreneur, who receded to a more supportive role in recent years, allowing new leaderships to emerge and collective and cooperative actions to take place, replacing in some cases, and complementing in others, the actions of symbolic figures, such as the 'anchor' entrepreneur. As can be seen in

Table 2.2, in the last 15 years four out of the five local institutions discussed had the maximum positive effectiveness on the cluster trajectory. The only one that decreased its influence were the local public policies – paradoxically, the first to have had a positive influence. This was not necessarily a sign of the ineffectiveness of the policy, but more a consequence of the ability of local entrepreneurs, by acting collectively, to defend their own interests, and create the conditions and new institutions that ensured that the cluster kept growing and faced the threats that emerged in its trajectory.

The final column of Table 2.2 shows the list of intangible factors, as defined by Venkataraman (2004), associated with each key event mapped in the cluster trajectory. It is important to note that, according to the dynamic identified, the intangible factors have a circular and virtuous relationship with the local institutional drivers. Thus the emergence and influence of a particular local institution, for instance, the law creating the industrial district, enabled the emergence of a particular set of intangible factors, for example, the emergence of informal forums for entrepreneurship and support for region-specific ideas to be created. But, the emergence of these two intangible factors also play an important role for other institutions to emerge or change, for instance, the gradual change in the cultural and social models, and the increasing role of the 'anchor' entrepreneur.

The same dynamic is associated with the decision of the 'anchor' entrepreneur to move to the industrial district, assuming the role of an institutional driver in the cluster development process. From that moment on he played the role of the executive leadership, an essential intangible factor, as suggested by Venkataraman (2004) and also emerged as a role model to be followed by other potential entrepreneurs. The emergence of these two additional intangible factors, on the other hand, strengthened the industrial district (a local public policy) and made an additional contribution for the cultural and social models to change. This, in turn, enhanced the conditions for the cluster to take off, and resulted in the emergence of a new intangible factor, that is, the support for focal points (the cluster itself) capable of producing novel ideas.

In a similar way, the creation of the Furniture Makers Association in 1978 was a result of the strengthening and positive influence of the existing institutional drivers and allowed for the remaining two intangible factors, the creation of safety nets and the gateways to large markets to emerge. As shown in Table 2.2, in the following events of the cluster trajectory the presence of the seven intangible factors helped to strengthen and reinforce the local institutional drivers and were, in turn, reinforced by them, which is confirmed by the maximum positive effect of all the local institutional drivers, except the public policies, and by the continuous presence of the seven intangible factors.

The analysis of the cluster trajectory also confirms the strategic and unique role of the executive leadership in ensuring that the other intangible factors emerge and work effectively, and so make the transformation of the region, from a traditional to a more entrepreneurial regime, viable. In fact the executive leadership of the 'anchor' entrepreneur is the third intangible factor identified in the cluster development process. All the other factors that emerged in the region were decisively influenced by actions and decisions of the 'anchor' entrepreneur in his role as the executive leader. Thus he emerges as a role model to be followed by potential entrepreneurs (Intangible 2), making an important contribution for the industrial district to change into a focal point capable of producing novel ideas (Intangible 1), playing a decisive role in the decision to create the Furniture Makers Association and the Furniture Trade Show (Intangibles 3, 5 and 6).

But the analysis of the Arapongas Furniture Makers cluster also shows that although the figure of the executive leadership is strategic and essential, at least in the initial stages of the regional transforming process, the individual's role can be gradually institutionalized, that is, assumed by institutions collectively created in order to ensure that the trajectory and the process can be sustained autonomously once the proper cultural, social and institutional environments are created. In the end this would be the final evidence of the positive transformation that took place in the region, that is, the continuity of the modernization process is no longer dependent of one individual, but relies on local models and institutions that support entrepreneurship.

The dynamic depicted and discussed in the previous paragraphs allows us to conjecture that, although the model proposed by Venkataraman (2004) links regional transformation to technical entrepreneurship, transformation could also take place as a consequence of traditional entrepreneurship, that is, entrepreneurship in mature industries. Moreover the case depicted here illustrates that the transformation process can be more effective if the local institutions create the conditions for the emergence of the intangible factors defined by Venkataraman (2004), which in turn strengthen the local institutions, creating a new kind of virtuous circle: one that enables the region to positively transform itself instead of keeping it trapped in traditional models that perpetuate social and economic backwardness.

CONCLUSION

We mentioned in this chapter's introductory section that there is a curious gap in the entrepreneurship literature. In the economics stream of entrepreneurship literature the central subject is the impact of entrepreneurship on regional economic growth, but no emphasis, or attention at all, is given to the individual entrepreneur, their firms and strategies. The entrepreneur is a 'given' or taken for granted. In the management stream of entrepreneurship literature the central subject is the individual entrepreneur, the process of venture creation, the entrepreneur strategies and firm performance, but no emphasis, or attention at all, is given to the macro-economic or regional impact of the entrepreneurial action. In this case the economic impact is a 'given', a normative assumption or is taken for granted.

The present study aimed to make an exploratory research to bring these two approaches together. We did this by studying the factors that led to the emergence and development of a furniture manufacturing cluster in the city of Arapongas, a medium sized city located in the countryside in the southern region of Brazil. With this purpose we studied simultaneously the trajectory, strategies and outcomes of one particular entrepreneur, and linked this trajectory to the cultural, social and economic transformation that the actions of this entrepreneur, and others that followed him, brought to the region. This particular entrepreneur, defined as the 'anchor' entrepreneur, was identified as the executive leadership whose innovative and proactive actions led to a path for other entrepreneurs, enabling the emergence and development of the cluster studied.

Through the actions of these entrepreneurs, or more simply through entrepreneurship, the city of Arapongas changed in the time span of three decades, from a stagnant agricultural economic based city, to one of the most industrialized cities in the state of Paraná, with a per capita industrial value added more than 50 per cent higher than in the main cities of the state and comparable to some of the most industrialized cities of the country. Starting with only a handful of small furniture companies, including the 'anchor' firm surveyed in this study, the city witnessed the emergence of an internationally competitive cluster of furniture manufacturers with more than 150 local and all family companies, generating annual gross revenues of more than US$ 800 million, from which more than US$ 40 million come from exports. The cluster generates more than 8000 direct jobs and 25 000 indirect jobs, allowing the city of Arapongas to have the lowest unemployment rate in the state of Paraná.

It is important to highlight that this transformation had a spontaneous component reflected by the decision of individual entrepreneurs to start their business by investing their money, accepting risk and so on. But it also had purposive components. This purposive component was expressed, first, by the local governmental policy which aimed at stimulating the industrial activity in the city back in the late 1960s. The second purposive component was the strategies of the entrepreneurs of the cluster that, since the beginning, resulted in collective and cooperative action by creating an association to promote training for workers and coordination of business and political collective actions. In a second phase, the Association of Furniture Manufacturers built a pavilion for furniture and equipment fairs, strengthening and disseminating the image and the brand of the local manufacturers, thus contributing to the consolidation and development of the cluster and, consequently, for the region as a whole.

We hope that beyond making a contribution to fill the gap in the entrepreneurship literature the present study makes other significant contributions. In the research field we illustrated the practical applicability of some novel and recent theories, like the effectuation theory proposed by Sarasvathy (2001), and the model proposed by Venkataraman (2004), who defines a set of intangible factors to foster regional transformation through entrepreneurship. We concluded that the analytical framework suggested in these two approaches seemed to be adequate for explaining the dynamics of the 'anchor' entrepreneur and cluster trajectories, and the relationship between them.

Most importantly, our study can bring important implications for policy making aimed at fostering entrepreneurship. We examined a case in which the actions of private entrepreneurs were able to radically transform and develop a region, increasing its income, reducing economic uncertainty, generating jobs and reducing inequality. All these outcomes are a high priority in developing countries that usually face a cruel combination of a drout of economic opportunities with high unemployment and social inequality. The promise of entrepreneurship could be really brilliant, as a tool for facing the severe problems of developing countries. One important lesson of the case studied and presented here is that although the presence of the government was of little importance during the development of the cluster, it was essential in the beginning by the decision of City Hall to donate land to the entrepreneurs and offering tax exemption. The policy makers should unleash local entrepreneurial forces and provide the appropriate support for the initial steps of the process. The evidence presented here leaves no room for doubt. When the entrepreneur starts to walk with their own feet they will walk all the way, with no additional help necessary.

We also analysed a case with no significant specificities. The industry of the cluster is a mature industrial sector, which utilizes a medium technology level, with no significant barriers to access. We are sure that in Brazil, or in many other developing countries, it wouldn't be difficult to identify other industries with similar characteristics, like shoe manufacturing, wearing apparel, food, food processing and many others, in which it is possible to identify a cluster with a similar history or, what is more promising, the existence of conditions that could lead a small and unknown group of entrepreneurs and could replicate the experience of Arapongas in other sectors. Good advice for local policy makers (or even for national policy makers) could be to find the local entrepreneur. With modest and proper support at the beginning of their ventures they can create new histories of success, like the one described here.

Although Venkataraman (2004) shows some concerns about the risk that non-technological entrepreneurship could inherit and the deficiencies of what he calls 'weak entrepreneurial force', and fail to promote a sustainable transformation process in the regions, we suggest that this process could be accomplished in two stages. In the first stage, the one depicted in our case study, the entrepreneurship in mature industrial sectors would be sufficient to move the region from a traditional and inefficient economic base to an industrial and a nationally, and eventually globally, integrated one, and in this process create the conditions for the intangible factors to emerge. In a second stage the industries in the region's cluster would have their technological content improved and complemented by other more technological based industries. This process is already taking place in the Brazilian furniture cluster studied here. Linked with the furniture industries, other industries are being attracted or locally developed, such as equipment, logistics and, more recently, design.

Finally, we are aware of the limitations imposed by the methodology used in this research, especially in terms of generalization. Since we choose to study a case of a traditional industry that can be found in many other regions in Brazil, or in many other countries, it would be necessary to replicate similar studies in order to the check the stability, reliability and generalization of the conclusions reported here. We challenge our colleagues interested in the field of entrepreneurship to do so.

NOTES

1. The Global Entrepreneurship Monitor is an annual research initiative conducted jointly by the Babson College (Boston, MA) and the London Business School (London) surveying the level of entrepreneurship activity in different countries. The research started in 1999 and has continued since then. The number of countries surveyed ranges from 30 to 40, and includes developed and developing countries.
2. Emphasis in the original.

REFERENCES

Acs, Z.J. and C. Armington (2004), 'Employment, growth and entrepreneurial acitivy in cities', discussion paper on entrepreneurship, growth and public policy, no. 1304, Max Planck Institute of Economics, Jena, Germany.

Acs, Z.J., D.B. Audretsch, P. Braunerhjelm and B. Carlson (2005a), 'Growth and entrepreneurship: an empirical assessment', Max Planck Institute of Economics, discussion paper on entrepreneurship, growth and public policy, no. 3205, Max Plank Institute of Economics, Jena, Germany.

Acs, Z.J., D.B. Audretsch, P. Braunerhjelm and B. Carlson (2005b), 'The missing link: the knowledge filter and entrepreneurship in endogenous growth', discussion paper on entrepreneurship, growth and public policy, no. 0805, Max Plank Institute of Economics, Jena, Germany.

Aloulou, W. and A. Fayolle (2005), 'A conceptual approach of entrepreneurial orientation within a small business context', *Journal of Enterprising Culture*, **13** (1), 21–45.

Alsos, G.A. and V. Kaikkonen (2004), 'Opportunities and prior knowledge: a study of experienced entrepreneurs, *Frontiers of Entrepreneurship Research*, Wellesley, MA: Babson College.

Ardichivili, A., R. Cardozo and S. Ray (2003), 'A theory of entrepreneurial opportunity identification and development', *Journal of Business Venturing*, **28**, 105–23.

Audretsch, D.B. and M. Fritsch (2002), 'Growth regimes over time and space', *Regional Studies*, **36** (2), 113–24.

Audretsch, D.B. and M. Keilbach (2004), 'Entrepreneurship capital and economic performance', discussion paper on entrepreneurship, growth and public policy, no. 0104, Max Planck Institute of Economics, Jena, Germany.

Audretsch, D.B. and R. Thurik (2004), 'A model of the entrepreneurial economy', discussion paper on entrepreneurship, growth and public policy, no. 1204, Max Planck Institute of Economics, Jena, Germany.

Audretsch, D.B., M. Carree and A.R. Thurik (2001), 'Does entrepreneurship reduce unemployment?', discussion paper no. TI 2001-074/3, Tinbergen Institute, Amsterdam, The Netherlands.

Baptista, R., V. Escária and P. Madruga (2005), 'Entrepreneurship, regional development and job creation: the case of Portugal', discussion paper on entrepreneurship, growth and public policy, no. 0605, Max Planck Institute of Economics, Jena, Germany.

Brown, T.E., P. Davidsson and J. Wiklund (2001), 'An operationalization of Stevenson's conceptualization of entrepreneurship as opportunity-based firm behavior', *Strategic Management Journal*, **22**, 953–68.

Carland, J.W., F. Hoy, W.R. Boulton and J.A.C. Carland (1984), 'Differentiating entrepreneurs from small business owners: a conceptualization', *Academy of Management Review*, **9** (1), 354–9.

Carree, M., A. Van Stel, R. Thurik and S. Wennekers (2000), 'Business ownership and economic growth in 23 OECD countries', discussion paper no. TI 2000-01/3, Tinbergen Institute, Amsterdam, The Netherlands.

Chiasson, M. and C. Saunders (2005), 'Reconciling diverse approaches to opportunity research using the structuration theory', *Journal of Business Venturing*, **20**, 747–67.

Christensen, C.M. (1997), *The Innovator's Dilemma*, Boston, MA: Harvard Business School Press.

Cooper, A.C. and W.C. Dunkelberg (1986), 'Entrepreneurship and paths to business ownership', *Strategic Management Journal*, **7** (1), 53–68.

Covin, J.G. and D.P. Slevin (1989), 'Strategic management of small firms in hostile and benign environments', *Strategic Management Journal*, **10** (1), 75–87.

Covin, J.G., K.M. Green and D.P. Slevin (2006), 'Strategic process effects on the entrepreneurial orientation-sales growth rate relationship', *Entrepreneurship Theory and Practice*, January, 7–81.

Dess, G.G., G.T. Lumpkin and J.G. Covin (1997), 'Entrepreneurial strategy making and firm performance: tests of contingency and configurational models', *Strategic Management Journal*, **18** (9), 677–95.

Ding, H.B. and P. Abetti (2001), 'The entrepreneurial success of Taiwan: lessons for rapidly industrializing countries', in M.D. Foo and P. Phan (eds), *Proceedings of the Conference on Technological Entrepreneurship in the Emerging Regions of the New Millennium*, June, Singapore: National University of Singapore, pp. 73–89.

Dosi, G. (1988), 'The nature of the innovative process', in G. Dosi, C. Freeman, R. Nelson, G. Silverberg and L. Soete (eds), *Technical Change and Economic Theory*, London and New York: Pinter Publishers, pp. 221–38.

Dutta, D.K. and Mary M. Crossan (2005), 'The nature of entrepreneurial opportunities: understanding the process using the 4I organizational learning framework', *Entrepreneurship Theory and Practice*, July, 425–49.

Eisenhardt, K.M. and C.B. Schoonhoven (1990), 'Organizational growth: linking founding team, strategy, environment, and growth among U.S. semiconductor ventures, 1978–1988', *Administrative Science Quarterly*, **35**, 504–29.

Fritsch, M. and P. Muller (2004), 'Regional growth regimes revisited: the case of West Germany', discussion paper on entrepreneurship, growth and public policy, no. 0404, Max Planck Institute of Economics, Jena, Germany.

Grilo, I. and R. Thurik (2005), 'Entrepreneurial engagement levels in the European Union', discussion paper no. 2905, Max Planck Institute of Economics, Jena, Germany.

Hayek, Friedrich (1984), *Individualism and Economic Order*, Chichago, IL: University of Chicago Press.

Hitt, M.A., R.D. Ireland, S.M. Camp and D.L. Sexton (2001), 'Strategic entrepreneurship: entrepreneurial strategies for wealth creation', *Strategic Management Journal*, **22**, 479–91.

Hornaday, J.A. and J. Aboud (1971), 'Characteristics of successful entrepreneurs', *Personnel Psychology*, **24** (2), 141–53.

Kirzner, Israel (1973), *Competition and Entrepreneurship*, Chicago, IL: University of Chicago Press.

Kunkel, J.H. (1971), 'Values and behavior in economic development', in Peter Kilby (ed.), *Entrepreneurship and Economic Development*, New York: Free Press, pp. 151–80.

Lichtenstein, B.B., K.J. Dooley and G.T. Lumpkin (2006), 'Measuring emergence in the dynamics of new venture creation', *Journal of Business Venturing*, **21**, 153–75.

Lumpkin, G.T. and G.G. Dess (1996), 'Clarifying the entrepreneurial orientation construct and linking it to performance', *Academy of Management Review*, **21** (1), 135–72.

Lumpkin, G.T. and B.B. Lichtenstein (2005), 'The role of organizational learning in the opportunity-recognition process', *Entrepreneurship Theory and Practice*, July, 451–72.

Lucas, R.E. (1988), 'On the mechanics of economic development', *Journal of Monetary Economics*, **22**, 3–39.

March, J.G. (1982), 'The technology of foolishness', in J.G. March and J.P. Olsen (eds), *Ambiguity and Choice in Organizations*, Bergen, Norway: Universitetesforlaget, pp. 69–81.

McClelland, D.C. (1967), *The Achieving Society*, New York: Free Press.

Miller, D. (1983), 'The correlates of entrepreneurship in three types of firms', *Management Science*, **29** (7), 770–91.

Miller, D. and P.H. Friesen (1982), 'The longitudinal analysis of organizations: a methodological perspective', *Management Science*, **28** (9), September, 1013–34.
Miner, J.B. (1996), *The Four Routes to Entrepreneurial Success*, San Francisco, CA: Barrett-Koehler Publishers.
Mintzberg, H. (1994), *The Rise and Fall of Strategic Planning*, New York: Free Press.
O'Gorman, C. and M. Kautonen (2001), 'Policies for new prosperity: promoting new agglomerations of knowledge intensive industries', in M.D Foo and P. Phan (eds), *Proceedings of the Conference on Technological Entrepreneurship in the Emerging Regions of the New Millennium*, June, National University of Singapore, Singapore.
Park, S. and Z.T. Bae (2004), 'New venture strategies in a developing country: identifying a typology and examining growth patterns through case studies', *Journal of Business Venturing*, **19**, 81–105.
Pettigrew, A.M. (1973), *The Politics of Organizational Decision-Making*, London: Tavistock.
Romer, P.M. (1994), 'The origins of endogenous growth', *Journal of Economic Perspectives*, **8** (1), 3–22.
Sarason, Y., T. Dean and J.F. Dillard (2006), 'Entrepreneurship as the nexus of individual and opportunity: a structuration view', *Journal of Business Venturing*, **21**, 286–305.
Sarasvathy, Saras D. (2001), 'Causation and effectuation: toward a theoretical shift from economic inevitability to entrepreneurial contingency', *Academy of Management Review*, **26** (2), 243–63.
Sarasvathy, Saras D. (2004), 'Making it happen: beyond theories of the firm to theories of firm design', *Entrepreneurship Theory and Practice*, Winter, 519–31.
Saxenian, A. (1994), *Regional Advantage*, Cambridge, MA: Harvard University Press.
Schumpeter, Joseph A. (1934), *The Theory of Economic Development*, Cambridge, MA: Harvard University Press.
Schollhammer, H. (1982), 'Internal corporate entrepreneurship', in C.A. Kent, D.L. Sexton and K.H. Vesper (eds), *Encyclopedia of Entrepreneurship*, Englewood Chiffs, NJ: Prentice Hall, pp. 209–23.
Shane, S. and S. Venkataraman (2000), 'The promise of entrepreneurship as a field or research', *Academy of Management Review*, **25** (1), 217–26.
Shepherd, D.A. and D.R. DeTienne (2005), 'Prior knowledge, potential financial reward, and opportunity identification', *Entrepreneurship, Theory and Practice*, January, pp. 91–112.
Solow, R. (1956), 'A contribution to the theory of economic growth', *Quarterly Journal of Economics*, **70**, 65–94.
Timmons, J.A. (1989), *The Entrepreneurial Mind*, Andover, UK: Brick House Publishing Company.
Van de Ven, A.H. and Rhonda M. Engleman (2004), 'Event- and outcome-driven explanations of entrepreneurship', *Journal of Business Venturing*, **19**, 343–58.
Van Stel, A. and D.J. Storey (2004), 'The link between births and job creation: is there a Upas Tree effect?', discussion paper on entrepreneurship, growth and public policy, no. 3304, Max Planck Institute of Economics, Jena, Germany.
Van Stel, A., M. Carree and R. Thurik (2004), 'The effect of entrepreneurship on national economic growth: an analysis using the GEM database', discussion paper on entrepreneurship, growth and public policy, no. 3404, Max Planck Institute of Economics, Jena, Germany.

Van Stel, A., M. Carree and R. Thurik (2005a), 'The effect of entrepreneurial activity on national economic growth', discussion paper on entrepreneurship, growth and public policy, no. 0405, Max Planck Institute of Economics, Jena, Germany.
Van Stel, M., R. Thurik, David Storey and S. Wennekers (2005b), 'From nascent to actual entrepreneurship: the effect of entry barriers', discussion paper on entrepreneurship, growth and public policy, no. 3505, Max Planck Institute of Economics, Jena, Germany.
Venkataraman, S. (1997), 'The distinctive domain of entrepreneurship research', in J. Katz (ed.), *Advances in Entrepreneurship, Firm Emergence and Growth*, vol. 3, Greenwich, CT: JAI Press, pp. 119–38.
Venkataraman, S. (2004), 'Regional transformation through technological entrepreneurship', *Journal of Business Venturing*, **19**, 153–67.
Vesper, K.H. (1979), 'Commentary', in D.E. Schendel and C.W. Hofer (eds), *Strategic Management*, Boston, MA: Little, Brown, pp. 332–8.
Weick, K.E. (1979), *The Social Psychology of Organizing*, Reading, MA: Addison-Wesley.
Wennekers, S., A. Van Stel, R. Thurik and P. Reynolds (2005), 'Nascent entrepreneurship and the level of economic development', discussion paper on entrepreneurship, growth and public policy, no. 1405, Max Planck Institute of Economics, Jena, Germany.
Wiklund, J. and D. Shepherd (2005), 'Entrepreneurial orientation and small business performance: a configurational approach', *Journal of Business Venturing*, **20**, 71–91.
Zahra, S.A. and J.G. Covin (1995), 'Contextual Influences on the corporate entrepreneurship-performance relationship: a longitudinal analysis', *Journal of Business Venturing*, **10**, 43–58.

3. Institutional transformation during the emergence of New York's Silicon Alley

Andaç T. Arıkan

INTRODUCTION

The importance of technology entrepreneurship in facilitating regional development is well known (Saxenian, 2000; Phan and Foo, 2004; Venkataraman, 2004). In fact many regional governments in the USA and across the globe aim to promote technology entrepreneurship through various policy initiatives, yet very few areas actually succeed in initiating entrepreneurship and even fewer regions turn into centers of large scale entrepreneurial activity such as Silicon Valley. In the face of such policy resistance (Sterman, 2001), an important question to ask is what is the process by which a region that lacks an entrepreneurial tradition initiates entrepreneurship and develops a favorable entrepreneurial infrastructure (Van de Ven, 1993).

Recent entrepreneurship research under the title of 'demand side perspective' has put a heavy emphasis on the influence of environmental factors on triggering entrepreneurial activity (Thornton, 1999). In contrast to the 'traits approach' that views entrepreneurial personality characteristics as the primary determinant of entrepreneurial outcomes (Brockhaus and Horwitz, 1986; Gartner, 1988), the demand side perspective holds that region specific environmental factors constitute an objective opportunity structure which gives particular regions differential advantage over others in terms of generating entrepreneurship. Based on this premise, researchers have studied the environmental characteristics of regions such as Silicon Valley that have been successful in stimulating large scale entrepreneurship. The factors that create a favorable environment for entrepreneurship were found to range from the availability of resources (for example, venture capital, land and facilities, technically skilled labor force, universities and research institutions) and support services (for example, lawyers, consultants, accountants and so on) to the presence of social networks, favorable government policies and a tolerant population that is receptive to entrepreneurship (Bruno and Tyebjee, 1982; Markusen et al., 1986; Schoonhoven and Eisenhardt, 1993; Gnyawali and Fogel, 1994). Based on this research, scholars have concluded that a change in a region's entrepreneurial profile requires a change in environmental conditions such that the new environment closely mimics those found in entrepreneurially successful regions.

While the aforementioned environmental factors are obviously important, institutional theorists have argued that there is more to regional entrepreneurial transformation (Romanelli, 1991; Van de Ven, 1993; Aldrich and Fiol, 1994; Aldrich, 1999). Institutional theory emphasizes the cultural and normative framework within which organizational populations are embedded (Scott, 1995). Entrepreneurial ventures leveraging an emergent technology are founded in the context of a local organizational community (Astley, 1985; Romanelli, 1989). In the course of the evolution of local organizational communities, processes of coercion, professional pressures and mimicry due to environmental uncertainties force local organizational populations to adopt similar organizational forms and business practices (DiMaggio and Powell, 1983). Over time these forms and practices become widely accepted norms that are taken for granted and organizations that are characterized by divergent forms or business practices are less likely to succeed (Hannan and Freeman, 1984). In other words, the local institutional environment determines appropriate ways of doing business and consequently creates selection pressures for new organizational foundings.

Aldrich and Fiol (1994) argue that entrepreneurial ventures leveraging a new technology require an entrepreneurial infrastructure (Van de Ven, 1993) that embodies social, economic and political factors that are most likely in conflict with the existing history dependent selection environment. That is why, in addition to problems associated with their nascent status, new entrepreneurial ventures in a non-entrepreneurial region are likely to suffer from lack of cognitive and sociopolitical legitimacy which makes the emergence of entrepreneurship in such regions particularly difficult. Cognitive legitimacy refers to the acceptance of a new kind of venture as a taken for granted feature of the environment. Cognitive legitimacy exists when an activity, a product, process or service becomes so familiar and well known that it is accepted as part of the sociocultural and organizational landscape (Aldrich, 1999). Defined this way, cognitive legitimacy relates to the extent to which information is spread about a new organizational form and thus to what extent external constituencies can assess its efficiency. Similarly, sociopolitical legitimacy refers to the acceptance by key stakeholders, the general public, key opinion leaders and government officials of a new venture as appropriate and right, given existing norms and laws.

The presence of these institutional constraints suggests that a change in a region's entrepreneurial profile requires a regional institutional transformation in addition to favorable environmental factors such as large local resource endowments. Regional institutional transformation is not a product solely of the actions of nascent entrepreneurs but rather a variety of different types of actors are implicated. As Nelson (1995, p. 78) contends,

> evolution of institutions relevant to a technology or industry may be a very complex process involving not only the actions of private firms competing with each other . . . but also organizations, like industry associations, technical societies, universities, courts, government agencies, legislatures, etc. In turn, the way these other organizations evolve and the things that they do may profoundly influence the nature of the firms and the organization of the industry.

Accordingly, a variety of different types of actors need to actively resist existing institutions (Oliver, 1991), and transform them through their strategic actions.

Even though the need for a regional institutional transformation for emergent local entrepreneurship is recognized, few studies have actually examined the process by which institutional transformation occurs during the early phases of regional entrepreneurial transformation. This chapter aims to examine the institutional transformation process that a region goes through on its way to becoming an entrepreneurial center. To this effect I provide a historical account of the transformation that the New York tri-state underwent during the late 1990s. I place particular emphasis on non-routine actions (that is, actions that are not necessarily in line with the routine and accepted ways of doing business) of pioneering actors from diverse walks of life and how those actions shaped the institutional environment in which Silicon Alley (New York's new media cluster) emerged. The case is particularly interesting since Silicon Alley's emergence coincides with (and is somewhat facilitated by) the resource munificent environment of the Internet boom years. The case discussion illustrates the influence of the movements in macro financial markets on the region's entrepreneurial development and its eventual demise. The case account is followed by a discussion of the lessons that can be drawn from the developmental history of this particular region.

THE ENTREPRENEURIAL TRANSFORMATION OF THE NEW YORK TRI-STATE REGION

New York City has been home to many established industries, such as financial services, entertainment and media; however before the early 1990s it has not been a technology entrepreneurship center (Florida and Kenney, 1988). The city's technology entrepreneurship profile started to change with the arrival of the Internet. First sparks of technology entrepreneurship started in Downtown Manhattan in the early to mid 1990s and later extended to the tri-state area. The region was given the name Silicon Alley (SA) after California's Silicon Valley. SA housed 'new media' companies that were mostly Internet content businesses aligned closely with advertising and entertainment sectors in the city. Starting from early 1995, SA grew at an unprecedented rate. By the end of the millennium, the region had created approximately 8500 new media businesses that generated over 250 000 positions and annual revenues of $17 billion (NYNMA Survey, 2000). It is the area's tremendous transformation within a matter of five years from an area with negligible levels of technology entrepreneurship to one of the fastest growing technology entrepreneurial centers of the 1990s that is examined in this section.

Case data reported below come primarily from an analysis of a total of 640 SA related articles that appeared between January 1995 and January 2001 in *Crain's New York*, an online and paper news outlet that covers the business life in New York and @NY, an online source that has covered SA exclusively during its growth years. Additional data come from books that examine the emergence and growth of SA (Kait and Weiss, 2001; Indergaard, 2004), New York New Media Industry Surveys (1996, 1997 and 2000), trade magazines, information from Internet sites, a case study on the New York Information Technology Center,[1] and informal conversations with industry participants during various networking events.

1987 Stock Market Crash

The history of SA can be traced back to an external shock to New York's local economy – the stock market crash in 1987 – which started shaping the regional context in which SA emerged. Downtown Manhattan had long been home to several stock exchanges including the New York Stock Exchange (NYSE) and many of the world's largest brokerage firms and banks. The city's status as one of the world's major financial capitals had mostly served downtown well, but it had also made the local economy highly dependent on the finance industry. Accordingly, the city's economy was hit hard when the stock market crashed. The crash prompted a major downsizing of the finance industry in Downtown Manhattan which continued into the early 1990s. Particularly affected by the crash were the real estate market and employment in the city.

The crash caused some of the area's most prominent corporate tenants to collapse and some others to migrate to cheaper and newer space outside of Manhattan leaving behind a large stock of empty office space. The accumulation of vacant office space accelerated further as new speculative properties that had begun construction before the downturn was completed. Between 1987 and 1989, nearly 12 msf (million square feet) of new office space came onto the market. Many tenants who moved into these properties were tenants of older buildings that were once celebrated for their architecture, but had become functionally obsolete. These buildings were in need of major renovation, but there was no incentive for real estate companies to undertake these renovations since there were no prospective business tenants.

The excess supply and diminished demand caused the office space vacancy rate to reach 20 per cent in 1994 – the highest in nearly 50 years. With over 21 msf empty office space, Downtown Manhattan had more vacant office space than the total building stock of many large cities. In the mean time, the city's residential space shortage remained unabated as vacancy rates for rental units hovered between 3–4 per cent. In a city that never sleeps the downtown area was considered a 9 to 5 business district and lacked amenities, such as bars, restaurants and entertainment venues, that would make the area an attractive residential area. Consequently, by the end of 1994, the real estate market in New York was characterized by a large pool of underutilized office space and a large unfulfilled demand for residential space.

The crash also caused the city to lose a lot of jobs. By the early 1990s New York City had already lost 300 000 jobs and was stuck in a recession. The loss of these jobs left a large group of New Yorkers unemployed. As a media and advertising capital, New York had always attracted creative talent; however the recession provided few job opportunities for new college graduates. In the early 1990s there was a group of young, talented, creative people in New York who found the notion of becoming a drone for a faceless corporation too dispiriting to consider. The following quote by Rufus Griscom, who co-founded Nerve.com – one of the most successful online magazines in SA – is representative of the mindset that many young college graduates who later became SA entrepreneurs had:

> I was a slacker after I graduated from Brown in 1991. I took a year off and was a ski instructor for a year. . . We had limited ambitions; we basically felt there were no attractive jobs. Maybe the best scenario was that you get to be a journalist and slave away on a terrible salary, not getting to write what you want, or you worked at a publishing house under similar circumstances, or you sold your soul and basically exchanged your twenties and thirties for a chunk of cash on Wall Street which I considered, begrudgingly. All the options seemed quite grim. The best possible option seemed to be to make coffee at a local diner and write, or to play in a rock band . . . there used to be a paradigm where you could either

make money or do something interesting. . . In fact not making money was a symbol of your integrity. (Kait and Weiss, 2001, p. 14)

The Arrival of the Internet and the Barriers to New Media Entrepreneurship in New York City

Another external shock that influenced New York's economy profoundly was the arrival of the Internet. The ability to post content on the web attracted the attention of a small group of multi-media artists, online publishers and software developers in New York who had impressive degrees in liberal arts but did not have attractive employment opportunities due to the ongoing economic recession. While traditional companies ignored the Internet, these individuals started to experiment with the new medium with great excitement and with the belief that something big was going to happen. These individuals, many of whom later became SA entrepreneurs, were perceived by others to be pioneering actors and accordingly described as bohemian, visionary, romantic, flamboyant, arrogant, cocky and irreverent.

The presence of these pioneering actors created a potential for the emergence of large scale Internet entrepreneurship in New York City. The city constituted a potentially favorable location for new media startups due to its status as a finance advertising and entertainment capital; however several hurdles stood before new media entrepreneurs.

First, the cost of doing business in the city was very high owing to New York being one of the most expensive cities in the world. In addition, government incentives for new media entrepreneurs were not in place since the Internet was a brand new technology and its commercial implications were yet unknown and its potential uncertain.

A second hurdle was that risk capital was not readily available to local new media startups. The city lacked an established network of angel investors since it did not have an entrepreneurial tradition such as the one that can be found in Silicon Valley. East Coast venture capitalists avoided early stage technology businesses and preferred to finance late stage manufacturing or service businesses as they could easily judge the market conditions for such businesses. West Coast venture capitalists that did have a tradition of financing technology businesses were mostly unaware of SA or did not understand the business models of New York new media firms. The venture capital industry in Silicon Valley had grown out of the personal computer industry which is based on proprietary technology. The venture capitalists would see a patented chip architecture that they understood and judged to be promising, have ideas about commercializing it and imagine carrying the company to an IPO. With content businesses there was no such clear product. The dominant business model was posting content on the web and making money either through advertisements or through subscription fees. Technology venture capitalists on the West Coast were not convinced that this business model was viable. To them New York Internet companies looked more like publishing businesses than technology businesses and they did not have experience investing in publishing businesses.

A third hurdle was the obsolete technology infrastructure in the city. The connectivity that new media businesses required was not offered in Manhattan office buildings. In fact even the basic heating, ventilation and air conditioning (HVAC) systems in most buildings were obsolete making them unattractive to new media entrepreneurs. Furthermore the ongoing recession made it unprofitable for office real estate firms to update these buildings.

A final hurdle for new media entrepreneurs was the challenge of attracting high quality employees. Early employees of SA were a handful of pioneering actors – young techies, artists and writers who were willing to labor 15 to 18 hour days in high tech sweatshops in exchange for a chance to be at the forefront of a media revolution. Their choice to work at new media companies was motivated by a romantic devotion to the Internet. Technical and managerial talent were scarce and available talent preferred working at traditional firms that offered better compensation, better work conditions and job security. The lack of legitimacy around the Internet made attracting high quality workers a challenge for new media firms.

Creation of an environment that is favorable for new media entrepreneurship required actions on behalf of a diverse set of actors including the local government, real estate firms, the workforce in the city and local financiers. But before contributions from these actors could be mobilized, the Internet pioneers needed to create social networks to coalesce. It is through these early networks that the pioneers of SA would find each other, share information about their activities and collaborate in their efforts to commercialize the new medium. Also important was the creation of SA specific media, particularly, since traditional media initially ignored the Internet and the actions of anyone who took it seriously. These early networks played a crucial role in the establishment of cognitive and sociopolitical legitimacy around the Internet.

Creation of Early Networks by Brokering Actors and Silicon Alley-Specific Media

Networking is important for any emerging industry, but for SA it was one of the defining characteristics. Alice Rodd O'Rourke, executive director of New York New Media Association said: 'I've worked in four industries, and I've never seen networking the way this industry networks. It's very purposeful: "Don't you want to know? Don't you want to share? Don't you want to be a part of this?" It's got a spark in it that is very exciting' (Kait and Weiss, 2001, p. 224). From small informal gatherings organized by early advocates of the Internet to large scale periodic events by industry associations, networking activities played a major role in building a community through social learning, creating the Internet celebrity culture and reinforcing the hype around the Internet. The networking scene in SA co-emerged with SA itself as a consequence of a few pioneering actor's non-routine actions.

The founding of the New York New Media Association (NYNMA)

In the summer of 1994 Mark Stahlman, an investment banker, and Brian Horey, a venture capitalist, founded the NYNMA. Their stated purpose was to build a community around new media in New York so that they 'could stop flying to California (that is, Silicon Valley) every other week to do business'. The NYNMA's roots go back to 1992 when Stahlman started his CyberSalon parties. CyberSalon was a small, monthly dinner party for people who were interested in computer related activities in New York, yet it constituted a trigger for the organization of larger gatherings. During the CyberSalon parties it became apparent that there were a lot of people using digital technologies but they were spread out in various businesses including advertising, graphic design, music and publishing. People from these different fields did not have a place to find each other and coalesce. NYNMA was founded to bring these actors together.

NYNMA's first action was the organization of CyberSuds – a monthly networking event for members of the new media community to get in touch with each other. The first CyberSuds was held at a downtown restaurant with only 15 people. Towards the end of the millennium the events were attracting close to 3000 people. Similarly, NYNMA's membership grew to over 8000 members making it the largest new media industry association in the world. With the growth in NYNMA came additional networking events, such as the Evening Panel Series, the Executive Roundtable Breakfast Series and finally Super Cybersuds – the largest tabletop trade show that provided new media companies the chance to market their products and services to potential business partners, customers and investors.

Another important action by NYNMA was the administration of new media surveys. In 1996 NYNMA partnered with Coopers and Lybrand and conducted the first large scale survey of the new media industry in New York. The survey was important because 'it let people working in their living rooms – or in someone's office that they managed to talk themselves into for free for six months – know that they are not alone, there are thirty-five thousand of them' (Kait and Weiss, 2001, p. 60). The surveys were an important social learning tool by making information about new media available to interested parties from different walks of life, and providing an identity for the collective. NYNMA repeated the surveys in 1997, 2000 and 2001, providing the most detailed statistical information on new media firms in the area.

CyberSlacker parties

Another set of early networking events were the 'CyberSlacker' parties that started in early 1995. The organizer was Jamie Levy, a 29 year old bleached blond who called herself 'the biggest bitch in SA' (www.weblab.org/benefit/lyrics.html). She was the founder of the world's first floppy disk-based electronic magazine, CyberRag. Her parties made Jamie Levy one of the most well-known characters in SA. Some even say that NYNMA's CyberSuds are modeled after Levy's CyberSlacker parties.

Cocktails with Courtney

Another brokering actor in SA was Courtney Pulitzer. In April 1997 she started writing a social column for @NY, chronicling what was happening in the industry. The column was one of the very first attempts to cover the emerging social life in SA. After a successful run of this column of about a year and a half, she started her own newsletter called 'The Cyber Scene'. By mid 2000 the newsletter had a readership of about 20 000. In late 1998, she started her signature event 'Cocktails with Courtney' addressing a desire to 'get together with friends in the industry'. The monthly parties were cocktail receptions where dot-com executives, top-level managers, analysts, programmers, investors and media professionals could meet in a friendly and elegant social networking environment. Cocktails with Courtney turned out to be an enormous success drawing on average 250–400 invited guests and, together with the Cyber Scene newsletter, made Pulitzer one of the most well known and highly connected figures in SA.

Silicon Alley party scene

In addition to periodic events, major SA companies, such as Pseudo, Razorfish and DoubleClick, started throwing their own parties. At the height of the Internet boom, there were as many as a dozen parties every single night of the week. These parties were characterized by lavishness and outrageousness. Large quantities of alcohol, drugs, go-go dancers, strippers, belly dancers and DJs showing porn flicks were not uncommon. Pseudo founder Josh Harris was famous for organizing parties in which SA socialites mixed with artists, musicians, dancers, freaks and hipster kids off the street just looking for something to do. His one month long, new millennium party project 'Quiet', which cost over a million dollars and attracted thousands of guests remains one of the most talked about parties in SA history. These parties were the major venues where social networks were formed and deals were made.

Bernardo's List

Another unproven entrepreneur, Bernardo Joselevich, emerged as the announcer of everybody else's parties. In late 1998 Joselevich was forwarding e-mails about SA events to a friend and his wife. The couple started forwarding Bernardo's List to their friends. Then those people contacted Bernardo to get his list directly and thus Bernardo's List started. By the end of 2000 Bernardo's weekly list was going to about 12 000 people including the top executives of SA. In time the list became 'the barometer of dot-com social patterns' (Industry Standard, 25 December 2000) and made Joselevich one of the early millionaires of SA.

The nature and function of networking events changed as SA evolved from a small group of evangelists to a widely recognized entrepreneurial center. Early events, such as the CyberSalon and CyberSlacker parties and early CyberSuds were very informal meetings attended by a handful of early true believers. The main goal was to create a community out of a small group of isolated, scattered people who were interested in new media. The meetings took place at small restaurants, bars or people's homes and served as sense-making occasions in which pioneers tried to figure out opportunities related to the Internet. As SA grew bigger, the scale and number of events increased dramatically. Networking became a way of life for all who were already or wanted to be a part of SA. Events became venues where deals were made. They were flooded with people who smelled money in new media and who wanted to meet with somebody – anybody in the industry. There came a point where some veterans became hesitant to attend the events because of the harassment by wannabe entrepreneurs.

Media played a role similar to that of networking activities in the development of SA. Given the initial lack of attention from established media companies, SA needed to create its own media. Three SA-specific media outlets emerged. They had a significant influence on the growth of SA by distributing information, bringing key actors together, and more importantly by defining what SA is and who the celebrities are, thereby providing the emerging new media industry with a distinct identity.

@NY

The first major SA-specific media outlet was @NY. Two local reporters, Jason Chervokas and Tom Watson, started @NY in September 1995 as an e-mail based newsletter. Chervokas commented on the non-routine nature of this action: 'We just kind of fell in love with the story of what was happening with a handful of Internet start-ups that sprung up for a bunch of accidental reasons. . . . no one thought it was a story . . . but it was fascinating to us . . . we knew something was happening' (Kait and Weiss, 2001, pp. 93–4). The two founders initially produced @NY as a weekly newsletter. It became a daily newsletter with an accompanying website after the company was sold to Internet.com in 1999. Siliconalleyjobs.com and siliconalleystocks.com, the two sister pages to @NY.com, became, respectively, the most widely used recruitment site in SA and the place where the public followed the progress of SA stocks. What started as an adventurous newsletter turned into one of the biggest online SA-specific information sources.

AlleyCat News

The second SA-specific media outlet was AlleyCat News. Anna Wheatley and Janet Stites, two writers from a local magazine, launched AlleyCat News in Fall 1996. While @NY was mostly news oriented, AlleyCat News intended to focus on financial aspects of SA and aimed to connect companies looking for funding with investors looking for opportunities. Eventually, it became a medium that brought two groups of key actors in SA together. The initially 16 page newsletter turned into a 100 page monthly magazine in 2000, making AlleyCat News one of the two biggest SA trade publications.

Silicon Alley Reporter

Even more influential than the above two outlets was Silicon Alley Reporter (SAR). In 1996 a 25 year old Jason McCabe Calacanis started SAR. At the time, he was writing a column for a magazine, covering the digital scene in New York. Unlike the founders of @NY, he was not a journalist but 'just a hack' as he said about himself (Kait and Weiss, 2001, p. 96). The *New Yorker* alluded to the non-routine nature of his actions: 'Where everybody saw twenty five-year olds in T-shirts fiddling unprofitably with monotonous graphic effects, he saw twenty-eight year old CEOs with IPOs and palm pilots. He saw a time when New York content companies would make Silicon Valley technology look dull. He saw glamour and money' (18 October 1999).

The first issue of SAR was a 16 page, double sided Xerox that Calacanis financed by putting $10 000 on his credit card. Initially, he did not have enough money to deliver the copies so he would put four to five hundred copies on a luggage cart and walk around SA dropping them off at company lobbies. He was out every night going from party to party, meeting people and then writing about them in SAR. In time the press built his persona as the 'insider'. In 1998, he started Silicon Alley Daily (SAD)

– the first daily e-mail newsletter dedicated to SA. By early 2000 SAR had become a 200 page magazine and was the biggest trade publication in SA. Similarly, SAD had over 35 000 subscribers making it the most read daily news source on SA. Calacanis also organized networking events. His first event was the Silicon Alley 97 Conference which was repeated in 1998, 1999, 2000 and 2001. Silicon Alley 2001 attracted over 2000 attendees. His other major event was the Silicon Alley Venture Capital Summit, which brought venture capitalists from East and West Coasts together with SA entrepreneurs.

Calacanis's strong personality, arrogance, and hostility towards other key figures in SA put him in the spotlight. SAR was considered by many to be opinionated and biased, yet it was embraced by SA mostly due to its celebrity making power. Frequent appearances in SAR and SAD bestowed a star quality to certain entrepreneurs. Realizing this, Calacanis started Silicon Alley Reporter 100 – an annual list of the top Internet players in SA – in 1997. Soon SAR 100 became the authority on who is 'happening' and who is yesterday's news. The accuracy or lack thereof of the list was less important than its ability to generate buzz. Clay Shirky of Site Specific commented:

> The signal innovation was the Silicon Alley 100: That was the moment. You hated yourself for looking at the list, you hated yourself for thinking you should be on the list, but you had to see – were you on the list? . . . You would see the list and think, how many people here do I know? It was a kind of measure of connectivity . . . The industry was too big for there not to be some arbiter, and since nobody was standing in his way, he [Calacanis] was going to be that arbiter. (Kait and Weiss, 2001, pp. 100–101)

Over time Calacanis came to be known as the medium mogul, the cheerleader of New York's new media scene, celebrity maker and bellwether of the industry. Venture capitalists and entrepreneurs alike used him as a human search engine for lucrative deals. People came to Calacanis because he was 'the connector' (*New Yorker*, 18 October 1999).

Early networking events and SA-specific media played a very important role in the emergence of large scale entrepreneurship in New York. A handful of pioneering actors from diverse walks of life took non-routine actions to bring the scattered Internet pioneers in the city together, which eventually facilitated the emergence of a community around the Internet and an associated identity for the emerging new media industry. The presence of this community made it possible for pioneering actors of various types (such as entrepreneurs, government officials, venture capitalists, real estate managers, media members and so on) to find each other, establish contexts for the creation of positive outcomes by collaborating, and sharing information about realized positive outcomes. Certain actions proved very influential in creating positive outcomes and helping the city pass some of the hurdles mentioned above. Below I discuss these actions.

Transformation of the Technological Infrastructure in New York City

Lower Manhattan Revitalization Plan (LMRP)

New York City entered the 1990s with a large stock of empty office space in Downtown Manhattan, a large demand for residential space and a troubled local economy. The city's mayor, Rudolf Giuliani, responded by establishing a task force that consisted of representatives from government, academic and private sector organizations to develop a plan to improve the economic conditions in Manhattan. The task force presented the LMRP in December 1994. The plan consisted of incentives for the conversion of older buildings into residential use and their renovation in order to entice tenants to renew and/or expand their space, and to attract new tenants. As a result of these incentives, real estate developers converted more than 5500 office units to residential units between 1995 and 2000. Consequently, the area's residential population started to go up, which in turn prompted the establishment of bars, restaurants and entertainment venues, thus transforming the area from a 9 to 5 business district to a 24 hour community. An unexpected consequence of this was that the amenities made downtown a very attractive place for the emerging new media community, because new media workers worked around the clock and needed a location that was alive for 24 hours. The amenities also fueled the emergence of the networking and party scene that established the hip SA culture.

LMRP triggered some entrepreneurial activity in Downtown Manhattan; yet the area's transformation was far from being complete. New media startups were scattered all over downtown without a visible symbol to represent the center of the emerging industry. The next action, the establishment of the New York Information Technology Center, was aimed at creating such a center.

New York Information Technology Center (NYITC)

Giuliani's task force had tentatively recommended the establishment of a technology center that would serve as the epicenter of the emerging technology district in Manhattan. But this project did not gain momentum until January 1995 when Carl Weisbrod, who previously worked at the city's Economic Development Corporation and advocated the establishment of a technology center downtown, became president of the Alliance for Downtown New York (ADNY).[2] Eventually, a request for proposals was issued to real estate developers.

Initially, 11 Broadway was chosen but the deal did not materialize because the co-owner of the building who had wanted the center suddenly died and the remaining partner preferred sticking to conventional rentals. Rudin Management, one of the biggest real estate firms in New York, chose not to bid on the project in the first round. In fact Rudin Management was one of the many large real estate firms that behaved in accordance with widely shared institutional beliefs that was not supportive of the emerging new media startups. Later on Weisbrod contacted Bill Rudin of Rudin Management Company in person and convinced him to undertake the NYITC project at 55 Broad Street, which was vacated by Drexel Burnham Lambert in 1990. The project started in June 1995 and the first tenant moved in December. 55 Broad Street provided many advantages to new media firms including low costs, a collaborative environment, flexible lease terms, 24 hour access, wiring required by technology businesses, credibility from working in a high profile office building and vicinity to Wall Street. By 1998 the building housed 75 leading new media firms and was nearly full.

The establishment of 55 Broad Street created major environmental changes for real estate firms and new media entrepreneurs. First, the building constituted a highly visible central location for the previously scattered new media businesses. In time it became the headquarters of SA providing legitimacy to the new media industry. Second, due to its status as the center of the technology district, it set the standard on technology infrastructure in the city buildings. In time high speed Internet access became one of the biggest selling points for major office buildings as real estate companies imitated 55 Broad Street and built connectivity into their new developments and refits. The action taken by Rudin Management resulted in the establishment of a benchmark for connectivity, and as more real estate developers adopted this benchmark, the expected standard level of connectivity that an average building needed to offer changed.

Despite these developments, increased numbers of new media businesses created an ever increasing demand for wired office space that was mostly unmet. The ADNY responded by taking another non-routine action – the Plug 'n' Go Program.

Plug 'n' Go Program

In 1997 the ADNY collaborated with the city's Economic Development Corporation and several downtown property owners to create Plug 'n' Go. It was a marketing effort that encouraged landlords to upgrade and rewire older buildings to provide Internet ready office space to emerging dot-coms in return for extensive marketing and advertising by the city and the ADNY. The program was a spectacular success. The initially six building program was extended to 15 buildings in 1998. Between 1997 and 2000

242 companies signed leases filling about 600 000 square feet of office space.

The LMRP, establishment of the NYITC and the Plug 'n' Go program, initiated and sped up the process of institutionalization of new real estate practices. As a result, the availability of technology infrastructure (a major weakness of New York in 1996) became one of the greatest strengths of New York City by 1998 (NYNMA Surveys, 1996, 1997). This transformation marked a major environmental change from Manhattan offering an obsolete technology infrastructure not adequate for Internet companies to Manhattan being one of the best wired cities in the nation. This change in turn increased Manhattan's attractiveness for new media firms and coupled with newly instituted tax reductions helped increase the rate of entrepreneurship in the city. Another factor that contributed to the emergence of large scale entrepreneurship in the city was the change in the equity financing environment.

Transformation of the Equity Financing Environment in New York City

The founding of Flatiron Partners and the emergence of a local venture capital market for new media

The founding of Flatiron Partners in 1996 by Fred Wilson and Jerry Colonna was the most important non-routine action related to the emergence of a local venture capital market for new media. Flatiron Partners was the first venture capital firm (VCF) to focus exclusively on New York new media startups. Wilson commented:

> I think what we saw in 1996 was the opportunity to create a venture capital firm that was focused on the Internet here in New York City. Looking back on it, it seems like that was pretty obvious. But at the time it was *not* obvious, and in fact when we announced that we were going to do this, people would say, what, you are going to be local – *and* you are going to focus on a single industry? That's never going to work. (Kait and Weiss, 2001, p. 144, emphases in the original)

Flatiron Partners gradually became the authority for investment strategies in SA after investing the $150 million committed by Chase and Softbank in prominent SA companies such as Starmedia Network and Street.com. In late 1998 Flatiron launched another $300 million fund aimed primarily at New York new media companies. The founding of Flatiron Partners constitutes the starting point of the emergence of a new collective behavior – investing in local new media firms. In four years more than two dozen new media VCFs imitating Flatiron's business model were founded in New York. Flatiron Partners played a major role in legitimizing investments in local new media companies.

Another important non-routine action was the New York City Investment Fund's decision in 1997 to invest $150 000 in a local new media startup – Starmedia Network. The New York City Investment Fund was created under the auspices of the New York City Partnership and Chamber of Commerce as a coalition of 60 individuals and companies, such as Citicorp, J.P. Morgan, Westinghouse and GE and media enterprises, such as CBS and NBC that represent the cream of New York's media and investment elite. Later the Fund announced that it set aside 10 per cent of its current capitalization ($6 million) for investments in SA. The fund was an introduction for its high powered investors to SA and constituted the trigger for a newly institutionalized collective behavior on behalf of institutional investors. Investing in SA was now acceptable.

In the meantime established New York VCFs, such as Prospect Street Partners, Sprout Group, Patricof and Co., Venrock Associates and Allen and Co. observed the positive outcomes to investing in new media and started investing in SA startups. Even commercial banks, such as Fleet Bank, that had avoided new media earlier started to reach out to the burgeoning sector. Investing in new media started by Flatiron Partners had become an institutionalized practice for the new and traditional VCFs in the area.

Seed stage funding and the emergence of angel investors and incubators

Seed funding is aimed at companies with a product under development but yet not fully operational. Institutional VCFs tend to avoid seed funding since the returns do not justify the efforts, or the time spent on the company. Thus, despite New York being a financial capital, small businesses had a hard time raising startup money. SA entrepreneurs were complaining that 'it was easier to raise $5 million in New York than it is to raise $500,000' (Internet.com, 6 August 1999).

Most seed funding is typically provided by 'angels' – high worth individuals with an interest in a particular industry, often because that is where they gained their wealth. Unlike Silicon Valley, New York lacked such people who understood technology businesses since the area lacked an entrepreneurial tradition. Towards the end of the 1990s, as the investments of a few pioneering angels paid handsomely through highly successful IPOs of their portfolio companies, others started imitating them. In time, a large group of angel investors who were willing to provide seed funding for SA companies emerged. Seed funding also became more institutionalized during early 1999. Wit Capital's $20 million Dawntrader Fund and Silicon Alley Partners, an independent VCF, both started to provide seed funding. The trend to provide seed funding also gave rise to the emergence of a new group of actors: Internet incubators. By the end of 2000 about 30 Internet incubators (for example, I-hatch Ventures, Business Incubation Group,

Doublespace, Efinanceworks, Gas Pedal Ventures and Global Bay) had sprung up in New York. In addition to these local developments SA was also able to attract the attention of West Coast VCFs.

Valley VCFs move into SA

The most important action aimed at attracting West Coast venture capital was the organization of the annual 'Alley to the Valley' conference that began in 1998 and was hosted by the New York City Economic Development Corporation. In four years 114 companies flew out to San Francisco to present their business ideas and seek funding. In the 1998 conference the goal was to educate the West Coast VCFs about businesses in the East Coast. By 2000 investing in SA was institutionalized so much so that the theme of the event was 'The Valley VCFs are moving to the Alley'. Famous West Coast VCFs, such as Draper Fisher Jurvetson, Impact Venture Partners, Bessemer Venture Partners and Accel Partners, all established a presence in New York.

The early actions listed above caused a once money starved SA to become one of the regions that attracted the highest amount of risk capital to early stage Internet ventures. While these actions were crucial in terms of bringing risk capital into SA, it was the strong stock market and a few record breaking SA IPOs (that is, idiosyncratic success stories) that reinforced these actions.

Successful IPOs – 'SA becoming real'

As the stock market embraced IPOs from very young companies that barely had any revenue, companies that traditionally needed second, third and fourth rounds of financing needed fewer numbers of rounds. This provided VCFs with fast and profitable exit opportunities and made the Internet a favorable investment opportunity. The first significant SA IPO was by DoubleClick. The day of DoubleClick's IPO – 20 February 1998 is known as 'the day SA became real'. The company raised $60 million in its IPO with a market capitalization of $518 million. On 11 November 1998 Earthweb did an IPO with a 248 per cent gain on its opening price making it the second biggest debut in IPO history. Just two days later TheGlobe.com went public with a 605 per cent rise from its opening price, setting a record as the fastest gainer in the history. These idiosyncratic success stories put New York on the IPO map. Content businesses were legitimized and financiers all around the nation became aware of SA. In 1998 and 1999 a total of 49 SA companies went public successfully. The stock market's embracing of IPOs by the very young SA companies presented VCFs with profitable exit opportunities and increased the legitimacy of new media as an investment opportunity.

A major consequence of the high flying IPOs was the resulting abundance of cash for Internet startups which turned the venture capital market into a buyers' market. SA entrepreneurs who were once willing to do anything to raise money were now being chased by VCFs. Consequently, they became arrogant and more demanding. VCFs that traditionally chose their portfolio companies based on a thorough evaluation of factors, such as cash out potential of the business, stage of development, the entrepreneurial team and soundness of the business model, started disregarding the traditional criteria and invested in unproven businesses with uncertain business models. They rationalized their investments primarily by referring to a 'new economy' in which traditional rules did not apply. Investment bankers' behaviors changed as well. A classic IPO candidate would be profitable, have a consistent track record of growth and earnings, and would have some predictability as to future growth. Most dot-coms that were made public by reputable investment banks did not have these characteristics. Just as VCFs were willing to invest in any company that can complete an IPO, investment banks were willing to sell any company that investors were willing to buy. Investment banks lent credibility to the sky high valuations of these companies by attaching their names and thus their reputations to these companies' IPOs. The outrageous IPO valuations even caused some observers to believe that the IPO market was operating based on the 'theory of greater fool'.

These developments resulted in two major changes for SA. First, an initially unknown SA became a legitimate investing opportunity for investors nationwide. Second, the newly gained legitimacy transformed SA from a struggling new media cluster to a full blown entrepreneurial center flushed with risk capital.

The favorable reaction of the public and private equity markets to SA influenced employment trends in the city significantly. Initially, people were hesitant to work at new media startups that did not offer good salaries or good working conditions. Those who did choose employment in SA firms were viewed as romantic risk takers – pioneering actors who were making choices not favored by the majority of the workforce in the city. This changed as new media businesses raised large sums of money and their working conditions improved. Consequently, talent started to flow into SA. Within three years new media employment in the city quadrupled. Around 30 per cent of SA employees came from Silicon Valley and other cities, such as Los Angeles, Seattle, Boston and Chicago. But the largest proportion of new media recruitment came from within the New York tri-state area. In time employee migration to new media became a big concern for traditional finance, advertising and entertainment firms as social learning amongst the workforce in New York caused many to quit their 'traditional' jobs for new media jobs. As a job in a new media startup became institutionalized, working for an Internet company became 'the hot thing to do' when once it was something the majority did not even consider.

The Decline and Collapse of Silicon Alley

The flood of a large amount of risk capital into SA led to many questionable deals as less savvy investors moved quickly to cash in on Internet mania. Companies that continually reported losses raised large amounts of capital. Crain's *New Yorker* article, titled 'Gold rush attracts new age opportunists', said:

> That an inexperienced twenty-something entrepreneur could raise $10 million from sophisticated investors is a story told many times over in SA. With unproven technologies and half-baked business models, a generation of technophiles has received vast sums of money because venture capitalists, who would have required exacting credentials of any fledgling manufacturer or retailer, have applied few of the standard tests when it comes to new media. Starry eyed investors have ignored the chaotic management styles, the constantly changing business plans and the astronomical levels of staff turnover. (*New Yorker*, 16 November 1998)

What was started by a handful of early true believers as a dedicated, passionate exploration of the Internet turned into a gold rush. As people observed the actions of early entrepreneurs, the successful IPOs and the overnight wealth that came with those IPOs, they imitated the pioneers to get a piece of the pie. The financing craze created a flood of firms founded on a different ideology than that of the early true believers. Kyle Shannon of Agency.com commented:

> . . . the gold rush probably started in '97 going into '98. . . . Because prior to that, the only reason people were quitting their day jobs and starting businesses was because they passionately believed in something. And after that, no one really gave a shit if the business was good or not – no one was passionate. It's like, 'Oh, he's doing a travel thing? I could do a travel thing – fuck it. Give it a different name, get a new URL, IPO baby, heah, whoo'. (Kait and Weiss, 2001, p. 155)

These new firms received large sums of money from public and private equity markets and then spent it on costly growth strategies that did not create performance outcomes, such as a steady stream of revenues, development of a proprietary technology or profits that markets traditionally favored. The typical business model was characterized by rapid, excessive growth fueled by external capital towards fast IPOs. Several factors made a rapid IPO very desirable for private SA firms. First, as more SA companies completed successful IPOs, private companies started facing immense pressure to go public to be able to stay competitive. Second, companies needed to go public to retain their customers and employees and attract new ones. A rash of successful Internet IPOs had heightened the value of stock options as incentives for employees and proceeds from offerings could help pay top dollar for digital talent that was scarce in New York. Also an IPO had PR value in that it helped companies stay in the public's mind and keep their customer base. Under these pressures fledgling Internet startups' ultimate goal became to establish a brand name as fast as possible to occupy an online niche and then go public. The way to do that was rapid growth. The munificent equity financing market, coupled with the tolerance for non-profitability, made raising money to pursue costly growth strategies the key activity for new media entrepreneurs. The level of spending far exceeded the ability of many of these startups to generate revenue. The following quote is very telling:

> There were companies that could have stopped growing, and could have created a profit, and would have done so based on good management. Whether highly experienced or relatively inexperienced, everybody bought into the proliferation idea: 'We are going to keep spending as long as someone is willing to fuel that growth'. (Alice Rodd O'Rourke of New York New Media Association in Kait and Weiss, 2001, p. 305)

With this mindset companies spent their venture capital and IPO dollars in excessive advertising, excessive hiring and acquisitions to achieve rapid growth. Spending on advertising became an increasingly larger part of a startup's budget as VCFs coached their portfolio companies to get big fast or else . . . Some companies were spending up to 70 per cent of their overall budget on marketing. In time marketing became an end in itself. Companies were also spending excessively to attract high tech talent. Scarcity of technical talent forced Internet companies to offer sky high salaries and stock options to lure talent away from Wall Street. Salary levels, especially for chief information officers, were increasing at 6 per cent to 8 per cent a year, twice the rate of pay increases in general. Another costly growth strategy pursued by new media companies was acquisitions. The explosion of dot-com mergers lifted the number of transactions involving New York companies up 42 per cent, from 730 in 1998. The second most active industry for mergers and acquisitions in New York was new media, with 108 announced deals in 1999. New York companies were involved in 11 per cent of all the mergers and acquisitions nationwide in 1999. Most deals were done by companies that were just trying to move fast, without assessing the gains from these deals. Many companies such as 24/7 Media, EarthWeb, Theglobe.com and DoubleClick went back to public markets for secondary offerings to finance acquisitions and internal expansion because they recognized the need to keep growing rapidly to justify their stratospheric valuations.

NASDAQ crash, the beginning of the end

The bitter moment for the Internet came at the beginning of April 2000 when NASDAQ started its nosedive. By 14 April the index had gone down to 3321 from its peek of 5048 on 10 March. The most important implication for SA was the abrupt change in the public and private equity markets. The IPO window for Internet businesses slammed shut. More than 20 SA companies that had filed to go public before the crash had to withdraw their IPOs. The closing of the IPO window meant that fledgling dot-coms with high burn rates had to find alternative sources for money. Most startups went back to their private investors for additional financing whereas some others started to look for strategic partners. Similar to the reaction of public markets, most VCFs cut seed stage funding significantly and adopted a triage approach to focus on their existing portfolio of companies rather than chase new deals. Of the 85 local deals closed in the third quarter of 2000, only 23 (27 per cent) were classified as seed stage or first round financing compared to 40 per cent in 1999. In choosing which portfolio companies to focus on VCFs started imposing more stringent standards. They started demanding that entrepreneurs have business strategies that showed a clear path to profitability, detailed spending plans and experienced management. Portfolio companies were pushed by VCFs and Wall Street alike to be at least cash flow positive by the end of 2001 as a benchmark of financial progress. As it got harder to raise capital, the market punished companies that were seeking additional money. Companies seeking first round funding at valuations of $10 million to $15 million were cut to $6 to $7 million. At later stages businesses previously valued at $50 million were now worth only $20 million.

Immediately after the NASDAQ crash, hiring in SA slowed down significantly and companies started laying off employees. By January 2001 5000 SA workers had already been laid off due to businesses folding or cutting back. The atmosphere of mistrust created by the layoffs, the chaotic work environment in SA and most importantly the devaluation of stock options due to the market correction resulted in many technology workers quitting the industry and going back to safe jobs in established companies. Employment in Internet companies became a risky choice that did not constitute the best career move.

As more and more Internet companies folded, doomsayers such as fuckedcompany.com, dotcomfailures.com and compost.com emerged to report the fates of failing dot-coms. First went the e-commerce sites, then the community sites and then the content sites. High flying interactive agencies that served these companies came next. Survivors had lost over 90 per cent of their value by the end of the year. Many SA companies that had been darlings of NASDAQ during the boom had been delisted from the index by the end of 2001 due to not being able to meet the trading requirements of the index. For example, TheGlobe.com, the holder of the record of the fastest gaining IPO in history, the company that is believed by some to have started the Internet boom, was delisted in April 2001 and folded in August 2001. Other delisted companies include highrollers, such as Bigstar, Interworld, Promotions.com, Opus 360, VitaminShoppe.com, Learn2.com and TheKnot.com.

In March 2001 Doubleclick removed the giant billboard on the rooftop at 22 Broadway that proclaimed 'DoubleClick Welcomes you to Silicon Alley'. The billboard had long served as the region's unofficial gateway. As SA kept shrinking, Silicon Alley Reporter folded in October 2001. The editor of the magazine, Jason McCabe Calacanis, said: 'The story is over, you can't have a magazine about unemployed people'. Days after the demise of Silicon Alley Reporter, the second trade magazine, AlleyCat News, which had gone bi-monthly in mid 2001, ended its five year run on 14 December 2001. The lavish party scene disappeared as well. Most companies stopped throwing parties and survivors became low-key events – pretzels instead of sushi, strictly cash bar and an office space instead of a famous downtown club. The events were also more focused, selective and mostly invitation only, designed to introduce key players to one another. The latest and final New York New Media Industry Survey conducted by the New York New Media Association and PriceWaterhouseCoopers in 2001 was only 12 pages compared to the 57, 59 and 52 pages of the 1996, 1997 and 2000 surveys, respectively. The number of respondents in the survey also dropped to 152 from 751 in 2000. Reflecting the demise of Silicon Alley, the 2001 survey was mostly about the sentiments of the industry insiders about the future of SA, and avoided providing any detailed statistics on the number of businesses or employment as in the previous surveys. On 24 March 2002, a *New York Times* article, titled 'Silicon Alley; a once evocative name falls victim to the bursting of the high-tech bubble', announced the death of SA. Finally, New York New Media Association, the trade association of SA, closed down its operation as of 13 December 2003.

CONCLUSIONS AND LESSONS LEARNED

Starting from the early 1990s to the end of the millennium, New York City transformed from an economically troubled city with no entrepreneurial tradition to a burgeoning entrepreneurial center. This rapid growth was followed by an even faster decline and the eventual collapse of the newly emerged new media cluster after the stock market crash of April 2000. Several explanations can be brought to bear on the emergence of this regional cluster. For example, population ecologists would view the case as one of the emergence of a new regional organizational population and would try to explain the growth of the cluster by recourse to the concept of population density (Aldrich, 1999; Baum and Oliver, 1996). Economic geographers would propose agglomeration economies as the primary force behind the growth of the cluster (Marshall, 1920; Storper, 1997). Complexity theorists would say that the emergence and the growth of the cluster are due to the amplification of accidental early events in the history of the region through increasing returns and positive feedback loops (Arthur, 1994; Chiles et al., 2004). Others would propose that the emergence of the cluster is associated with the presence of a financial bubble in the stock market and associated herd behavior on behalf of investors (Kait and Weiss, 2001; Mills, 2002). One or more of these processes may indeed be in place but the case discussion above points to another dimension of regional entrepreneurial change and that is institutional transformation.

The emergence of the new media cluster in New York City was accompanied by major changes in beliefs and institutionalized practices of different types of actors. For example, governmental institutions that were previously doubtful about the potential of Internet entrepreneurship changed and provided full support to new media through various types of incentives. Real estate developers that initially did not prefer to upgrade their office buildings changed and directed their efforts to providing fully equipped office space to new media entrepreneurs. Venture capitalists who did not view Internet companies as favorable investment opportunities changed and established funds directed particularly at early stage Internet startups. The workforce in the area initially did not view employment in Internet startups as a favorable career choice but later on herds of qualified workers quit their traditional jobs to work for Internet companies. Traditional media that previously ignored the Internet later on spoke about a 'new economy' and the new ways of doing business that the Internet created. Support service providers, such as accounting firms, law firms and consulting firms, changed their typical business practices and created a new 'pay by equity' system to cater to the specific needs of new media startups. Similarly, venture capitalists and investment banks changed their criteria for screening investments and underwriting IPOs, respectively. After the NASDAQ crash, a deinstitutionalization process took place such that the newly institutionalized practices during the emergence of the cluster became unacceptable. These changes on behalf of different types of actor groups did not occur independently of each other but rather co-evolved as each change became a trigger for other changes on behalf of interdependent actors' practices. Several lessons can be drawn from the case with respect to the institutional transformation process that is associated with these changes.

First, institutional transformation involves many different types of actors in addition to entrepreneurs. The case account above suggests that entrepreneurship does not take place in a vacuum, but rather a variety of interdependent actors, including local socialites, media institutions, real estate developers, financiers (including venture capitalists, angel investors and institutional investors), local workforce and government agencies, are implicated in the process of entrepreneurial transformation. Commercialization of a new technology requires entrepreneurs to mobilize contributions from a variety of actors (for example, financing from financiers, labor from the local workforce, coverage of positive outcomes from media institutions, facilities with appropriate infrastructure from real estate developers, incentives from the government and so on). However these contributions are not readily available since what the entrepreneurs would like these different actors to do are most likely in conflict with accepted norms of doing business in the region. In other words, lack of cognitive and sociopolitical legitimacy around the new technology makes it difficult for entrepreneurs to mobilize required contributions.

An important factor in the initiation of institutional transformation is the presence of individuals from all these subgroups who are willing to take risks by engaging in actions that are not justified by the selection environment associated with currently institutionalized practices. These are pioneering individuals who see potential (that others do not yet see) associated with the emerging technology and are willing to take non-routine actions in an effort to initiate change. Accordingly, they constitute the micro foundations for macro transformation. These individuals are likely to have the characteristics, such as need for achievement, internal locus of control, high risk taking propensity and tolerance for ambiguity, that traits approach researchers have identified (Brockhaus and Horwitz, 1986). In the language of evolutionary economics, these individuals are the actors that introduce variation into a constantly evolving economic system and start shaping the selection environment in accordance with the needs of the newly emergent technology (Lambooy and Boschma, 2001). While the presence of such individuals is crucial for an entrepreneurial transformation, it is by no means sufficient.

The second important factor in the initiation of institutional transformation is the presence or creation of a medium where these pioneering individuals can come together and create an early community around the new technology. In the absence of such a community, pioneering individuals remain as lonely, delinquent actors or dreamers who favor actions that are not in line with institutionally accepted practices. The establishment of an early community provides pioneers with a distinct identity which in turn isolates them from traditional industries and initiates the process of legitimization of their non-routine actions (Garud and Rappa, 1994). Of particular importance for the creation of an early community are networking events and the presence of actors who organize them. These actors are typically highly connected individuals or institutions who use their connections to establish ties between pioneering actors so that they can mobilize each other's support. In this regard the presence of an active social scene and venues, such as bars, restaurants and so on, in the region where pioneers can meet each other, exchange ideas and establish ties is crucial for institutional transformation. The case discussion illustrates how the early networking activities of actors such as Jamie Levy, Mark Stahlman, Jason McCabe Calacanis and Courtney Pulitzer, and the vibrant social scene in Downtown Manhattan proved to be influential for the creation of an early new media community in SA.

The third important factor in the initiation of institutional transformation is the creation of media outlets that will distribute information about the emerging community, their activities or outcomes and the opportunities associated with nascent entrepreneurship to the world outside the early community. Traditional media may ignore the early community or otherwise present a negative picture of the actions of its members. Therefore creation of new outlets by pioneers may be required. Media portrayals of key actors, businesses and their behaviors and outcomes affect the meaning making process by which the public constructs the social image or identity of an emerging community. A favorable social image created by the media generates support for participation in the activities of the early community members and helps establish cognitive legitimacy. Furthermore distribution of information through media outlets facilitates others' adoption of the non-routine actions of pioneers through social learning and hence extends the early community by the addition of new members. The more actors adopt the non-routine actions of pioneers, the bigger the early community becomes and the more likely the non-routine actions of its members are to become institutionalized.

The very early adoption of non-routine practices may be facilitated purely by an interest in the new technology or even by people having nothing better to do, as illustrated by Rufus Griscom's quote earlier. However large scale adoption requires the presence of largely publicized, idiosyncratic success stories. For example, news about SA had been circulating through the new media community in New York but everybody else in and outside New York heard about SA after the record breaking IPOs of a few SA companies, such as DoubleClick, Earthweb and TheGlobe.com. Idiosyncratic success stories provide proof that the non-routine actions of pioneers who were acting against institutionalized norms were indeed justified. In other words, the early community's beliefs about the potential of the new technology were more than just a romantic pursuit of a non-realizable dream but rather the pursuit of a vision that will lead to profitable business establishments. Furthermore these types of success stories are likely to generate wide coverage by traditional media which helps dissemination of information to wider audiences beyond the region. Consequently, idiosyncratic success stories contribute in a major way to the establishment of cognitive and sociopolitical legitimacy.

At this juncture it would be in order to discuss the role of financial markets on the emergence (and decline) of SA in particular and regional entrepreneurial transformation in general. The explosive growth of SA coincides with what is referred to as the dot-com bubble (Ljungqvist and Wilhelm, 2003). Whether SA would have grown so big (and as fast) in the absence of a strong stock market is a question whose answer will not be known. Perhaps the growth of the cluster would have been much slower and the cluster more sustainable. Regardless, the role of the stock market bubble on the area's entrepreneurial transformation (particularly with respect to the decline and eventual collapse of the cluster) cannot be ignored. The positive investor sentiment associated with the strong stock (and IPO) market of the late 1990s influenced the institutional transformation around the growth of SA in major ways.

First, it increased the availability of risk capital to the newly emerging entrepreneurial ventures and provided successful exit opportunities for risk capital providers. Investors commit money to venture capital funds only if VCFs can provide returns that are higher than the opportunity cost of capital (Gompers and Lerner, 1999). Given that VCFs' returns materialize when they exit their investments, the status of exit markets is central to the operation of the venture capital industry. VCFs typically exit their investments through IPOs, acquisitions, buybacks, liquidations or write-offs. Amongst these options, IPOs provide the highest return (Soja and Reyes, 1990). The higher returns that IPOs generate make them the desired exit mode for VCFs. In fact VCFs consider alternative exit modes only when they judge that the company will not be able to do an IPO (Gompers and Lerner, 1999). Given that the most profitable exit mode for VCFs is IPOs, hot IPO markets – periods during which IPOs generate extremely high returns (Ritter and Welch, 2002) – are associated with high levels of new commitments to venture capital funds. In addition to the large amounts of new commitments to venture capital funds, hot IPO markets create an environment in which VCFs can exit lesser quality investments through a successful IPO due to positive investor sentiment (Lowry, 2003) and the associated increased receptivity of the stock market to lower quality firms (Ritter and Welch, 2002).

A favorable response from financial markets in return signals opportunities and provides financial incentives to potential entrepreneurs. Consequently, an entrepreneurial boom may start during hot IPO markets. A caveat is that firms founded during such booms are not necessarily high quality firms. For example, Florida and Kenney (1990) argue that entrepreneurial booms are associated with a 'startup mania' in which entrepreneurs with low quality ideas rush to the marketplace with copycat product designs to cash in on the latest technology fad. Similarly, Barnett et al. (2003) provide evidence that entrepreneurial booms are associated with a more lenient entry selection environment and thus the entry of lower quality firms. Their evidence suggests that firms that enter during entrepreneurial booms experience higher death rates.

The hot IPO market during the Internet boom helped SA generate the idiosyncratic success stories that helped legitimize the emerging cluster; on the other hand, it may be argued that it caused the institutionalization of unproductive practices. Venture capitalists and investment bankers abandoning the traditional criteria to screen investments and underwriting IPOs are examples. These changes in financial markets in return facilitated the institutionalization of a business model whereby SA firms would aim to raise large sums of money from public and private equity markets and then spend it on costly growth strategies that did not create performance outcomes, such as a steady stream of revenues, development of a proprietary technology or profits that markets traditionally favored. The newly institutionalized business model for new media startups was characterized by rapid, excessive growth fueled by external capital towards fast IPOs. Accordingly, when the hot IPO market came to an end with the stock market crash and the newly institutionalized practices were no more acceptable, firms founded based on these practices started to fail. The SA example thus demonstrates the extent to which macro financial markets can be influential in terms of determining a region's path of entrepreneurial transformation. A strong stock market may speed up the process of regional growth to a great extent by channeling much needed investment money to entrepreneurial startups. On the other hand, temporary speculative bubbles, especially if they last a long time, may systematically legitimize unproductive business practices in a self-reinforcing manner which may result in regional growth that is superficial and non-sustainable.

From a regional policy making perspective, the case of SA points to the importance of the creation of buzz and an early community around emerging technologies. The generation of new technologies in a region is a major policy problem in itself and several solutions, such as providing training to create a qualified local workforce and establishing local universities and research institutions, have been previously proposed (McQuaid, 2002). In the case of SA the technology (that is, the Internet) was not created locally but its commercialization took place at a regional level due to the successful transformation of the institutional environment in the city. New York was privileged as a metropolis and enjoyed urbanization economies due to the presence of many established industries, such as finance, entertainment and media. Accordingly, it housed many individuals that had the potential to be pioneers during the commercialization of an emergent technology. The fact that the economic recession prior to the emergence of the Internet left most such individuals unemployed also helped. For regions that do not enjoy urbanization economies a major priority should be generating diversity in the region so that pioneers who are willing to go against institutional norms would emerge. This could be achieved by making the region an attractive place to live for a diverse population. A nice physical environment, high quality education and medical facilities, entertainment venues where a social life can flourish and superior means of transport and communication are prerequisites. Beyond these infrastructural requirements, local governments need to keep a close watch on emerging technologies and facilitate the emergence of early communities around promising new technologies. This may be achieved primarily through establishment of mentoring programs, organization of networking events and creation of outlets where outcomes of pioneers can be publicized. In addition government incentives should not be directed solely at entrepreneurs but also other actors whose contributions are needed for successful entrepreneurship. For example, government incentives were key in the establishment of the NYITC and the Plug 'n' go Program which facilitated the transformation of the technology infrastructure in New York City office buildings.

More generally, developmental policies should be custom designed to meet a region's particular needs and path dependent historical development (Lambooy and Boschma, 2001). This contrasts with the currently predominant approach of trying to find optimizing policies that have universal applicability. Local policies should co-evolve within the specific context of the regional system as policy makers improve their understanding of the self-organizing mechanisms at work in the region through a trial and error process, and implement incremental policies along the way that are aimed at short-term changes rather than long-term, pre-defined targets.

NOTES

1. http://www.ite.poly.edu/55case/begin.htm.
2. ADNY is a commercial organization that collaborates with the local government and other organizations in the area to provide Manhattan's historic financial district with a premier physical and economic environment.

REFERENCES

Aldrich, H.E. (1999), *Organizations Evolving*, Thousand Oaks, CA: Sage Publications.

Aldrich, H.E. and C.M. Fiol (1994), 'Fools rush in? The institutional context of industry creation', *Academy of Management Review*, **19**, 645–70.

Arthur, W.B. (1994), *Increasing Returns and Path Dependence in the Economy*, Ann Arbor, MI: Michigan University Press.

Astley, W.G. (1985), 'The two ecologies: population and community perspectives on organizational evolution', *Administrative Science Quarterly*, **30**, 224–41.

Barnett, W.P., A.N. Swanson and O. Sorenson (2003), 'Asymmetric selection among organizations', *Industrial and Corporate Change*, **12**, 673–95.

Baum, J.A.C. and C. Oliver (1996), 'Toward an institutional ecology of organizational founding', *Academy of Management Journal*, **39**, 1378–427.

Brockhaus, R.H. and P.S. Horwitz (1986), 'The psychology of the entrepreneur', in D.L. Sexton and R.W. Similor (eds), *The Art and Science of Entrepreneurship*, Cambridge, MA: Ballinger.

Bruno, A., V. and T.T. Tyebjee (1982), 'The environment for entrepreneurship', in C. Kent, D. Saxton and K. Vesper (eds), *Encyclopedia of Entrepreneurship*, Englewood Cliffs, NJ: Prentice Hall.

Chiles, T.H., A.D. Meyer and T.J. Hench (2004), 'Organizational emergence: the origin and transformation of Branson, Missouri's musical theatres', *Organization Science*, **15**, 499–519.

DiMaggio, P.J. and W.W. Powell (1983), 'The iron cage revisited: institutional isomorphism and collective rationality in organizational fields', *American Sociological Review*, **48**, 147–60.

Florida, R. and M. Kenney (1988), 'Venture capital and high technology entrepreneurship', *Journal of Business Venturing*, **3**, 301–19.

Florida, R. and M. Kenney (1990), 'Silicon Valley and Route 128 won't save us', *California Management Review*, **33**, 68–85.

Gartner, W.B. (1988), '"Who is an entrepreneur?" is the wrong question', *American Journal of Small Business*, Spring, 11–32.

Garud, R. and M.A. Rappa (1994), 'A socio-cognitive model of technology evolution: the case of cochlear implants', *Organization Science*, **5**, 344–62.

Gnyawali, D.R. and D.S. Fogel (1994), 'Environments for entrepreneurship development: key dimensions and research implications', *Entrepreneurship Theory and Practice*, **18**, 43–62.

Gompers, P.A. and J. Lerner (1999), *The Venture Capital Cycle*, Cambridge, MA: MIT Press.

Hannan, M.T. and J. Freeman (1984), 'Structural inertia and organizational change', *Ameican Sociological Review*, **49**, 149–64.

Indergaard, M. (2004), *Silicon Alley, the Rise and Fall of a New Media District*, New York: Routledge.
Kait, C. and S. Weiss (2001), *Digital Hustlers*, New York: Regan Books.
Lambooy, J.G. and R.A. Boschma (2001), 'Evolutionary economics and regional policy', *The Annals of Regional Science*, **35**, 113–31.
Ljungqvist, A. and W.J. Wilhelm (2003), 'IPO pricing in the dot-com bubble', *Journal of Finance*, **58**, 723–52.
Lowry, M. (2003), 'Why does IPO volume fluctuate so much?', *Journal of Financial Economics*, **67**, 3–40.
Markusen, A., P. Hall and A. Glassmeier (1986), *High-Tech America: The What, How and Why of the Sunrise Industries*, Boston, MA: Allen & Irvin.
Marshall, A. (1920), *Principles of Economics*, 8th edn, London: Macmillan.
McQuaid, R.W. (2002), 'Entrepreneurship and ICT industries: support from regional and local policies', *Regional Studies*, **36**, 909–19.
Mills, Q.D. (2002), *Buy, Lie, and Sell High: How Investors Lost Out on Enron and the Internet Bubble*, Upper Saddle River, NJ: Pearson Education.
Nelson, R.R. (1995), 'Recent evolutionary theorizing about economic change', *Journal of Economic Literature*, **33**, 48–90.
New York New Media Association Survey (NYNMA), 1996, 1997, 2000, New York New Media Association, New York: Coopers & Lybrand.
Oliver, C. (1991), 'Strategic responses to institutional processes', *Academy of Management Review*, **16**, 145–79.
Phan, P.H. and M.D. Foo (2004), 'Technological entrepreneurship in emerging regions', *Journal of Business Venturing*, **19**, 1–5.
Ritter, J.R. and I. Welch (2002), 'A review of IPO activity, pricing, and allocations', *Journal of Finance*, **57**, 1795–828.
Romanelli, E. (1989), 'Organization birth and population variety: a community perspective on origins', in L.L. Cummings and B. Staw (eds), *Research in Organizational Behavior*, **11**, Greenwich, CT: JAI, pp. 211–46.
Romanelli, E. (1991), 'The evolution of new organizational forms', *Annual Review of Sociology*, **17**, 79–103.
Saxenian, A. (2000), *Regional Advantage: Culture and Competition in Silicon Valley and Route 128*, Cambridge, MA: Harvard University Press.
Schoonhoven, C.B. and K.M. Eisenhardt (1993), 'Entrepreneurial environments: incubator region effects on the birth of new technology firms', in M.W. Lawless and L. Gomez-Mejia (eds), *High Technology Venturing*, Greenwich, CT: JAI Press.
Scott, W. (1995), *Institutions and Organizations*, Newburry Park, CA: Sage Publications.
Soja, T.A. and J.E. Reyes (1990), *Investment Benchmarks: Venture Capital*, Needham, MA: Venture Economics Inc.
Sterman, J.D. (2001), 'System dynamics modeling: tools for learning in a complex world', *California Management Review*, **43**, 8–211.
Storper, M. (1997), *The Regional World*, New York: Guilford Press.
Thornton, P.H. (1999), 'The sociology of entrepreneurship', *Annual Review of Sociology*, **25**, 19–46.
Van de Ven, A.H. (1993), 'The development of an infrastructure for entrepreneurship', *Journal of Business Venturing*, **8**, 211–30.
Venkataraman, S. (2004), 'Regional transformation through technological entrepreneurship', *Journal of Business Venturing*, **19**, 153–67.

PART 2

Government and non-governmental organization influences on entrepreneurship in emerging regions

4. Institutional entrepreneurship in the emerging regional economies of the Western Balkans

Denise Fletcher, Robert Huggins and Lenny Koh

INTRODUCTION

As evidenced in the monitoring and review efforts that have begun to feature in entrepreneurship inquiry (Sarasvathy, 2000; *Entrepreneurship Theory and Practice*, 2001; *Journal of Management*, 2003; Steyaert and Hjorth, 2003), significant progress has been made in identifying the range of theoretical resources for understanding how and why entrepreneurial activities 'come about' in various contexts. This work takes account of the characteristics of entrepreneurial activity in particular types of personalities, people, teams, cultures, neighbourhoods, communities, organizations, industries and economies of the world. Each emphasis has been important for facilitating a wide range of conceptual and methodological approaches to investigating entrepreneurial people, policies and practices. Each has been responsive to a range of disciplines to theorize this activity, indicating that entrepreneurship research has its intellectual roots in a variety of social science disciplines as illustrated by Swedberg (2000).

This diversity of practice is further acknowledged in inquiries that investigate how entrepreneurial activities occur in different local or regional economic, social and cultural contexts. Here research efforts concentrate on examining the relationship between entrepreneurship and its expression in different social, economic milieu (Hjorth and Johannisson, 2003), communities (Johannisson, 1990), industrial districts (Amin, 1994; Pyke et al., 1990) and regional networks (Butler and Hanson, 1991). But over the last ten years or so the local or regional embeddedness of entrepreneurial activity has been overshadowed by the more recent 'opportunity discovery' line of inquiry in entrepreneurship research. The opportunity recognition frameworks (including Shane, 2000; Shane and Venkataraman, 2000; Lumpkin et al., 2003) collectively offer a range of concepts (that is, networks, experience, ideas sharing, prior knowledge of markets, entrepreneurial alertness) that characterize what is at the core of entrepreneurial activity. However much of this inquiry tends to overemphasize the actual opportunity recognition or discovery processes rather than the personal, product, service, organizational, industry, regional or institutional transformations that are brought about because of opportunity enactments. Also it is rare to see empirical studies that examine the patterning of inter-organizational relationships in a particular community, locality or region and which converge over time to create what Van de Ven (1993) refers to as an infrastructure of entrepreneurship. For this reason in this chapter we examine the processes through which resources are enacted and mobilized by a network of organizations and local actors as they strive to achieve regional economic interests. In so doing, we draw attention to the transformative effects of local (entrepreneurial) action strategies – strategies which converge and help to diffuse new (and transform old) institutional arrangements and contribute to regional emergence.

The empirical research reported here examines the Balkans part of Europe. In particular we investigate the Western Balkans region, which the European Union (EU) designates as encompassing Croatia, Bosnia, Serbia-Montenegro, Kosova, Macedonia and Albania. For our study we draw upon three of these (Serbia-Montenegro, Kosovo and Macedonia), all of which were republics in the former Yugoslavia. This region is politically distinctive because, in addition to having been devastated by war, ethnic tension and out-migration, it is also facing severe post-socialist economic restructuring and is struggling to achieve cultural and ethnic stability.

For theoretical purposes the notion of institutional entrepreneurship is utilized. By this we mean the enactment and patterning of inter-organizational relations that enable the transformation of existing or emergence of new socio-economic institutions. This conceptualization of institutional entrepreneurship is distinctive on several levels. First, institutional theory has not been widely applied in empirical studies of emerging regions (Hoskisson et al., 2000). Second, although institutional theory is quite often concerned with organizational fields (or the aggregation of organizations that constitute 'a recognized area of life' (DiMaggio and Powell, 1983, p. 148), the ways in which institutions are transformed at a local level through individual acts of agency are under-examined (Lawrence and Phillips, 2004). However, rather than focus on special individuals as the key agents of institutional entrepreneurship, we draw attention to the cumulative effect of local action strategies in enacting entrepreneurship. By focusing on the patterning of inter-organizational relations we can also examine the relationship between entrepreneurship and regional emergence.

In the following section we discuss the relationship between entrepreneurship and the emergence of regions. We then relate this discussion on regional emergence to the notion of institutional entrepreneurship, and add case material and some empirical problems and issues (engagement with problems in the world). A review of concepts from entrepreneurship and institutional theory allows the formulation of research questions, which enable the study of the Western Balkans as an inter-organizational (regional) field. Following an outline of the research methodology, an analysis of the data is undertaken in order to examine the inter-organizational patterning of institutional entrepreneurship. This analysis evaluates patterns of activity in the small business and government or policy making community. The chapter concludes with a discussion about the nature of institutional entrepreneurship in the emergent Western Balkans region.

ENTREPRENEURSHIP, REGIONAL DEVELOPMENT AND EMERGING ECONOMIES

International initiatives such as the Global Entrepreneurship Monitor project (Minniti et al., 2006) argue that there are direct links between entrepreneurship and long-term economic growth and development (Fritsch and Mueller, 2004) – links that are usually measured by the number of new businesses created. It is not surprising, therefore, that entrepreneurship is argued to be important for emergent economies. Emergent economies are defined by Hoskisson et al. (2000) as low income, rapid growth economies (that is, China, India, Asia, Latin American, Africa, the Middle East and the former countries of the Eastern bloc or Soviet Union) that are using economic liberalization as their primary engine of growth. Whilst we acknowledge that some inquirers are critical of the discriminatory effects of entrepreneurial discourses and the eulogizing ideologies they embody (Nodoushani and Nodoushani, 1999; Ogbor, 2000), it is not the intention in this chapter to challenge the positive correlation between entrepreneurship and regional emergence. What is important, instead, is to consider how entrepreneurship can be stimulated and facilitated within (emergent) regions. This is of specific interest for economic development practitioners and policy makers. It is also important for entrepreneurship inquiry and theory development, particularly if theory development can arise from engagement with problems in the world alerting us to research opportunities hitherto unanticipated (Kilduff, 2006).

The role of entrepreneurship as a lever of economic development has become increasingly tied to the concept of so-called ‘clusters’ of regionally interdependent firms and institutions (Feldman et al., 2005). Porter (1990, 1998), the chief architect of cluster theory, argues that along with improvements in innovation and productivity, clusters are a key means of triggering new business formation. Therefore, within the regional development framework, clusters are seen to serve as a spawning ground for new entrepreneurs who are linked by a strong professional culture and are likely to depend on a local network of colleagues, suppliers, clients, servicing firms and financial organizations (Huggins, 2000). Although this is undoubtedly true in a number of regions around the world (such as Silicon Valley, Boston's Route 128 and the Italian industrial districts), it is less clear if, and how, such clusters can be catalyzed elsewhere (Martin and Sunley, 2003). This is partly due to political, cultural and ethnic diversities across regions. But also communities, localities and regions have specific sets of institutions (that is, practices, rules, referents and understandings) that affect firm formation and cluster sustainability.

In view of this there has been a shift in the investigation and implementation of regional development from foreign investment attraction, which is increasingly seen as 'foot loose' and difficult to regionally embed (Dicken, 2003), to a focus on indigenous development through supporting and nurturing the establishment of an entrepreneurial culture and society. The underlying perspective is that regional development requires a long-term and 'velvet glove' approach, based on creating and providing a cultural and attitudinal environment receptive to entrepreneurship and business creation (Mueller and Thomas, 2001; Hayton et al., 2002). However this indigenous or 'grass roots' entrepreneurial development is often more difficult to stimulate than mainstream business support in that it requires an ability to enhance and develop the capabilities and confidence of potential entrepreneurs (Von Bargen et al., 2003).

When we turn to entrepreneurship and regional development issues in emergent or peripheral regions, the challenges are even greater (Cecora, 1999; Benneworth, 2004). To explain this it is helpful to turn to institutional theory where a distinction is often made between mature and emergent organizational fields. Organizational fields are characterized by aggregations or networks of organizations that collectively constitute a recognizable area of life (DiMaggio and Powell, 1983). To transfer this to a regional context, we might propose that regional organizational fields are made up of industrial actors, regulatory agencies, consumers, small businesses, government bodies and other international or foreign agencies. In mature organizational fields the patterning of inter-organizational relations is familiar and established over time as repeated interactions give rise to shared norms and common understandings. In emergent organizational fields, or what Trist (1983) and Hardy (1994) refer to as 'under-organized domains', there is a diversity of actors but their interactions and activities are fairly uncoordinated. This is partly because their macro economic and political institutional structures are often in flux (either because of radical political reform and/or intense institutional change). Also cultural norms, values and practices are being transformed as social structures are in transition. In emerging economies this means that there is greater potentiality for 'regional becoming' in the sense that new social, economic and cultural boundaries are being renegotiated as institutional rules and practices are being (re)constructed. We turn now to the presentation of some contextual data on the empirical situation in the Western Balkans. This is done not only to illustrate the emergent nature of various institutional structures in the Western Balkans but also, as referred to earlier, to ensure that the theoretical analysis undertaken here engages directly with particular challenges that are specific to that region of the world.

THE WESTERN BALKANS REGION

This study focuses on three former Yugoslav countries of the Western Balkans (FYR-Macedonia, Serbia-Montenegro and Kosovo). This European region has been selected because it is one of the least evolved in the transition process. For example, many of the Western Balkan countries are late starters in the transition process towards a market economy with Bulgaria, Romania and Croatia expected to be in the 2010 wave of EU enlargement (Stone and Syrri, 2003). Also in transition reports (EBRD, 2004) it is frequently cited that the countries of South East Europe have an average index score of 2.6 (with an index range from 1, indicating no progress, to 4.3, indicating standards similar to advanced industrial economies). This is low compared to Central Eastern Europe (CEE) or the East Balkans where the overall score averages at 3.8. The main reason for this low transition indicator is that this region is characterized by weak macroeconomic and business environment indicators compared to the rest of Europe. This is expressed in Table 4.1.

Despite the challenges associated with macroeconomic restructuring, there has been significant progress since 2000 reflected in economic growth averaging around 4 per cent per annum and inflation having stabilized close to single digit figures. Even so unemployment rates which are over 30 per cent remain very high and this, along with low per capita incomes around US$ 2500, constitute a source of economic and social tension. In addition, the contribution of manufacturing to the Gross Domestic Product (GDP) has declined by over 20 percentage points to 23–28 per cent due to the winding down of obsolete industrial plants in favour of services (particularly trade) that have come to account for almost 55–60 per cent of GDP. The contribution of agriculture ranges between 13–17 per cent. Fast industrial decline, overall downsizing of the public sector and a still weak private sector account for the inability to absorb excess labour.

Table 4.1 Economic and financial indicators of countries in South East Europe

Indicators for 2004	EU	SEE	ALB	BiH	BU	CR	Fyrom	SiM	Kosovo	RO
GDP growth	1.8	5.4	6.0	3.2	4.3	4.3	2.9	8.6	3.7	4.9
Inflation	1.7	5.5	2.5	0.2	2.3	2.2	1.1	13.8	1.1	15.4
Govt. balance (% GDP)	−3.0	−3.4	−4.5	−0.2	−0.4	−6.3	−1.6	−3.2	−4.9	−2.4
Current account (% GDP)	0.5	−12.8	−7.6	−17.4	−8.4	−6.1	−9.6	−14.5	−41.5	−5.8
Foreign debt (% GDP)	75.0	49.0	23.1	34.9	65.6	81.8	41.3	61.7	n/a	34.6
FDI (% GDP)		4.7	2.9	5.4	7.1	6.9	2.1	4.2	1.1	3.2
Interest rate	2.5	12.1	4.6	4.5	5.2	4.8	13.9	17.6	14.1	20.0
GDP per capita $	32000	3152	1942	1849	2531	6518	2704	2576	1318	2624
Unemployment rate	11.0	29.9	15.0	44.0	13.5	18.7	37.9	31.0	42.0	7.2

Notes: Country names: ALB (Albania), BiH (Bosnia-Herzegovina), BU (Bulgaria), CR (Croatia), Fyrom (FYR-Macedonia), RO (Romania), SEE (South East Europe), SiM (Serbia and Montenegro).

Sources: IMF (1999–2006); EBRD (2004).

In Serbia-Montenegro the macroeconomic outlook has not yet secured low and stable inflation rates while weak export performance, with exports constituting only 13.8 per cent of GDP, contributes to excessive external deficits around 14 per cent of GDP. Rising Foreign Direct Investment (FDI), as a result of the privatization process, is a hopeful sign. In FYR-Macedonia significant progress has been achieved after the establishment of a Stabilization Association Agreement (SAA) with the EU. However the economy is believed to be growing below potential sustaining a very high unemployment rate at around 38 per cent. The external deficit is still very high. This is linked to the fact that the productive structure of the economy continues to be dislocated and the country has not yet established significant areas of international comparative economic advantage. This is a major drawback given that it is a very small but quite open economy. In the case of Kosovo, the economy is dependent on external support and there is limited space for independent policy making. With income per capita around US$ 1300, Kosovo is the poorest region with around 40 per cent of the population below the poverty line and unemployment at 42 per cent.

Exports cover only 5 per cent of total imports and the external deficit of around 41 per cent of GDP is covered by foreign aid. High past immigration has resulted in non-Kosovo communities accounting for 25 per cent of the population contributing around US$ 350 million in remittances per annum. This should imply a significant increase in the future population should the region achieve economic and political stability. The potential for fast progress in the next five years is quite significant since the economy is starting from a very low level and the process of institutional change could result in significant productivity gains (IMF, 1999–2006).

The region of the Western Balkans is an emergent organizational field. At the level of business activity there is evidence of economic dualism with sectors of the economy dominated by a small number of relatively modern firms and a substantial number of technologically underdeveloped micro and small enterprises (OECD, 2005a–c). For example, over 99 per cent of registered companies in the region are SMEs with micro companies constituting around 93 per cent of the total. They account for around 45 per cent of economic activity and 47 per cent of employment in the private sector. The density of small firms in relation to the population is rather small, with figures showing that there are 14 small firms per 1000 inhabitants, compared to 53 for the EU-19 where SMEs also generate a much higher share of private employment at 66 per cent. Regarding sector allocation of enterprises, over 50 per cent engage in wholesale and retail trade followed by manufacturing, construction and transport. About one-quarter of all enterprises are sole proprietors.

At the societal level aspirations for EU membership constitute the driving force and hope of the peoples of the region. This is a vital factor, since EU membership will be dependent on the development of institutional arrangements that conform to the EU's political, economic and regulatory frameworks. The ability to modernize fast, therefore, becomes a key challenge for this region. Further challenges are the development of the legal and political institutional environment for business, inefficient tax systems, progress in the development of business laws on bankruptcy and liquidation, competition law, transparency of public procurement processes and corporate governance, stakeholder rights, minimization of economic dualism and a pervasive unofficial economy (Development Researchers Network, 2005; Jefferson Institute, 2006). Furthermore, from this analysis of the macroeconomic and institutional environment, the regional economy of South East Europe more broadly is opening up to global competition at a much faster rate than to the modernization of its economic and institutional structures. This is indicated by the high external deficits and reliance on foreign capital and aid inflows in order to sustain growth and low inflation. Also in this region the emphasis is on market neutral policies (rather than interventionist industrial policies to support specific sectors) whereby the economy is opened up to global competition to determine its true comparative advantage via the global market.

THE EMERGENT NATURE OF THE WESTERN BALKANS REGION: IMPLICATIONS FOR REGIONAL DEVELOPMENT AND ENTREPRENEURSHIP THEORY

This emphasis on market neutral policies and exposure to global competition is important for regional development in that it creates an 'open' or 'uncoupled' organizational field (Greenwood and Hinings, 1996) in which local actors are brought into contact or association with ideas, practices, resources and institutional referents from other settings. This openness, these authors argue, can stimulate the cross fertilization of ideas between various actors and the enactment of new organizational or regional arrangements. In so doing, creative and innovative practices are enabled. This contrasts with what is sometimes called 'tightly closed' organizational fields, a characterization that, in the context of the former communist economies (such as the Western Balkan situation discussed here), refers to the particular (command) practices of resource mobilization that produced certain economic and social institutional arrangements and which continually tied human agency to the state. In this situation there is limited exposure to alternative cultural, economic and social norms, values, practices or other institutional referents. This has the effect of circumscribing opportunities for organizational creativity, possibly driving it into the informal or black economy.

When considering the role that entrepreneurship plays in regional emergence, it is insufficient, therefore, to focus on the advantages of 'top down' or 'bottom up' approaches for regional development. What is needed is a more sophisticated theoretical understanding that recognizes the intertwined relationship between regional emergence, regional culture and regional organization. To develop this theoretical understanding we might promulgate discourses about the need to create an 'entrepreneurial society' or a cultural and attitudinal environment receptive to entrepreneurship and business creation (Mueller and Thomas, 2001; Hayton et al., 2002). We might also argue for 'economic gardening' forms of development (Hamilton-Pennell, 2004) that help to create an 'infrastructure of entrepreneurship' (Van de Ven, 1993) or mobilize entrepreneurial activity at a grass roots level (that is, small business owners). And these are not trivial or straightforward activities, especially in the former communist societies where individuals have been used to a different sort of relationship with the state than is the case in market economies.

What is more important, therefore, is to develop an understanding of entrepreneurship and regional emergence that addresses two particular (and inter-related) challenges. The first is taking account of the inter-relationship between the industrial actors, regulatory agencies, consumers, small businesses, government bodies and other international or foreign agencies that regional organizational fields constitute – and which are instrumental in entrepreneurial and regional 'becoming'. The second is developing an institutional theorization of the notion of entrepreneurship that is appropriate for a regional context. And it is to these issues that we now turn.

THEORETICAL FRAMEWORK AND RESEARCH QUESTIONS

The term institutional entrepreneurship is usually used to describe organized actors who leverage support and acceptance for new institutional arrangements to serve an interest they value (Dorado, 2005 citing DiMaggio, 1988; Rao, 1998; Beckert, 1999). The notion combines conceptual insights from entrepreneurship and institutional theory. From the entrepreneurship standpoint this conceptual interface is enabled because of the interest in how organized actors leverage resources, identify opportunities and gain legitimacy or support for new activities, innovations, product and service transformations. The interface with institutional theory occurs because, in the process of bringing about social, economic, cultural or political transformations, such actors and activities change or subvert institutional rules and referents, and therefore engender institutional change. As Dorado (2005) comments, although there is a long tradition of institutional research (citing Selznick, 1949, 1957; Stinchcombe, 1968), this theory has received fresh impetus with the interest in how actors gain support and acceptance for institutional change projects (Fligstein, 1996, 1997; Rao et al., 2000).

Ideas from entrepreneurship and institutional theory have been applied in a variety of contexts. These range from industry creation (Aldrich and Fiol, 1994), family business groups in the Association of Southeast Asian Nations (ASEAN) (Carney and Gedajlovic, 2002), cultural entrepreneurship in nineteenth-century Boston (DiMaggio, 1982), micro finance organizations in Bolivia (Dorado, 2001), development of common technological standards (Garud et. al., 2002) and transformation processes in Norwegian fisheries (Holm, 1995). Also studies directly addressing institutional entrepreneurship include institutional management and organizational change in a transition context (de Holan and Phillips, 2002), the whaling industry on Canada's west coast (Lawrence and Phillips, 2004), Chinese entrepreneurship (Yang, 2004), adoption of new technologies such as the 'Kodak moment' (Munir and Phillips, 2005), the legitimation of American sovereignty (Steinman, 2005) and Association to Advance Collegiate Schools of Business (AACSB) agencies (Durand and McGuire, 2005). Importantly, Dorado (2005) isolates institutional entrepreneurship as a particular profile of institutional change.

The study of institutional entrepreneurship and regional emergence is still quite rare (Hoskisson et al., 2000), and for this reason this chapter considers the enactment and patterning of inter-organizational relations that are stimulated by entrepreneurial discourses, and which contribute to the becomingness of a social and economic (regional) reality. This theoretical stance is important for two reasons.

First, it builds upon the analytical tradition put forward by Berger and Luckmann (1966) and built upon by Giddens (1984, 1994) and Sztompka (1993) in his sociology of social change in which inquirers are encouraged to consider linkages between organizations, agencies, institutions and the various actors that constitute (and contribute) to the patterning or shaping of these relations. Also, in relating to this analytical tradition, we can work towards an integration or fusion of structure and agency, operation and action (Sztompka, 1993) – an undertaking which is important for bridging the various levels of analysis that are often singularly privileged in entrepreneurship inquiry and which, it is argued, lead to the fragmentation or reduction of entrepreneurial activity into separate categories or units of analysis (Busenitz et al., 2003).

Second, drawing attention to the linkages between organizations, agencies and institutions within a regional organizational field as they attempt to develop an infrastructure of entrepreneurship (Van de Ven, 1993) is important because it is unusual to see empirical studies of this kind. Most regional studies addressing entrepreneurship focus on either policy driven ('top down') or 'grass roots entrepreneurship' ('bottom up) strategies for stimulating a cultural and attitudinal environment or infrastructure that is receptive to entrepreneurship. Very few focus on the inter-organizational relationality of such strategies. Also, given the situation in the Western Balkans and its emphasis on market neutral policies and exposure to global competition, it is important to take account of the extent to which local actors are brought into contact or association with ideas, practices, resources and institutional referents from other settings. This was discussed earlier in terms of 'open' and 'closed' organizational fields. It is also important to note that organizational fields can be 'too open', resulting in a combination of resource exchange activity, environmental turbulence and lack of institutionalization in an organizational field conspiring to generate uncertainty and ambiguity (Duncan, 1972), which inhibits creativity. These two key challenges give rise to the first research question:

1. How do patterns of activity and relationship between industrial actors, regulatory agencies, consumers, small businesses, government bodies and other international or foreign agencies converge to bring about entrepreneurial activity?

Through institutional theory it is possible to draw attention to the ways in which local (entrepreneurial) action strategies help to diffuse new (and transform) old institutional arrangements. This is important for extending our entrepreneurship inquiry of opportunity 'discovery' beyond micro, personal, behavioural and situational understandings to theories that explain organizational, industry, regional and institutional transformations. However, unlike Dorado (2005), we are less concerned with whether institutional entrepreneurship is more or less possible according to a particular mix of agency, resources and opportunities in (opaque, transparent or hazy) organizational fields. Instead we follow Maguire et al.'s study (2004) in identifying those actors that are better able to engage in institutional entrepreneurship in an emerging context. We also draw attention to the processes through which institutional entrepreneurs promote the adoption of new institutional practices associated with entrepreneurship. In taking this emphasis, consideration can be given to the cultural discourses and emergent strategies that influence the creation of new institutions (Lawrence and Phillips, 2004). These interests are expressed in the second research question:

2. What are the processes through which resources are enacted and mobilized by a network of organizations and local actors as they strive to achieve regional economic interests?

In examining the transformative effects brought about by institutional entrepreneurs, we highlight how particular institutional practices are being transformed as a result of the activities of local actors and economic organizations. We draw attention to how entrepreneurs create connections between 'sites' kept separate by existing institutions (Yang, 2004). Also, taking note of Fligstein and McAdam's (1995) views about the role that institutional entrepreneurs play in helping to create and maintain collective identities, we consider their contribution to the development of a sense of regional identity and becoming (and the role in legitimizing entrepreneurship). This takes account of what Beckert (1999) refers to as the 'institutional disembedding' that goes hand in hand with the emergence of new institutional structures (thus creating spaces of uncertainty that become filled by activities of entrepreneurs). These theoretical concerns give rise to the third research question:

3. In what ways do the local (entrepreneurial) action strategies converge and help to diffuse new (and transform old) institutional arrangements?

FIELDWORK AND DATA COLLECTION

The fieldwork for this study occurred between 2004 and 2006. The data collection process involved three core phases. The first phase involved secondary data collection and evaluation of the extensive reports on transition processes. This data enabled a mapping of the various actors who were economically involved in the region. These groupings emerged as: the small business sector (mostly micro small firms); policy makers; intermediaries (including donor agencies) and the research community (including universities). Some groupings were involved at a national level in policy developments or initiatives. Others had a more local role. However they were all involved in institutional capacity building relating to enterprise and small business initiatives in one of the three countries.

Having identified these groupings, the second phase of the data collection involved interviews and focus group consultations with a selection of these actors. These took place over a 14-month period during 2005–6. Interviewees and focus group members came from a variety of institutions, including foreign donor or international agencies (34 per cent), research institutes (23 per cent), universities (20 per cent) and government bodies (23 per cent), their key activities being higher education, research and development, consultancy, advice and guidance. A list of the organizational actors interviewed during the fieldwork is outlined in Table 4.2. A total of 30 interviews were undertaken with each of these groupings within the organizational field. Where possible interviews were recorded and later transcribed. When recording was not possible (as there was some sensitivity to this), extensive notes were taken by the researchers.

The third phase of data collection involved a survey, administered by both face to face and telephone means, with 60 small business owners (20 in each country). The choice of the SMEs and research performers interviewed was made by the national experts that formed part of the focus groups. They were tasked with developing a list of SMEs that was to some extent representative of the industrial structure of each of the three economies (each list containing up to 100 firms) and from this ensuring that a mix of a minimum

Table 4.2 Organizational characteristics of actors interviewed during research investigation (2004–7)

Institution	Role/Activity
Serbia	
Serbian Chamber of Commerce	Representative of the business community
University of Novi Sad	Higher education institution
Mihajlo Pupin Institute	Research Institute
Belgrade Regional Centre for Development of Small and Medium Enterprises and Entrepreneurship	Local enterprise centre
Kruševac Regional Centre for Development of Small and Medium Enterprises and Entrepreneurship	Local enterprise centre
Kragujevac Regional Agency	Regional development agency
Kosovo	
Riinvest	Not-for-profit economic development research centre
Comtel	Infrastructure development company
Comtrade	ICT company
Faculty of Machinery, University of Pristina	Academic research department
Faculty of Construction, University of Pristina	Academic research department
City Group	Local development company
UK – Pharma	
Faculty of Agriculture, University of Pristina	Academic research department
Kosovo Veterinary and Food Agency and Consumer (KVFA)	Government agency
Former Yugoslav Republic of Macedonia	
Institute for Agriculture	National research institute
United Nations Development Programme (UNDP)	International donor initiative
Agency for Agricultural Development	Government agency
South Eastern Enterprise Development	Regional enterprise agency
CDS	International training, consultancy and research company
Union of Economic Chambers	Umbrella body of local chambers
Foundation for sustainable development of the region of Ovce Pole (FOROP)	Not-for-profit development foundation
Agency for Promotion of Entrepreneurship (APE)	National entrepreneurship agency
Apparel Technology Centre	Textile research centre

of 20 SMEs were interviewed. This approach meant that representativeness across the three economies as a whole is maintained. A range of micro, small and medium sized enterprises, with an average of 1.1 million euro turnover were interviewed. Most SMEs are relatively new – the youngest was just established in the last three years whilst the oldest were established slightly over ten years ago. Nearly 80 per cent of the SMEs operate in the manufacturing sector (IT production, food and beverage and electrical engineering are the main sectors covered) and about 20 per cent are in the service sector.

Each organizational group was asked a mixture of questions relating to the inter-organizational relationships and interactions they had with other groups. In addition they were asked what the barriers were, if any, to coordinating with other local, national or regional agencies or bodies. Finally, they were asked about their general perceptions of other institutional groupings in the region. From these focus group discussions, two institutional entrepreneurs were selected for closer analysis. These two agencies were frequently mentioned by local actors as being 'the main drivers of entrepreneurship in the region'. In the following sections some of the key findings from the research with these various groupings are presented. In selecting four groups (policy makers, small business owners, research institutions and international aid donors), our intention is to identify those actors that are more or less able to engage in institutional entrepreneurship in an emerging field. We also identify the local action strategies and highlight the resources or discourses that enable some actors to engage in entrepreneurial activities.

INSTITUTIONAL ENTREPRENEURSHIP IN THE WESTERN BALKANS

As can be seen from the survey of local, regional and national entrepreneurial initiatives, there are a variety of agents involved in stimulating entrepreneurial activity (industrial actors, regulatory agencies, small businesses, government bodies and other international or foreign agencies). By mapping the various initiatives underway in the region it became apparent that there is an emergent institutional, regulatory and education framework supporting entrepreneurial activities. In general, however, the institutional framework in FYR-Macedonia appears to be more evolved than in Serbia and Kosovo. For instance, in FYR-Macedonia the Ministry of Economy established the Agency for Promotion of Entrepreneurship in 2004. The task of this agency is to implement the national strategy for SMEs and to ensure their involvement in the legislative process. Also, in 2004, the National Council for Competitiveness and Entrepreneurship aimed to bring together government organizations and business representatives. The ongoing decentralization process (Ohrid Agreement) in FYR-Macedonia implies that in addition to local SME offices the municipalities will also become active participants in local economic development. At the national level, in addition to the Economic Chamber of the Republic, which focuses more on representing larger companies, various chambers were created in 2004 (that is, the Trade Chamber, the Industrial Chamber and the Services Chamber) – set up by SMEs. Their task is to promote competitiveness of their members, improve the business environment and promote export growth.

The policy making and SME representation framework in Serbia-Montenegro is similar, although less evolved, to that encountered in FYR-Macedonia, with the Ministry of Economy and the National Agency for the Development of SMEs standing at national level and implementing policy at regional level offices. The major policy document is the 'Strategy for Development of Small and Medium Size Enterprises and Entrepreneurship in the Republic of Serbia – 2003–2008'. The SMEs are primarily represented by the Chamber of Commerce, which also has organizations at a regional level. The National Employment Service has started to develop a network of Centres for Entrepreneurial Development aiming towards stimulation of self-employment through education, training and information. Regarding the availability of skills, the National Agency for Development of SMEs has implemented the Scheme for Educational and Consultancy Stimulation, providing access to existing and potential entrepreneurs of consulting and training services at lower than market prices. With the help of the European Union the Euro-Info Correspondence Centre (EICC) has been founded, as a part of the international EICC network, offering services of business liaison and consultancy regarding EU issues relating to SMEs. Various international donors assist towards the establishment of Agencies for Local Economic Development.

The current status of Kosovo has not yet allowed for the emergence of a clear institutional framework for SMEs and innovation promotion. Initiatives are fragmented and the institutional set-up is at a relatively initial stage, hence no concise analysis can be performed at this stage. The Ministries for Trade and Industry and the Ministry of Education, Science and Technology are the major government agents for SME and R&D policies. Supporting offices are the Investment Promotion Office, the Export Stimulation Office and at the local level the Municipal Business Centers and the Business Service Centers. The Kosovo Chamber of Commerce is the major SME representative organization that also engages in some activities stimulating R&D and providing business information. The Riinvest Institute for Development Research undertakes research projects on most fields relating to development issues, mostly working in projects financed by international donors.

ENTREPRENEURIAL PROCESSES

Although policy frameworks are emerging at differing rates in all three countries, they are all still relatively weaker when it comes to the advanced support needs of enterprises, especially relating to innovation issues and the development of skills. A significantly large proportion of small firms surveyed (average of 84 per cent) have never collaborated with an external research organization as a means of innovating. This deficit can be seen in the low take-up percentage and frequency (average of 15 per cent) in patenting inventions, licensing technology, joint venture and other research and development. More positively, some 57 per cent of the sample firms frequently operate their own internal new product development. In general new product (89 per cent), new market (82 per cent), new process (80 per cent) and supply chain (63 per cent) developments are very much internally operated. Many small firms commented that they viewed the overall institutional framework as fragmented, overlapping and complex. It was further identified that the current use of intermediary organizations to facilitate knowledge-based interactions is extremely limited. Some 70 per cent of respondents have not used any business intermediary. The most utilized intermediaries are international and donor organizations. The least utilized are business associations and chambers of commerce. This is outlined in Table 4.3 and indicates that at present existing economic institutions and business intermediaries are not significantly utilized by small businesses. These results highlight the low level of trust these firms have towards existing intermediaries – as high as 92 per cent of the responding small firms have doubts concerning non-firm institutions in their region.

Slightly more than half of the respondents (51 per cent) perceived that competition or secrecy is the main barrier to accessing external knowledge.

Table 4.3 Facilitation of SME interaction through intermediary organizations

Intermediary organizations facilitating interactions in relation to research, development or other information appropriation	Very often	Quite often	Very infrequently	Never
Business intermediaries	5%	13%	12%	70%
Chambers of commerce	2%	6%	29%	63%
Business associations	7%	19%	13%	61%
International organizations/donors	17%	18%	24%	41%
Government agencies	5%	6%	26%	63%

This indicates the possibility of a lack of trust and minimal collaboration between firms and other institutions. The main barriers to accessing external knowledge are a lack of specialized institutions, a lack of relevant knowledge, the financial burden and the overall business environment. Some 58 per cent of the firms also considered that enhanced intervention could be achieved through establishing a platform for the region's research and knowledge generating sector, and 57 per cent stated that access to more relevant finance and funding could better stimulate knowledge transfer.

In addition, in terms of local policy support for the private sector level, the small businesses surveyed reported that the business advisory services, incubators and clustering projects are largely inadequate and needed a more focused approach, moving from the provision of simple services (business plans, off the shelf information) towards technical advice, product market analysis and targeted information with relevant case studies and exchange of experience. This support would be more beneficial for enhancing and developing the capabilities and confidence of would-be entrepreneurs (Von Bargen et al., 2003). In total, 68 per cent of the small businesses surveyed would prefer intervention through the development of more and better intermediary organizations to link the research and business sectors.

This evidence is helpful for addressing the first research question, as it leads us to the conclusion that there are extensive patterns of activity and relationships between industrial actors, regulatory agencies, small businesses, government bodies and other international or foreign agencies. These patterns are purposefully directed at stimulation of entrepreneurial activity in the region. However, in spite of this diversity of organizational actors, their interactions and activities are fairly uncoordinated. This means that 'sites' or practices previously kept separate by existing (cultural and political) institutions are still being maintained (Yang, 2004). An example of this that emerged during the fieldwork are the activities of senior staff in universities who, it was claimed, utilize their senior positions to secure access to new business opportunities.

The Western Balkans region is, therefore, very much an under-organized or emergent inter-organizational domain (Trist, 1983; Hardy, 1994). The institutional framework for policy making relating to small business enterprise and R&D development is still to fully evolve and inter-organizational cooperation between policy makers and the business community is fairly rudimentary. There is no clear strategic conceptualization at the central level or coordination with regional and local level agents. The lack of coordination and networking between government organizations, research performers and the small business community is identified as a significant barrier. At present, a number of international initiatives are characterized by a high degree of duplication and overlapping mechanisms and processes, which may be reducing their effectiveness. The education and training system is not yet geared towards the needs of the emergent economy and there are limited institutional resources for skills development support and technology transfer. This is because the macro economic and political institutional structures are in flux and interactions between the various actors have not had the time to evolve into a set of shared norms or common understandings. In this sense the 'openness' of the Western Balkans region, with its exposure to global competition and diversity of organizational activity, is what contributes to environmental and institutional uncertainty and ambiguity (Duncan, 1972). But in the longer term it is as new social, economic and cultural boundaries are being renegotiated and institutional rules and practices are being (re)constructed that entrepreneurial activities can take hold.

LOCAL ACTION STRATEGIES: INSTITUTIONAL ENTREPRENEURSHIP AND THE ROLE OF INTERNATIONAL DONOR AGENCIES

Throughout the research process it became apparent that the main impetus for entrepreneurial development in the region is coming from donor agencies who act as intermediaries in the region, promoting and legitimizing entrepreneurship. These are not entrepreneurs in the traditional sense of alert individuals, teams or families that have identified business opportunities from the inertia or disequilibria of particular (transforming) markets. Instead, the role that intermediaries or brokers play is in the Kirznerian sense of bringing people together to enable economic transformation. In this we address the third research question, which is examining the local action strategies adopted by institutional entrepreneurs as they gain support for and legitimization of entrepreneurship projects. In order to further analyse this role we have selected two organizations for closer analysis here. These two organizations have been purposefully selected because, although they have similar aims and aspirations for entrepreneurial development in the region, they give different accounts to the success and barriers found there. One is a public body responsible for the generation of education and training and support programmes for enterprise at a regional and national level. This organization is given a fictitious name, the European Agency (EA), to protect our interviewee. The second is a private organization from Germany that provides technical assistance to small business development in the region (GTZ).

EA is responsible for enterprise development initiatives in the Balkans, including non-financial and financial support for SME development; FDI; export related programmes; vocational education and training; new curricula and occupational standards across sectors and in schools. The main thrust of enterprise development (which encompasses all programmes above) is through the EU Charter for SMEs, as a result of which there is a focus on larger issues (access to capital, finance and enterprise support). However the agency tends to use the term 'enterprise' rather than 'entrepreneurship', because in our interviewee's words, 'entrepreneurship is more limited than enterprise'. In building regional enterprise support centres, their approach is to promote a 'stakeholder model of regional development' (involving local government and developing local training programmes of centres through research and development programmes and partnership building activities. EA's aim is to work with local and national government, involving various agencies and building institutional capacity.

GTZ is a German holding company responsible for technical cooperation. It provides technical knowledge in 120 countries throughout the world and holds bilateral relations with each country in the Balkans. The organization is an implementation agency for programs of the German government. However they are also privately funded. GTZ's focus is on start-ups, which according to our interviewee is due to three key reasons: for regional economic development; promoting entrepreneurial spirit in a country; and promoting cooperation (between local and national agencies). For example, in Macedonia, GTZ have worked with the national agency for entrepreneurship promotion (government agency) and they have been involved in the creation of six regional centres that have also been supported by the British know-how fund.

GTZ target three client target market groups: the unemployed, business support and the media. The business support market is fairly standard in that they offer training programmes on business planning, export and so forth. Targeting support for the unemployed is deemed to be important for encouraging a culture change with regard to entrepreneurship. The GTZ interviewee comments:

> Because of the bad negative image of entrepreneurship in Macedonia, we needed to change this. There is a lot of negative energy in media. 'Collective apathy to start a business' especially in Eastern Macedonia, [we] need to encourage young people to become self employed. Not just to blame everything on the government. There is a lack of entrepreneurial spirit. People here are risk averse. They would rather be employed for 200 euros rather than take the risk of running their own business. Working for [an] established company gives you better status.

When asked about where the main impetus for entrepreneurship is coming from, the EA interviewee stated that:

> Entrepreneurial activities are beginning to occur although this is still largely unstructured and uncoordinated at a local level. [In his view] . . . enterprise in

> the region has been forced by the EU charter and the Thessalonica agenda. This is the main driver or lever. But now, the countries are slowly coming around to the need for enterprise development/SME and entrepreneurship. The donors are driving this agenda. It is a donor-led intervention. . . . much of the focus [now] is on the Minster of economics and labour; curricula; business advice for would-be and existing entrepreneurship. But in reality, it is a donor-led intervention (although taking into account government strategy). We are pushing an agenda but meeting obstacles and constraints.

According to the EA interviewee, donor-led intervention is not fully effective because of old institutionalized practices, such as the allocation of financial recourse and human resources, a lack of political will and a lack of inter- and intra-ministerial and government agency cooperation. In practice, this translates to the dominance of central planning economic, top down decision making from Belgrade, no local initiatives and the feeling of 'having no control over one's own destiny' – all of which relate to the traditional relationship that individuals have with the state in the former communist countries. When questioned further about which institutions they were transforming, the response was: 'economic planning, government, and culture'. He explains further:

> The registration issue has been sorted, it is easier now to register a business. And there is a culture of business; and a lot of people want to do something, but there are constraints surrounding credit. The government is not keen because there is a lack of understanding of the importance of enterprise. Also the governments are weak. Administrative arrangements are weak. There is regional good will. But it is all talk and there is an 'implementation vacuum'. There is a lack of people, lack of skill at the government level. And very weak civil administration system, people there are not relevant. There is not a culture of risk taking and failure. It still holds with the people that the central government needs to take care of us. Not trying to do things for themselves. Also, private business is seen as 'negative, dirty and corrupt'.

One of the strategies EA have undertaken to change the public perception of entrepreneurship was taking out a logo on a bus '*biznis mali – sta mu fali*?' (translated, this means 'a small business is just a small large business, so how about it?'). At GTZ institutional change is engendered by the development of training programmes directly with journalists from the media. They do this by promoting successful business stories for use in the press that act as benchmarks or role models for young people. In Macedonia the word 'entrepreneur' has negative connotations.

In the words of our informant 'it is a bad word; in the past it is associated with the construction industry – someone who is manipulating things. It was also associated with semi legal activities'. Implementing training programmes directly for journalists, it is envisaged, will bring about institutional change at the level of culture and changing peoples' perceptions of what entrepreneurship is concerned with.

Other institutionalized practices that GTZ were targeting to transform were the activities of the public sector Chambers of Commerce. Before transition it was commented that these chambers did not want to cooperate with the donor agencies or new business support agencies. This was explained as a 'lack of confidence' and 'conflicts between business and government'. But some success has been achieved whereby they are now willing to establish partnership agreements with business support agencies. Similar success has been achieved with the Ministry of Economy (the national policy makers for promoting entrepreneurship). When asked how this success was achieved, the GTZ interviewee reported that:

> It is difficult to change their mentality but through positive practice it has now changed. We have used other methods – not telling them what to do. Not direct influence but involving them in practices (i.e. trade fair for export, in the past they only wanted to go there to get a trip to Europe but now they are thinking more strategically). We try to encourage them to learn from example. It is not ideal yet but things are changing.

When commenting on the institutional practices that were more difficult to change, the GTZ interviewee referred to universities as 'being more difficult', especially where there is a 'closed inner circle of good people, professorial elites who also have associations with political parties; they also tend to be the older generation who make things difficult for the younger generation'. A 'degree of arrogance' was cited as an explanation. Some of these people are entrepreneurs themselves with many links with private companies but it was felt that these are 'difficult institutions to change or involve'. Other institutions that were reported to be difficult to change are monopolized banks where it is difficult for foreign investors to invest. Also certain interest groups and particular privatized companies were signalled out as difficult to work with. These companies tend to be the ones that were previously state owned, becoming privatized early in the transformation process and favouring those individuals that were well placed with resources and information to benefit from privatization. Our informant commented that 'although they are private on paper . . . in practice they operate as before and it is difficult to change their attitude'. However it was also noted that, in his view, these firms are now losing their competitive edge and the disembedding of these institutionalized practices (citing Beckert, 1999) would occur through market forces (rendering these firms uncompetitive).

Addressing the role of institutional entrepreneurs in enhancing a regional identity for the Western Balkans, the GTZ interviewee commented that:

> The SEE region is very artificial. There is no regional identity. There is only 1 per cent trade between Albania and Macedonia. The region was imposed by external factors. We are not proud of the Balkans label – it has connotations of bad and negative image because of existing cultures. E. Balkans escaped this and W. Balkans cannot now catch up with the rest of Europe. It is difficult to develop a regional identity when each country does not have their own identity. There is such a lack of self confidence in each country and so we cannot work together to cooperate until they have built their own identity and overcome nationalistic movements which are still strong.

Similarly, in relation to the extent to which their activities are helping to create a sense of regional identity, the EA interviewee stated that:

> The region does not function as a region except as a regional free trade agreement. The Western Balkans region is purely an administrative region following the EU definition. And this is because of politics. The Western Balkans cannot and will not in the foreseeable future (because of war, ethnic divisions, politics/instability in the region and the special status of Kosovo) function as a region.

DISCUSSION

An infrastructure of entrepreneurship is beginning to emerge in the Western Balkans, with the key actors engaging in this infrastructure development being agencies of the European Union, international donor organizations, national ministries of economy, chambers of trade and commerce, regional centres for small business development, universities, and private and public research institutes that use local labour but which are funded or supported from outside the Balkans region. As a result, it is argued that the Western Balkans is an 'open' organizational field. This is because of the wide possibilities for local actors to come into contact or association with ideas, practices, resources and institutional referents from other settings. In addition, there are national education and vocational training programmes for fostering entrepreneurial attitudes and culture. These are important for promoting the adoption of new institutional practices associated with entrepreneurship. But, also, we would argue that entrepreneurial discourses are being drawn upon to influence institutional change in the region.

This change includes the development of an international support base, the development of a local base of trainers and skills development of people working in international organizations who will, over time as donor organizations move out of the area, build on the skills to develop their own businesses or go into consultancy and so on. Other areas include the implementation of local support strategies, the rolling out of entrepreneurship programmes from the large towns more widely into the region, changing the culture of entrepreneurship and developing support for cooperative ways of working. As such, entrepreneurship is closely associated with (and promoted by) discourses of regional development, economic prosperity, individual well-being and inter-organizational cooperation and coordination. However, as shown in the earlier fieldwork, some of the support is still rather uncoordinated and many efforts to develop coordination are driven 'top down'.

It is possible to identify the variety of ways in which institutional entrepreneurs promote the adoption of new institutional practices associated with entrepreneurship. Donor organizations have a significant role to play in legitimizing entrepreneurship (as can be seen in the work with the media institutions and the unemployed). Also they are active in creating connections between 'sites' kept separate by existing institutions as Yang (2004) referred to (that is, the attention to promoting inter-organizational coordination and cooperation between chambers of commerce and other agencies). However the region is still very much in transition with many barriers and blockages preventing entrepreneurship from being institutionalized. The main areas of institutional disembedding that are still ongoing are with universities, some specialized interest groups and the older privatized companies. There is also, in spite of the efforts to promote cooperation and coordination, only limited success in promoting a sense of regional identity for the Western Balkans – a role which is identified by Fligstein and McAdam (1995) as important for institutional entrepreneurs. As such, a regional identity is still very underdeveloped and unlikely to emerge until national identities are more firmly rooted.

Within this study we have related to a set of sociological ideas which enable the study of linkages between organizations, agencies, institutions and the various actors that constitute (and contribute) to the patterning or shaping of these relations. A theoretical framing for the study has been derived by linking institutional theory and entrepreneurship. However, in contrast to traditional ways of conceptualizing institutional entrepreneurship, we offer a different perspective. And we propose this theoretical understanding by linking directly to the empirical regional field of the Western Balkans. Entrepreneurship is broadly occurring within the region as a result of the resource mobilization of government bodies, donor agencies, small businesses and various research institutes. We claim that entrepreneurial practices are occurring in the three countries studied through the collective activities of small businesses, policy makers, research communities and donor agencies. However the institutionalization of these practices is slow to take effect because the coordinations between the various groupings are still fairly fragmented and have yet to converge. This is perhaps what might be expected in the early stages of entrepreneurship infrastructure development (Van de Ven, 1993). However we also illustrate how the major entrepreneurial impetus within the region is coming from the international donor agencies who are activating particular local intervention strategies to bring about entrepreneurship at a faster pace. Much of this local intervention work is directed at changing existing institutions relating to government and universities and the cultural perceptions of the local population who generally perceive entrepreneurship in a negative way. In general, therefore, we find that entrepreneurial activities play a significant role in the institutional change and 'social becoming' (Sztompka, 1993, p. 17) of emergent regions.

Based on our analysis, however, we offer an adapted theoretical understanding of the relationship between entrepreneurship and institutional change. The conceptualization of institutional entrepreneurship offered here highlights the enactment and patterning of inter-organizational relations that enable transformation of existing or emergence of new economic institutions. This is distinctive because it draws attention to the cumulative effects of particular patterns of organizational activity that over time contribute to entrepreneurship in a regional economy.

CONCLUSION

Following our review of entrepreneurial activity in the emergent regions of the Western Balkans, we argue that institutional entrepreneurship is the overarching profile of institutional change in an emerging region. This is because institutional entrepreneurship is constituted through a variety of convening and partaking inter-organizational relationships that converge over time. This relates partly to the fact that, in emerging regions, the transformation challenges are too complex for any single grouping of organizations to deal with. As a result, different groupings of actors practice entrepreneurship and contribute to institutional change in different ways. Small business owners tend to act autonomously and 'partake' in institutional change by creating a business to generate a livelihood or income for the family or household. Institutional change and regional development occurs slowly and incrementally through their countless and semi-autonomous exchange and resource mobilization activities. Intermediaries, chambers of commerce and international donor organizations also engage in 'convening' activities, jump-starting processes to overcome particular problems. But all of these constitute institutional entrepreneurship – that is, processes (and not special individuals) that converge to enable actors to identify, recognize and enact novel product, service, industry or personal and institutional transformations. Institutional entrepreneurship, therefore, refers to the inter-organizational processes through which resources are enacted by a network of local actors as they mobilize support for and acceptance of new institutional arrangements.

ACKNOWLEDGEMENT

We would like to thank George Anastasiadis at the South East European Research Center who kindly provided us with significant support during the data collection process.

REFERENCES

Aldrich, H.E. and C.M. Fiol (1994), '"Fools rush in"? The institutional context of industry creation', *Academy of Management Review*, **19** (4), 645–70.

Amin, A. (1994), 'The difficult transition from informal economy to Marshallian industrial district', *Area*, **26** (1), 13–24.

Beckert, J. (1999), 'Agency, entrepreneurs and institutional change: the role of strategic choices and institutionalised practices in organizations', *Organisation Studies*, **20**, 777–99.

Benneworth, P. (2004), 'In what sense "regional development?": entrepreneurship, underdevelopment and strong tradition in the periphery', *Entrepreneurship and Regional Development*, **16** (6), 439–58.

Berger, P. and T. Luckmann (1966), *The Social Construction of Reality*, New York: Anchor Books.

Busenitz, L.W., G.P. West, D. Shepherd, T. Nelson, G.N. Chandler and A. Zacharakis (2003), 'Entrepreneurship research in emergence: past trends and future directions', *Journal of Management*, **29** (3), 285–308.

Butler, J.E. and G.S. Hanson (1991), 'Network evolution, entrepreneurial success and regional development', *Entrepreneurship and Regional Development*, **3**, 1–16.

Carney, M. and E. Gedajlovic (2002), 'The co-evolution of institutional environments and organisational strategies: the rise of the family business groups in the ASEAN region', *Organisation Studies*, **23** (1), 1–32.

Cecora, J. (1999), *Cultivating Grass-roots for Regional Development in a Globalising Economy: Innovation and Entrepreneurship in Organised Markets*, Aldershot, UK: Ashgate.

de Holan, M. and N. Phillips (2002), 'Managing in transition: a case study of institutional management and organizational change', *Journal of Management Inquiry*, **11** (1), 68–83.

Development Researchers' Network (2005), 'Study on the linkages between economic development of the countries of the Western Balkans and progress in institutional reform', Development Researchers' Network, Rome, Italy.

Dicken, P. (2003), *Global Shift, Fourth Edition: Reshaping the Global Economic Map in the 21st Century*, New York: Guildford Press.
DiMaggio, P. (1982), 'Cultural entrepreneurship in nineteenth-century Boston: the creation of an organizational base for high culture in America', *Media Culture Society*, **4**, 33–50.
DiMaggio, P.J. (1988), 'Interest and agency in institutional theory', in L. Zucker (ed.), *Institutional Patterns and Organizations: Culture and Environment*, Cambridge, MA: Ballinger, pp. 3–22.
DiMaggio, P.J. (1992), 'Nadel's paradox revisited: relational and cultural aspects of organizational structure', in N. Nohria and R.G. Eccles (eds), *Networks and Organizations: Structure Form and Action*, Boston, MA: Harvard Business School Press, pp. 118–42.
DiMaggio, P.J. and W.W. Powell (1983), 'The iron cage revisited: institution isomorphism and collective rationality in organisational fields', *American Sociological Review*, **48**, 147–60.
Dorado, S. (2001), 'Social entrepreneurship: the process of creation of microfinance organizations in Bolivia', PhD thesis, McGill University, Canada.
Dorado, S. (2005), 'Institutional entrepreneurship', *Organisation Studies*, **26** (3), 385–414.
Duncan, R.B. (1972), 'Characteristics of organizational environments and perceived environmental uncertainty', *Administrative Science Quarterly*, **17** (3), 313–27.
Durand, R. and J. McGuire (2005), 'Legitimating agencies in the face of selection: the case of ASCSB', *Organisation Studies*, **26** (2), 165–96.
EBRD (2004), *Transition Report*, various issues, 1999 to 2004, London: EBRD Publications.
Entrepreneurship Theory and Practice (2001), special issue, **24** (4).
Feldman M.P., J. Francis and J. Bercovitz (2005), 'Creating a cluster while building a firm: entrepreneurs and the formation of industrial clusters', *Regional Studies*, **39** (1), 129–41.
Fligstein, N. (1996), 'Markets as politics: a political cultural approach to market institutions', *American Sociological Review*, **61**, 656–73.
Fligstein, N. (1997), 'Social skill and institutional theory', *American Behavioural Scientist*, **40** (4), 397–405.
Fligstein, N. and D. McAdam (1995), 'A political-cultural approach to the problem of strategic action', unpublished paper.
Fritsch, M. and P. Mueller (2004), 'Effects of new business formation on regional development over time', *Regional Studies*, **38** (8), 961–75.
Garud, R., J. Sanjay and K. Arun (2002), 'Institutional entrepreneurship in the sponsorship of common technological standards: the case of Sun Microsystems and Java', *Academy of Management Journal*, **45** (1), 196–214.
Giddens, A. (1984), *The Constitution of Society*, Cambridge, UK: Polity Press.
Giddens, A. (1994), *Elements of the Theory of Structuration in the Polity Reader in Social Theory*, Cambridge, UK: Polity Press.
Greenwood, R. and C.R. Hinings (1996), 'Understanding radical organizational change: bringing together the old and the new institutionalism', *Academy of Management Review*, **21** (4), 1022–54.
Hamilton-Pennell, C. (2004), 'CI for small businesses: the city of Littleton's economic gardening program', *Competitive Intelligence Magazine*, **7**, 13–15.
Hardy, C. (1994), 'Under-organized interorganizational domains: the case of refugee systems', *Journal of Applied Behavioural Sciences*, **30**, 278–96.

Hayton, J.C., G. George and S.A. Zahra (2002), 'National culture and entrepreneurship: a review of behavioural research', *Entrepreneurship Theory and Practice*, **26**, 33–52.
Hjorth, D. and B. Johannisson (2003), 'Conceptualising the opening phase of regional development as the enactment of a "collective identity" ', *Concepts and Transformations*, **8** (1), 69–92.
Holm, P. (1995), 'The dynamics of institutionalization: transformation processes in Norwegian fisheries', *Administrative Science Quarterly*, **40** (3), 398–422.
Hoskisson, R.E., L. Eden, C. Ming and M. Wright (2000), 'Strategy in emerging economies', *Academy of Management Journal*, **43** (3), 249–67.
Huggins, R. (2000), *The Business of Networks: Inter-firm Interaction, Institutional Policy and the TEC Experiment*, Aldershot, UK: Ashgate.
IMF (1999–2006), 'Staff paper reports' for Serbia-Montenegro, FYR-Macedonia and Kosovo over the period 1999–2006.
Jefferson Institute (2006), 'Competitiveness of the Serbian economy', Jefferson Institute, Belgrade, Serbia.
Johannisson, B. (1990), 'Community entrepreneurship – cases and conceptualisation', *Entrepreneurship and Regional Development*, **2**, 71–88.
Journal of Management (2003), special issue, **29** (3).
Kilduff, M. (2006), 'Editors comments: publishing theory', *Academy of Management Review*, **31** (2), 252–5.
Lawrence, T.B. and N. Phillips (2004), 'From Moby Dick to Free Willy: macro-cultural discourse and institutional entrepreneurship in emerging institutional fields', *Organization*, **11** (5), 689–711.
Lumpkin, G.T., G.E. Hills and R.C. Strader (2003), 'Opportunity recognition', in H.P. Welsch (ed.), *Entrepreneurship: The Way Ahead*, New York: Routledge, pp. 73–90.
Maguire, S., C. Hardy and T.B. Lawrence (2004), 'Institutional entrepreneurship in emerging fields: HIV/AIDs treatment advocacy in Canada', *Academy of Management journal*, **47** (5), 657–79.
Martin, R. and P. Sunley (2003), 'Deconstructing clusters: chaotic concept or policy panacea?', *Journal of Economic Geography*, **3** (1), 5–35.
Minniti, M., W.D. Bygrave and E. Autio (2006), *Global Entrepreneurship Monitor: 2005 Executive Report*, Wellesley, MA and London: Babson College and London Business School.
Mueller, S. and A.S. Thomas (2001), 'Culture and entrepreneurial potential: a nine country study of locus of control and innovativeness', *Journal of Business Venturing*, **16**, 51–75.
Munir, K.A. and N. Phillips (2005), 'The birth of the "Kodak moment": institutional entrepreneurship and the adoption of new technologies', *Organisation Studies*, **26** (11), 1665.
Nodoushani, O. and P.A. Nodoushani (1999), 'A deconstructionist theory of entrepreneurship', *American Business Review*, **17** (1), 45–9.
Ogbor, J.O. (2000), 'Mythicising and reification in entrepreneurial discourse: ideology-critique of entrepreneurial studies', *Journal of Management Studies*, **37** (5), 605–35.
OECD (2005a), Stability Pact, EBRD, Investment Compact for SEE 'Enterprise Policy Performance Assessment: Serbia-Montenegro'.
OECD (2005b), Stability Pact, EBRD, Investment Compact for SEE 'Enterprise Policy Performance Assessment: FYR-Macedonia'.

OECD (2005c), Stability Pact, EBRD, Investment Compact for SEE 'Enterprise Policy Performance Assessment: Kosovo'.
Porter, M. (1990), *The Competitive Advantage of Nations*, London: Macmillan.
Porter, M. (1998), *On Competition*, Harvard, MA: Harvard Business School Press.
Pyke, F., G. Becattini and W. Sengenberger (1990), *Industrial Districts and Inter-firm Cooperation*, Geneva, Switzerland: International Institute for Labour Studies.
Rao, H. (1998), '"Caveat emptor": the construction of nonprofit consumer watchdog organizations', *American Journal of Sociology*, **103** (4), 912–61.
Rao, H., C. Morrill and N.Z. Mayer (2000), 'Power plays: how social movements and collective action create new organizational forms', *Research in Organisational Behaviour*, **22**, 237–81.
Sarasvathy, S.D. (2000), 'Report on the seminar on research perspectives in entrepreneurship' (1997), *Journal of Business Venturing*, **15** (1), 1–58.
Selznick, P. (1949), *TVA and the Grass Roots*, Berkeley, CA: University of California Press.
Selznick, P. (1957), *Leadership in Administration: Sociological Perspective*, Berkeley, CA: University of California Press.
Shane, S. (2000), 'Prior knowledge and the discovery of entrepreneurial opportunities', *Organization Science*, **2**, 448–69.
Shane, S. and S. Venkataraman (2000), 'The promise of entrepreneurship as a field of research', *Academy of Management Review*, **25** (1), 217–26.
Steinman, E.W. (2005), 'Legitimizing American Indian Sovereignty: mobilising the constitutive power of law through institutional entrepreneurship', *Law and Society Review*, **39** (4), 759–92.
Steyaert, C. and D. Hjorth (eds) (2003), *New Movements in Entrepreneurship*, Cheltenham, UK and Northampton, MA, USA: Edward Elgar.
Stinchcombe, A. (1968), 'Social structure and organizations', in J.G. March (ed.), *Handbook of Organizations*, Chicago, IL: Rand McNally, pp. 142–93.
Stone, D.A. and D. Syrri (eds) (2003), *Integrating the Western Balkans into Europe: the aftermath of the Greek EU Presidency*, Thessaloniki, Greece: South East European Research Centre.
Swedberg, R. (ed.) (2000), *Entrepreneurship: The Social Science View*, Oxford, UK: Oxford Management Readers.
Sztompka, P. (1993), *The Sociology of Social Change*, Oxford, UK: Blackwell.
Trist, E. (1983), 'Referent organizations and the development of interorganizational domains', *Human Relations*, **36**, 269–84.
Van De Ven, H. (1993), 'The development of an infrastructure for entrepreneurship', *Journal of Business Venturing*, **8** (3), 211–30.
Von Bargen P., D. Freedman and E.R. Pages (2003), 'The rise of the entrepreneurial society', *Economic Development Quarterly*, **17** (4), 315–24.
Yang, K. (2004), 'Institutional holes and entrepreneurship in China', *Sociological Review*, **52** (3), 371–89.

5. The role of government in the formation of late emerging entrepreneurial clusters of India

Kavil Ramachandran and Sougata Ray

INTRODUCTION

Developing countries, eager to catch up on industrialization, have identified industries based on information technology (IT) as a major growth driver. It is in this context that the recent interest in industrial clusters as a means to accelerate regional development, particularly after the boom in the IT industry in Silicon Valley, has to be viewed (Sturgeon, 2003). At the same time, efforts by several states and countries to recreate Silicon Valleys have had mixed results. Why does it happen so? Are such clusters human made? Is their growth a natural phenomenon or can it be catalyzed? If so, how and how much? Knowledge drawn from manufacturing industries or other service industries provides only partial answers. This is true in the Indian context as well. Unfortunately, though many scholars have tried to explain why India could emerge as an IT superpower, little has been said on how and why successful IT clusters have been formed in some cities such as Hyderabad and what needs to be done to sustain and replicate the success. This chapter attempts to provide some initial explanation to this complex process.

While most studies on clusters reveal the history and factors that assisted the formation of clusters, a key challenge, often not discussed is whether it is possible to compress the time normally and naturally taken to form clusters and, if so, how. This insight is particularly important now since time has become a critical success factor for both regions and entrepreneurs to be winners. The experience of Hyderabad and Kolkata clusters provide some insight into this area.

The chapter, based both on primary and secondary data, discusses a theoretical model on cluster formation based on case studies of three cities in India; one already established (Bangalore), another in the early growth stage (Hyderabad) and the third, an established industrial and commercial city trying to revive its fortunes through IT (Kolkata). Our enquiries through these case studies bring out four sets of factors as key to the formation of IT clusters. While the role of industry attractiveness and factor conditions are well known, the role of a pool of local entrepreneurs is less discussed. We find that the extent to which the state plays the role of an entrepreneur in attracting and encouraging entrepreneurship and thus synergizing the other three factors is equally if not more important, and is particularly elaborated on in this chapter.

INDUSTRIAL CLUSTERS AND INDIAN IT INDUSTRY

Interest in industrial clusters goes back to Marshall's (1920) pioneering work, which still remains one of the most important pieces in this area. Marshall attributed clustering tendency to the external economies that firms derive from a neighborhood location. Marshall's arguments that firms in clusters specialized through splitting up of activities (which are now called links on the value chain) and benefited from economies, laid the foundation for further studies in this area. He argued that the reasons that clustering occur included economies in labor supply and information sharing from locating firms in proximity to each other. In fact he assumed the existence of cooperation and competition emerging in the process, both good for building overall competitiveness of locations and firms. There has since been general consensus in describing a cluster as a geographic concentration of several inter-connected firms belonging to a particular field. Some of the key features of a cluster are information spillover across firms, development of a skilled labor market and specialization and division of labor among enterprises (Sonobe and Otsuka, 2003).

Though Marshall observed the phenomenon of clusters, it did not attract academic attention for the next five decades, primarily because of the growing interest in vertical integration as the key growth model (Rocha, 2004). While some of the conceptual work can be traced to Italian clusters, there are several others covering different industries (both modern and traditional) in industrialized and less industrialized areas. There are three broad streams of studies on clusters that appeared in the past two decades or so. These cover regional development, economics and sociology, but not entrepreneurship. Italy's clusters have been a major interest of study, particularly in the context of regional development. For instance, Becattini's (1990) study emphasized the role of clusters in improving firm efficiency and also regional development.

Michael Porter (1990, 1998a and b, 2000) gave a new impetus to the continuing but often isolated interest in industrial clusters. Porter looked at the phenomenon more comprehensively, building on his arguments made in the context of his discussion on the competitive advantage of nations. He argued that his 'diamond' influences creation of clusters. This involved the co-existence of the following: factor conditions, demand conditions, the existence of related industries and a local context for firm strategy, such as rules and incentives.

Krugman (1991) provided further insight into clusters by looking at the phenomenon from yet another angle. He argued that the central reason for clustering is the benefit of increasing returns to scale, either at the local or national level. This would include positive economies for the firm and also diseconomies for moving from an existing location, so that those existing continue to be in a location, while attracting new firms, resulting in cluster formation. In effect, Krugman's major contribution is in terms of identifying the causes for cluster formation.

A close look at the arguments of Porter and Krugman show that clusters are formed because of the perceived internal and external economic benefits. The process is accelerated if the industry is growing rapidly due to positive demand conditions. In addition, Porter has emphasized the role of government as a facilitator (Porter, 1998a).

A totally different, but complementary, approach to understanding clusters came from sociologists who brought to the fore the role of social networking. Building on Granovetter's (1985) original thesis, Polanyi (1994) and Storper (1997) and later Keeble and Wilkinson (2000) demonstrated the process and effect of knowledge spillovers and innovation across firms in the same location. Interest in strategic alliances and inter-firm networking that grew in recent years (Lahiri, 2004) also provided a general back drop to the keen interest in networking as a facilitator of cluster formation. Saxenian (1994) had noted that these are more important than external economies of an economic and technical nature. Rocha (2004), after a review of relevant literature, concluded that informal personal contact networks are necessary for knowledge transfer and innovations to be geographically concentrated.

Over the years a number of additional thoughts on the concept and scope of industrial clusters have emerged. Camagni's (1991) work on innovation networks found that information exchange across people located in a cluster is more significant than business contacts among them. Van Dijk's (1999) work on the typology of industrial clusters provided new insight to such dimensions relevant in the context of IT clusters. Brenner's (2004) study of clusters added further insight to this dimension. He noted that a shared cultural background and a strong social network developed by the entrepreneurs and employees of these clusters are key features of clusters. These recent findings are particularly relevant for our discussion here.

In essence, the dynamism and economic success of numerous clusters operating in countries, such as Italy, Germany and Japan, in diverse industries, such as car, leather, textile, jewelry and optical frames, has been well documented. In spite of some inter-firm variability, the firms in a cluster show striking similarities in the way they are structured, behave and perform as they are more or less governed by the same policy, institutional, competitive, technology and socio-economic environment (Porter, 1990). The constellation of these forces together creates an environment that becomes a source of constraints, contingencies, problems and opportunities that affect the terms on which a firm transacts business and derives and sustains competitiveness. These can cover the entire value chain or parts of it, within the same country or abroad. In that sense, the arrival of the Internet and powerful communication facilities have enabled the creation of virtual aggregation by clusters of firms located in different geographical points.

A closer look at all these materials would show that most researchers have discussed the influence of certain factors in the process of the formation of clusters. While this process is interactive, it is not clear how the cluster formation process can be accelerated beyond offering incentives. This is primarily because most scholars have looked at the phenomenon from their respective disciplines, and none from the angle of entrepreneurship. We believe that this is a major gap in understanding clusters.

Our anecdotal analysis of cluster formation in Bangalore and Hyderabad shows that the existing literature cannot explain why these two clusters are behaving the way they are. The rapid growth of Hyderabad is amazing. It was never considered to have the factor conditions for growing IT. It never had a pool of IT entrepreneurs to network with. It was not a natural destination for IT given its rather hot climate. We found that the literature does not recognize any distinctive role for a government that behaves more like an entrepreneur than a bureaucracy. Also, the role of an entrepreneurial government in creating a cluster from scratch has not yet been studied.

Our aim is to make an exploratory study of the role of entrepreneurial leadership in creating and growing clusters. We have chosen the IT industry because of its rapid growth in general and the efforts of varying degrees made by different state governments in India to promote it.

GROWTH OF THE IT INDUSTRY IN INDIA

Unlike most developing nations, India is demonstrating a high degree of competitiveness in knowledge intensive software development and IT Enabled Services (ITEs). The Indian IT industry has been experiencing rapid growth since the early 1990s and has become a major player in the Indian economy. There has been international recognition of the potential of Indian IT firms, which have become suppliers to a large number of Fortune 500 firms. The software industry[1] in India grossed an annual revenue of Rs. 708 billion (US$ 15.6 billion) during 2003–4, up from Rs. 382 billion (US$ 8.4 billion) in 2000–1, registering an average growth of about 30 per cent in rupee terms. More than 8000 firms, located in cities like Bangalore, Chennai, Hyderabad, Kolkata, New Delhi and Pune have been providing a range of software services, mostly targeted at foreign customers. The software exports from India grew from Rs. 283 billion (US$ 6.2 billion) in 2000–1 to Rs. 555 billion (US$ 12.2 billion) during 2003–4, registering an average growth of more than 30 per cent in dollar terms.

The growth of the Indian IT industry has seen clear phases, with both Indian entrepreneurial firms and MNCs contributing significantly to developing the industry. The initial deregulation of the hardware industry encouraged many hardware firms to emerge in India. Later the majority of these hardware firms diversified into software businesses. The IT industry has responded to the various markets that have grown in different time spans and has tried to cater to in-house software services, software exports and software product firms. The software capabilities were initially developed through import substitution and with a full-fledged orientation to the domestic IT market. Gradually however, software exports started and became the buzzword of the industry in the 1990s.

While the software industry has been growing rapidly in recent years, centers of excellence have already emerged in India, and the geographical spread of the industry in India has changed considerably in the last decade. The share of southern Indian software exports grew from 25 per cent in 1991–2 to more than 40 per cent in 2003–4. Though Chennai had emerged as a software center early on, it was soon overtaken by Bangalore with Hyderabad, Pune and more recently Kolkata gaining ground to catch up. The Bangalore cluster is the largest in terms of sales and exports. It also houses the most sophisticated firms. Compared to Bangalore's software exports of Rs. 146 billion in 2004, Hyderabad and Kolkata exports amounted to Rs. 56 billion and Rs. 22 billion, respectively. However the number of firms in both Bangalore and Hyderabad were about the same (*Business Standard*, 2003). To promote rapid growth of the IT industry, central as well as many state governments have created software technology parks in which the necessary infrastructure is readily available. Notable examples include Bangalore's Electronic city and Hyderabad's HITEC city, which offer not only office space and communications links, but several social amenities as well. The software parks located in these cities have also played an important role in allowing firms to develop clusters.

In this chapter we discuss how IT clusters have been formed in three selected Indian cities. We have highlighted the significant contribution that state governments make in shaping the formation of IT clusters in the capital cities of three major states in India, namely Karnataka, Andhra Pradesh and West Bengal. Karnataka houses Bangalore, which has emerged as the 'Silicon Valley' of Asia. Hyderabad in the neighboring Andhra Pradesh is fast catching up, and is rated as the next 'Silicon Valley' (Zwingle, 2002). Kolkata (erstwhile Calcutta), the British capital of the Indian subcontinent, which was once the most industrially and commercially developed city in India, has been trying to revive its fortunes by aggressively attracting investment in IT and ITES industries.

We shall now discuss the three case studies in detail to understand the three clusters in greater detail.

CASE I: FORMATION OF THE IT CLUSTER IN BANGALORE

Bangalore houses the most prominent IT cluster in India. From a mere 13 software firms in 1991–2, the city now has a pool of over 1200 firms working in areas, such as computer chip design, systems software and communications software, and employing over 100 000 IT professionals. Compared with other locations in India, Bangalore has a high-end technology/industry concentration, such as very large scale integration (VLSI) and telecommunications services, and a higher degree of MNC presence with over 200 foreign firms in operation. It is ranked fourth as a global hub of technological innovation, behind San Francisco and Austin in the USA and the Taiwanese capital, Taipei. But why and how did Bangalore emerge as the leading hub of Indian IT industry?

Early Factor Advantage

There are several factors – historical, geographical, economic, cultural and political – that have contributed to the emergence of Bangalore as the dominant IT cluster in India. Bangalore has been fortunate to be rich in the supply of both economic and non-economic factors, and as Ramachandran (1986) noted had the key ingredients required to make it a preferred choice of business location.

Skilled labor

Among the advanced factors, highly skilled labor plays a very important role in the development of any industry (Hanna, 1994; Porter, 1990). As the software industry makes intensive use of human capital, the availability of highly skilled technical and managerial personnel becomes the key location factor. Bangalore has a large, highly skilled IT talent pool available at a relatively low cost, thanks to the historical development of the city's educational, research and industrial infrastructure. Karnataka has one of the strongest educational infrastructures, both at undergraduate engineering and postgraduate levels. Of the 67 engineering colleges in the state, 26 are located in and around Bangalore, including the Indian Institute of Information Technology. In addition, entrepreneurial and managerial talents are encouraged by institutions such as the Indian Institute of Management Bangalore.

Research institutions

There has been early localization of science and technology related research and training institutions, as Bangalore is considered an ideal place – in terms of climate and infrastructure – to conduct scientific research in sensitive areas like defense and electronics as well as fundamental research in science. The seed was sown with the establishment of the elite Indian Institute of Science, a number of the largest and most prestigious public sector enterprises in fields, such as electronics, aeronautics, earth moving equipment and machine tools, and research organizations, such as the Indian Space Research Organization and the Defence Research and Development Organization. They were established partly for climatic reasons and partly for technical and economic reasons. These enterprises created a good pool of technical personnel in Bangalore, an important condition in developing the IT industry anywhere.

Social network

Over the years the pool of technical and managerial talent has grown, making Bangalore one of the richest social networks for IT. Many of the graduates who migrated to the USA for higher education and jobs form part of the social network that nurtures the local software industry.

Investments made by government and other public institutions for specialized infrastructure, such as technological and management institutions, and educational programs and investments by companies – in training programs, infrastructure, centers of excellence, testing laboratories and so on – have contributed to the development of this network and Bangalore as a city.

Economic infrastructure

Bangalore has had a good network of roads and airport connectivity for a long time thanks to the city's attractiveness for setting up industrial, research and defense establishments. Even during the pre-independence period, it was called the 'Garden City' of India, with an excellent infrastructure. Bangalore started experiencing power shortages only in recent years when the demand for power grew rapidly and without adequate measures taken by the government to enhance the supply. A similar situation exists regarding growing traffic congestion on the roads.

Social infrastructure

Bangalore has been identified as an ideal place to live with an abundance of rich social infrastructure. This includes a moderate climate, good housing, parks and educational facilities.

Growth in Local Entrepreneurship

Bangalore became home to three categories of firms. Initially, it was the MNCs, starting with the establishment of Texas Instruments in 1984 as a 100 per cent export oriented firm. The early days of growth of IT saw a few more MNCs, such as Digital Equipment, IBM and HP choosing to be in Bangalore.

However the rapid growth of indigenous entrepreneurs of firms, such as Wipro, Infosys and Microland, whose promoters had no family business background, has been remarkable. Many IT executives set up their ventures in Bangalore either because they belonged to the city or the state or because they had developed strong social networks there. The new generation of entrepreneurs were inspired by the success of those around them and the opportunities they had exploited as managers. The virtuous circle continued to attract more and more start-ups, while the city continued to have branches of existing firms from other cities.

Attractiveness of the Industry

The success of most Indian software firms comes from serving foreign customers, especially in the USA (Kapur and Ramamurti, 2001). Given the small size of the domestic market, Indian firms had to be export oriented and were dependent on the growth of the export market (Chakraborty and Dutta, 2002). Since the 1970s outsourcing of software development activity by firms in developed economies became a trend due to the huge increase in software costs, increasing demand for complex information systems applications, rapid obsolescence rates of the IT infrastructure and inadequate supply of IT personnel. As a result, IT work is today distributed globally on the basis of cost, location of customer sites and expertise, and is largely independent of the IT company's country of origin (Salzman, 2000). As a result, many large and medium sized US firms focused their attention on more valuable and creative projects (Arora et al., 2001). Differences in time zones allow work to be carried out by Indian teams on a 24 hour basis, shortening cycle times and improving productivity and service quality. The social network connecting people of Indian origin in the USA, often working in Silicon Valley, with engineers and managers in India has also played an important role in exploiting this advantage (Kapur and Ramamurti, 2001). Being members of a vibrant cluster, Bangalore based firms could tap the network and capitalize on the US demand more than others.

This is clearly visible from the fact that Bangalore remained the most favored destination for both Indians setting up ventures based in India or working for the Indian subsidiaries of foreign multinationals (Heeks, 1999; Taeube, 2004). In 2000 71 of the 75 multinationals located in Bangalore had executives of Indian origin returning from abroad as heads (Ghemwat, 2000). Overseas Indians, who returned to start new companies or supply venture capital, have fueled new venture formation. In addition, overseas companies opened software centers in India to strengthen interaction between their organizations and Indian suppliers. By 2001 several MNCs had R&D subsidiaries also in Bangalore.

Benefiting from Cluster Advantage

The basic principle of clustering for manufacturing industries applies to IT, the major difference being the virtual nature of the activity. In terms of spatial concentration traditional manufacturing firms and knowledge based firms follow the same pattern. Success of a cluster depends on the attractiveness of the specific location for firms in the related industries and the ecosystem that gets evolved. A cluster of independent and informally linked software firms and institutions located in the same city has definitely allowed firms to exploit advantages in efficiency, effectiveness and flexibility (Porter, 1998a, 2000). This is fine with Bangalore, which has witnessed close relationships between local industry and major research universities and institutions in the area, reasonably active venture capital industry, some degree of inter-firm cooperation, tolerance for spin-offs, and nurturing of firms largely outside the purview of large, ponderous, bureaucratic firms and financial institutions. The process of synergy building is steady, and its speed depends on the growth of the industry and the benefits that individual firms can derive from the cluster.

The Bangalore cluster has affected the development of software firms in three broad ways as argued by Porter (1998a). First, by increasing the productivity of companies based in the area; second, by driving the direction and pace of innovation, which underpins future productivity growth; and third, by stimulating the formation of new businesses, which expands and strengthens the cluster itself. Companies have been able to operate more productively in sourcing hardware and software, accessing information, technology and local institutions, coordinating with related companies, and measuring and motivating improvement. Software firms have been able to tap into an existing pool of specialized and experienced employees, thereby lowering their search and transaction costs in recruiting. It has been easier to attract talented people from other locations because the cluster of firms signal opportunity and reduce the risk of relocation for employees. However there is not enough evidence available to suggest that formal inter-firm kiretsu type linkages (Tyrini, 1994) exist among IT firms in Bangalore, whereby they gain from disintegration of the value chain and use of efficient networks of market transactions (Scott, 1988; Storper, 1997).

Naturally, there has been an accumulation of extensive market, technical and competitive information within the cluster, and the local firms have preferred access to it. In addition, personal and professional relationships, old boys' networks built in technical colleges and universities, and community ties have fostered trust and facilitated the flow of information. The Bangalore cluster has also developed an unmatched reputation as an industrial location, which in turn has benefited the firms located in the city dealing with global buyers and suppliers. Beyond reputation, the city based firms have often profited from a variety of joint marketing mechanisms, such as company referrals, trade fairs, trade magazines and marketing delegations.

Bangalore's cluster of high quality IT companies made it easy for overseas companies to meet many potential vendors in a single trip, which allowed them to multi-source or switch vendors if necessary. This also led to a high degree of local rivalry, which was very motivating. Peer pressure amplified competitive pressure among other Bangalore based firms, even among non-competing or indirectly competing companies. Pride and the desire to look good in the local community had spurred professionals to attempt to out do one another. The state government that promoted IT in a big way during the latter half of the 1990s also fueled rivalry. As a result, there has been a rapid increase in the number of IT companies in Bangalore, thereby intensifying rivalry.

The diffusion of technology, facilitated by the entry of multinational firms from the USA, has been helping firms move up the value curve. At the same time, Indian software firms like Infosys and Wipro opened offices in the USA or acquired US companies, to better serve their clients on high-end projects and to have listening posts in Silicon Valley. Thus physical distance was bridged by the strengthening of cross-national, intra-firm networks and by inter-firm social networks among Indians and overseas Indians.

Clustering in Bangalore has played a vital role in innovation and value upgradation in the same manner as envisaged by Porter (1998a). Because most sophisticated buyers in the world have been part of this cluster, local firms usually have a better window of the market than isolated competitors do. Small and medium size companies continued to get opportunities to grow as the bigger companies moved into larger projects, vacating space for the smaller ones. Being located in the same place and being part of the network, these players got the first crack at outsourcing orders vacated by the big ones. The ongoing relationships with other entities within the cluster have also helped companies learn early about evolving technology, hardware and software availability, service and marketing concepts and so on. Such learning was facilitated by the ease in making site visits and frequent face to face contact. In addition, serial entrepreneurs tended to choose their familiar and a comfortable location for their subsequent ventures, as was found in Cambridge, UK, by Vyakarnam and Myint (2005).

Local Government – Indifference to Intermittent Active Facilitation

The role of the state and the local governments, albeit small, was also important in developing the industry. The 1971 policy of the government attracted substantial investments to locations around Bangalore. In 1997 Karnataka became the first state to come out with a specific policy for the promotion of IT, which included a number of tax concessions and other benefits to IT investors.

In recent years, like most fast movers, Bangalore started facing the pinch on two counts. On the one hand, the rapid growth of the population rising on the wave of IT success impaired the livability of Bangalore with dramatically escalated real estate prices, congestion, poor road conditions and other overloads on the local infrastructure. On the other hand, many other state governments started to aggressively promote IT and provide better infrastructure and policy support to attract investments. This provided alternative locational choices for software firms looking for new investment opportunities in India.

In the absence of government initiatives, the leading software firms started early on to invest heavily to overcome the bottlenecks in physical infrastructure. For example, Infosys had to install its own telecommunications system, stock 11 tons of back up batteries to keep the computers running, 4000 gallons of diesel fuel to power its generators during power cuts, operate its own sewage treatment plant to reuse water as a remedy to erratic water supply and run a fleet of buses to transport its employees (Ghemwat, 2000). Shortage of public infrastructure forced several Bangalore based firms to start moving to new locations for expansion of their operations. With the rapid erosion of the relative factor advantage, the role of local government has increasingly become critical to sustain the prominence of the Bangalore based IT cluster.

Recent initiatives by the state government have started paying dividends again, with several new IT based investments of high quality flowing in. For instance, the state government has set up a new university with excellent facilities exclusively for engineering education. There is also greater attention now paid to decongest the city.

Learning from the Bangalore Cluster

Overall, it is observed that factors such as historically and accidentally created human resources, proactive policies to attract and allow multinational firms to exploit human resources, technology leverage by local firms, emergence of a new class of entrepreneur and their link with the epicenter of IT in Silicon Valley, and to some extent the facilitating role of the state government have played crucial roles in the emergence of the Bangalore cluster. A closer look at these factors shows that the competitiveness of a location for cluster formation depends on the confluence of a number of variables as shown in Figure 5.1.

These factors are quite in line with the existing literature on the competitiveness of location (Porter, 1990, 1998a).

Industry attractiveness

Influenced by the global demand condition and industry structure, rapid growth in the industry expectedly emerges as a critical factor that creates the necessary customer pull and momentum for any firm to flourish, and consequently any location to grow rich. This is a fundamental requirement for any location to flourish.

Factor supply

Bangalore, with its developed pool of physical, technical and economic infrastructure, fulfills the key dimensions of supply of factors in the formation of a cluster. Lessons from the experience of regional development in a number of countries, including relocation strategies of firms in mature industries to low wage zones such as Malaysia and Thailand, provide insight into the continuing attractiveness of Bangalore. Entrepreneurs and managers look for a combination of factors when choosing their location.

Knowledge based industries are influenced by a different set of location factors compared to traditional manufacturing based industries

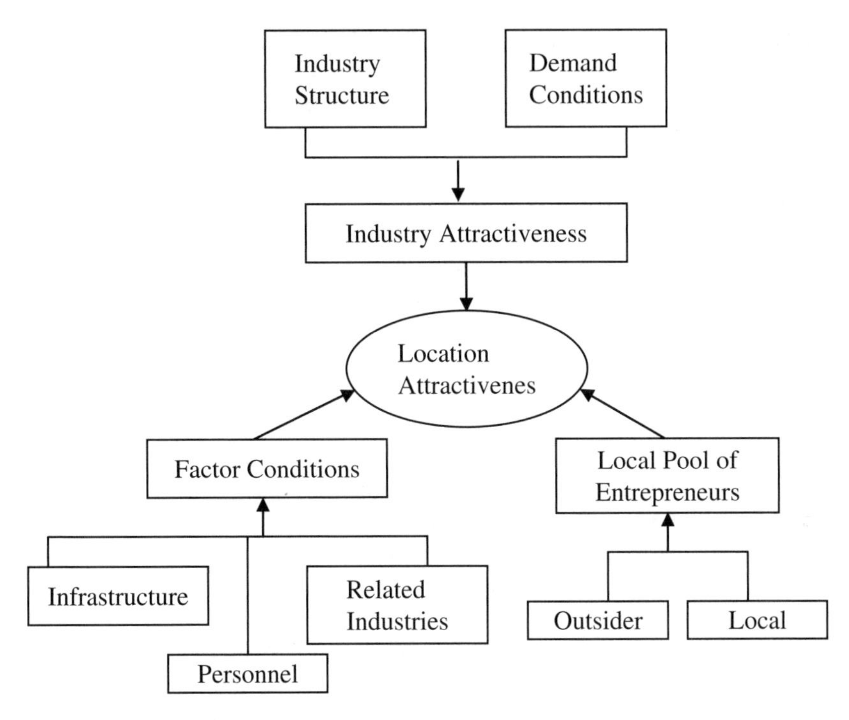

Figure 5.1 Natural growth drivers of clusters

(Ramachandran, 1986; Saxenian, 2000). Regional development indeed depends on the development of economic activities and income generated in the region. Ramachandran (1986, 1989) has shown a shift in the mix of location factors from economic to social as a region develops its economic infrastructure, in line with the Maslowian (1954) hierarchy of needs argument. According to Ramachandran, as the supply of physical infrastructure reaches a reasonable level, the need of entrepreneurs and managers to have fairly high quality non-economic factors (pleasant environment, social infrastructure) would go up. This is particularly so in knowledge driven industries, where employees, the key source of value addition, are highly mobile. The kind of economic infrastructure would not be the same for manufacturing and non-manufacturing industries. For instance, the quantity and quality of communications infrastructure of IT industry is much higher compared to manufacturing.

Local pool of entrepreneurs

The third factor is the supply of a local pool of entrepreneurs; both at the early stage to kick start the growth of the cluster and at a later stage to sustain the momentum. For entrepreneurs, 'home proximity' is the most important location factor (Ramachandran, 1986), particularly in societies such as India with strong cultural bonds among family and community members. In case the relative factor advantage of the cluster goes down temporarily, a capital flight is bound to happen. However the local entrepreneurs are more likely to continue and work for the revival of the cluster advantage because of their greater bond with the location (Stuart and Sorenson, 2003). In addition, in dealing with a turbulent industry, such as IT, the location of the business in a place where one's social and professional networks are present becomes important (Johannisson, 1998). This is particularly so for start-ups with limited internal resources (Siegel et al., 1993), but with the need to have a growing pool of resources (Barney, 2002). This is evident in the case of the Bangalore cluster where companies like Infosys and Wipro have taken a very active role in supporting the Karnataka government to revive factor conditions and to some extent arrest the flight of capital to other locations, such as Chennai and Hyderabad. The presence of this important factor, that is, a local pool of entrepreneurs shaping the competitiveness of a location has not received enough attention in the extant literature.

However a fundamental question that remains is the replicality of the 'Bangalore Strategy' elsewhere in India. Since the competitive, technological and regulatory environments are not the same, it will be interesting to see the extent to which clusters like Hyderabad and Kolkata have followed the same strategy as Bangalore.

CASE II: FORMATION OF CLUSTER AS A LATE MOVER – CASE OF HYDERABAD

Our second case study is Hyderabad, which marked its beginning with the establishment of the Software Technology Parks of India (STPI) in the years 1991–2. After a slow start it started growing in the mid 1990s and has since grown rapidly to have 1288 firms with software exports of Rs. 185.82 billion in 2006–7 (STPI, 2007). The top ten firms contributed more than 55 per cent of the exports from Hyderabad. The industry has grown particularly rapidly in the last three years. About one-third of the registered units are foreign companies contributing about 30 per cent of the total investments. Out of 174 new firms registered with STPI, Hyderabad in the F.Y. 2006–7, about 70 firms are in the process of commencing their operations. Once these firms become operational, the formation of the IT cluster in the city will get a major boost. In the past few years, the Andhra Pradesh government has been able to attract a number of leading foreign multinational firms and Indian firms to set up operations in Hyderabad by providing better infrastructure, concessions, hidden subsidies and local demand through e-governance projects. It has simultaneously backed local IT firms, such as Satyam Computers and InfoTech Enterprises, to grow and flourish. As a result, Satyam has emerged as the biggest exporter from the city, and one of the 'big four' in IT industry in the country.

We shall next discuss the major differences between Bangalore and Hyderabad in the context of the IT clusters.

Factor Conditions in the Mid 1990s

Hyderabad was often considered as a 'hot, dusty and dirty' city, with crowded and badly maintained roads. Power supply was erratic and civic services of poor quality. It had a few engineering colleges, but not of very high quality. Discipline in academic institutions was nothing to boast about. Graduates in large numbers used to migrate to other places in search of jobs.

There were a few advanced defense research institutions located on the outskirts of the city, away from the public glare. In addition, the city had a few large public sector enterprises, manufacturing electrical and electronic goods, located on the outskirts. It had four major industrial estates which had a number of small and medium size pharmaceutical and engineering manufacturing units.

In short, Hyderabad was often called a 'big village' with a laid back attitude to life. Politicians used to be busy trying to retain their seats and government highly bureaucratic and corrupt. It grew at a snail's pace and nobody seemed to be concerned. Nobody, then, would have imagined in their wildest dreams such a transformation of the city within such a short period of time.

Entrepreneurial Leadership of State

In this section we present an analysis of how the Andhra Pradesh state government led by the then Chief Minister Chandrababu Naidu (1996–2004) made a giant stride to shape an IT cluster in and around Hyderabad. Naidu started his career as an entrepreneur and left it with his family to run when he joined politics. In 1996, when he became the Chief Minister, he had spotted the emerging entrepreneurial opportunities in knowledge intensive industries. Though Hyderabad was already known for its cluster of pharmaceutical firms, he believed that the potential in IT was greater. Early on, he realized that he had to make a major change in the mindset of people in the government and outside. His entrepreneurial vision and the clarity of the strategy are evident from the number of initiatives Naidu had taken to position Hyderabad as the knowledge hub of India. This included not only upgrading economic and social infrastructure, but also providing pioneering leadership in e-governance in India. This not only pushed up IT orientation in the society, but also opened up business opportunities for several companies, including Microsoft.

As someone with proven entrepreneurial capabilities, he had realized the need to stir up the administrative machinery from its slumber and provide high quality service to give confidence to prospective investors. With a team of chosen civil servants who shared his dream and with proven managerial capabilities, he started pushing administrators to think and act differently. He demonstrated that accountability and performance can be improved even in the government machinery. For instance, he improved the management information system (MIS) in the government. Follow up became routine. He privatized and outsourced a number of services that until then were carried out directly by the government, without retrenching people. For instance, every road in Hyderabad got cleaned before sunrise every day. The government secretariat started functioning in time and wore a professional look.

The state government partnered with private companies to promote IT literacy too. This included campaigning and facilitating the creation of a number of pioneering international quality institutions in the area of IT, biotechnology, life sciences, business management and insurance, all in emerging growth areas. All of them are designed to be of international repute. The creation of a finance district to house a number of institutions and agencies is expected to lead to a cluster of financial organizations in Hyderabad. The entrepreneurial leadership is as much reflected in these initiatives as it is in brand building around them. A number of foreign delegations and dignitaries were invited to the HITEC city in the last few years and several road shows and seminars were held in various countries, including the USA, UK and Germany, to give visibility and build image for the city. Naidu, as the chief minister, took a personal interest in negotiating with a number of multinational firms. All these formed the building blocks of a grand strategy to catch up with the development process, which is knowledge driven in the emerging context. These steps are far superior to anything happening in any other Indian state.

The entrepreneurial leadership of the state is reflected in a number of other ways too. For instance, Andhra Pradesh was one of the first states in India to formulate an IT policy in 1999. Recognizing the emerging global opportunity in the IT Enabled Services sector (ITES), Andhra Pradesh formulated a separate ITES policy in 2002. The Chief Minister demonstrated superior entrepreneurial leadership compared to his counterparts in other states. When the Indian School of Business (ISB) start up team was deciding which location to choose for the prestigious institution among Karnataka, Tamil Nadu and Maharashtra, the Chief Minister personally persuaded the high powered team to visit Hyderabad and give him a chance to make a presentation, though his state was not on the shortlist. The visiting team's experience in Hyderabad contrasted with those in other cities. While one chief minister (CM) wanted a quota of seats reserved for applicants from there, another CM made them wait for 45 minutes and showed indifference at the meeting. In Hyderabad the Chief Minister not only individually greeted the members of the team and garlanded them, but also made a power point presentation, and personally handed out plates at dinner. Many investors have experienced the same pattern of behavior from him both before and after making investments. Interestingly, three years after the setting up of ISB the entire area, lying on the outskirts of the city, has suddenly become an IT hub.

The current strategy of Hyderabad to encourage the setting up of a number of high quality private schools for primary and secondary education is relevant in this context. However some of the other initiatives that make sense in this context include the creation of a number of parks, botanical gardens and other ecotourism centers, a major urban afforestation program all over the city, the promotion of tourism in a large and concerted way and building the image of Hyderabad as the first choice in where to live. These are all strategic variables forming part of a grand strategy based on the emerging industrial location factors. These strategies are synergistic with the current efforts to promote tourism in the whole of Andhra Pradesh, including Hyderabad. There is a definite and conscious effort to make Hyderabad and surroundings and the state as a whole attractive for the knowledge and service driven industries to grow. It is such qualities of entrepreneurial leadership that provide the necessary fuel to push the state to a higher production function and enter a new and higher paradigm, as explained by Schumpeter (1949).

The initiatives have started paying off as can be seen from a survey conducted by NASSCOM (2002) on city competitiveness. Hyderabad was ranked one ahead of Chennai (three) and Bangalore (six). The competitiveness assessment was carried out based on three parameters: availability and quality of infrastructure, availability and cost effectiveness of personnel and policy support.

In essence, the Hyderabad case proves that it is possible for a late mover to catch up if it follows the principles of identifying entrepreneurial opportunities (Ramachandran, 2003; Shane, 2005) and successfully executes them (Bossidy and Charan, 2002). It is not easy for a sleepy government system to wake up and act like this suddenly.

Cluster Literature and Entrepreneurial Leadership of State

Our earlier discussion on business clusters showed that so far no notable work has been done on the creation of a cluster within a short period of time, when the factor conditions are unfavorable. Even Porter's (1998b) discussion on government role in cluster growth did not envisage a situation where the local government plays the role of an entrepreneur in identifying an attractive opportunity and exploiting it. After a review of the role so far played by governments across the world, Porter finds a role for the government '. . . in ensuring that appropriate factor conditions are present as well as setting a context that encourages upgrading through appropriate policies . . .' (p. 11). Here also, implicitly Porter talks about a managerial role that involves a reactive response, and not a proactive entrepreneurial approach. Later, in the same paper, Porter finds 'an important role for government in facilitating the upgrading of clusters' (p. 12). Governments are expected to be social engineers to foster private entrepreneurship, and not to be entrepreneurs themselves (Chang, 1995).

However, state governments in China have started showing entrepreneurial vision to enhance national or regional competitiveness (Pereira, 2004). It is to be remembered that the power of the state to effect changes is unlimited in a totally controlled economy such as China. Singapore is another country that has demonstrated entrepreneurial leadership at the national level, during its process of transformation from a poor, newly independent country to a rich, city nation. In fact it was forced to develop the economy by attracting overseas investments (Rastin, 2003).

Of course, Singapore did not try to create a cluster, but behaved entrepreneurially. As noted by Schein (1996), the local government demonstrated honesty and trustworthiness in its transactions with MNCs. This pattern of behavior was demonstrated in Hyderabad when Naidu convinced Bill Gates to open Microsoft's research center in that city. There are two differences between Singapore and Hyderabad. First, Singapore enforced political autocracy like China to ensure stability and the rule of law whereas Hyderabad has functioned as a democracy through its history. Second, unlike Hyderabad, Singapore had a culturally fluid society, where people were willing to adapt their own cultural norms and values to the changing environment (Choy, 1987). Demonstrating entrepreneurial leadership in a tradition bound, bureaucratic, democratic government set up requires not only vision but also capabilities that are noticed in path breaking entrepreneurs.

In essence, the existing cluster literature is silent about any entrepreneurial dimension in cluster creation. Our major contribution is in identifying and discussing this role. We notice that it is possible to create a cluster from scratch if the government provides entrepreneurial leadership in spotting an emerging opportunity and exploiting it. Effective entrepreneurial leadership involves the creation of an organization with an integrated approach to converting information into ideas and solutions (Marquis, 1969; Utterback, 1971). Such organizations, whether private, public or government, highlight the importance of an explicit and well articulated strategy that enables managers to have an integrated vision of where to go (Miller, 1983). The commitment of an entrepreneurial organization to seize opportunities is revolutionary rather than evolutionary and the processes are tuned to it (Stevenson and Gumpert, 1985). Stevenson and Jarillo (1990) had observed that an entrepreneurial organization pursues opportunities independent of existing resources. They follow the strategic intent arguments of Hamel and Prahalad (1989). McGrath and MacMillan (2000) have identified three sets of practices that are affected in this process. These are work climate, seeking and realizing opportunities, and quick problem solving with people at work.

Success of such leadership initiatives in organizations depends on the presence of a supportive leadership that builds commitment and enthusiasm by creating a shared sense of purpose and meaning in the organization (Roberts, 1984). Devarajan et al. (2003) conclude that effective entrepreneurial leadership would create a shared passion in the organization.

CASE III: FORMATION OF CLUSTER AS A LATE MOVER – CASE OF KOLKATA

During the mid 1980s the first set of software firms were set up in Kolkata, the capital of West Bengal, a state ruled by the communist government for about the last three decades. At the peak of trade unionism in the state and amidst large scale opposition to computerization, IT companies in Kolkata received virtually no government support; instead, they faced a very hostile environment in the early years. Though the state government set up the first Software Technology Park of Kolkata (STPK) in 1992–3 soon after economic liberalization, it did not pay enough attention to it for the next several years. As a result, the progress made during the major part of the 1990s was abysmally slow with only 39 registered units and 30 operating firms and Rs. 2 billion exports in 1999.

During the last five years Kolkata has emerged as an important IT cluster in India with more than 400 registered units and 215 operating firms employing over 25 000 professionals and exports of Rs. 22 billion in 2004. The STPK has grown particularly rapidly in the last three years (growing at the rate of 70 per cent), earning the distinction of the fastest growing STP in India. Unlike many other Indian cities Kolkata has always had an inherent advantage in areas such as talent pool, reliable supply of power, and quality and cost of living, which are among the key enablers of knowledge based industries like IT. The bottlenecks used to be in the areas of civic infrastructure, work culture, hostile industrial climate, investor unfriendly image of the local government, and absence of policy and investor support. However, in the last three years, the state government has been able to attract a number of leading foreign and Indian IT firms to invest in Kolkata by laying the red carpet, removing all the regulatory hurdles with alarming speed and presenting an IT investment friendly face to the investors. The government has also initiated a project to set up by 2007 India's largest IT hub, with the facilities to become a modern IT hub, not only in India, but in the whole of Southeast Asia in 500 acres of land near Kolkata airport. In addition, more than 150 firms registered with STPK are expected to start their operations soon. Once these firms become operational and the new electronic city project is complete the formation of the IT cluster in the city will get a major boost.

Entrepreneurial Leadership of State

In this section we present an analysis of how the West Bengal state government led by Chief Minister Buddhadeb Bhattacharjee along with his colleagues in the IT department has been able to take giant strides in shaping an IT cluster in and around Kolkata. The state was a late starter in promoting itself among the IT players. West Bengal's share in the country's IT services export was only 3–4 per cent in 2002. The state hired the services of McKinsey & Company to draw up the IT vision and road map for the state. Accordingly, the state government envisioned the state to become among the top three IT states in the country by 2010, contributing 15–20 per cent of the country's total IT revenue. The focus on the long term would be on high value added IT work, developed through a leadership at corporate, intellectual and government levels.

Since then the government has been adopting a three-pronged approach towards achieving the target and has made significant progress in the last three years. It has provided legal, administrative and physical infrastructure to potential players who wanted to set up shop in the state. Following Naidu's strategy in Hyderabad, the government has taken up several active e-governance initiatives to spread the use of IT in departments with maximum citizen interface.

The entrepreneurial vision and the clarity of the strategy are evident from the number of initiatives the state government has taken to effect the resurgence of Kolkata as the intellectual capital of India. The city has had a number of pioneering high quality institutions of excellence for decades in the areas of engineering, science and technology, statistics, management, medicine and insurance. However these institutions started declining a few decades ago and almost lost their pre-eminent positions. The entrepreneurial leadership of the state is reflected in reviving these institutions and also encouraging new institutions to be set up by private initiatives. The government has promoted a new university, under which 65 private engineering colleges and business schools have been established. It has started promoting industry academic meeting to bridge the gap between the supply of talent and recruitment.

In addition, one of the main issues that needed to be tackled was the image of the city to place Kolkata firmly on the IT and ITES map. To convert Kolkata as a destination for IT and ITES firms, the state government has taken a number of steps to erase the negative image of the city, and has been making a lot of promotional efforts, including road shows abroad.

The Chief Minister has taken a personal interest in wooing big Indian industrial houses and multinational firms. A seasoned politician in India's communist party, Buddhadeb Bhattacharjee became Chief Minister of West Bengal in 2000. Since then he has been on a mission to win investment and change perceptions of the state and its capital, Kolkata. After a late start, the state government has been actively wooing technology companies and discouraging unions from striking. In an interview with *Far Eastern Economic Review* (2004), he said:

> It's a very competitive world. Therefore I have to perform or perish. We must get rid of 'red-tape-ism' and bureaucratic bungling. In the past, we committed certain mistakes in the trade unions. Sometimes their behaviour was beyond our control. Now we say, look, we won't allow this sort of agitation or intimidation. Labourers have to share it, otherwise the industry will collapse and you'll lose jobs. We won't allow any irresponsible behaviour or activities in the name of trade unions.

Though West Bengal faced a different set of challenges in setting up an IT cluster in Kolkata, many of the steps resemble that of the Andhra Pradesh AP government a few years ago. For instance, the West Bengal government formulated an IT policy in 2001. Recognizing the emerging global opportunity it announced a separate ITES policy in 2002. Similarly, quite like the Andhra Pradesh government, the West Bengal government has given major emphasis on e-governance projects which would not only help improve administration, but also create business opportunities for the IT companies. The state has created a single window system to facilitate IT investments and a separate organization serves as an interface between the private sector and the government on e-governance projects.

Both Hyderabad and Kolkata have seen the local governments playing an active and direct role in shaping the IT cluster; still there are some differences in approaches adopted by the West Bengal government vis-à-vis the Andhra Pradesh government as Kolkata has been facing a different set of challenges compared to those faced by Hyderabad. Kolkata has been carrying a legacy of over a hundred years of rapid industrialization punctuated by recent decades of labor unrest, industrial disputes, and flight of capital and talent. While over 20 per cent of the top IT talent in India have their origin in West Bengal, only a handful of firms are based out of Kolkata. The West Bengal government faces a dual challenge not only to win the confidence of investors but also to woo the migrant Bengali IT professionals working elsewhere to return to the state. These two are highly interconnected – success in one will rub onto another. No doubt a lot of resources have been deployed to invite existing national and international IT and ITES players to set up their operations in the state, and successfully so. However Kolkata's biggest challenge perhaps lies in establishing a few large IT players developed from local entrepreneurship. Some of the home grown entrepreneurs need to capitalize on the entry of big players and push themselves adequately to catapult themselves into the big league as happened in Bangalore earlier and more recently in Hyderabad.

DISCUSSION

Obviously, the strategy followed by late entrants like Hyderabad and Kolkata are not the same as that of Bangalore. We shall discuss the differences and the key variables that contribute to the late entrants' success.

As is clear from Table 5.1, Bangalore alone had the preparedness to become a cluster in terms of the dimensions discussed in the literature. Bangalore had a pool of relevant factors, organically grown over a period of time. That configuration enabled firms to exploit opportunities in an attractively growing industry. Though the local government support fluctuated over a period of time, the cluster had a lot of inherent strengths to grow on its own, disregarding major government support. Bangalore has the characteristics of a healthy cluster as discussed in Porter's diamond.

However both Hyderabad and Bangalore provide contrasting pictures (Table 5.1).

The Hyderabad cluster did not have a pool of factor resources to start with, but had an entrepreneurial government, which created and strengthened the pool on all fronts. A similar approach is adopted in Kolkata too.

Table 5.1 At the beginning of the cluster formation

	Bangalore	Hyderabad	Kolkata
Early factor advantage			
(a) Skilled people	High	Medium	Medium
(b) Research Institutions	High	Low	Low
(c) Social network	High	Low	Low
(d) Economic infrastructure	High	Medium	Medium
(e) Social infrastructure	High	Low	Low
Growth in local entrepreneurship	High	Medium	Low
Attractiveness of the industry	High	High	High
Benefiting from cluster advantage	High	Low	Low
Local government support	Medium	High	High

In a growing industry, such as IT, attractive demand conditions very often facilitate entry of new players, that too when entry barriers are not very strong. It would not have been possible for late entrants like Hyderabad or Kolkata to make the place attractive for IT investment had the IT industry been showing signs of maturity.

The role of the state as an entrepreneur and strategist in shaping the formation of IT clusters for late starters has been the most important revelation of the case studies of Hyderabad and Kolkata. In an organizational context whether it is a firm or a government, building and sustaining competitive advantage revolves around the quality of leadership. For innovations to occur, in a tradition bound, bureaucratic, mature organization such as a government in this context, the leadership has to possess additional qualities. The leader must have entrepreneurial leadership qualities. Not only has the leader to create a work climate for innovative practices to flourish, but also to orchestrate the seeking and realizing of opportunities to grow the business (McGrath and MacMillan, 2000). The leader has to support and encourage hands on practices that involve problem solving with people at work. In addition, an entrepreneurial leader has to allocate resources, attention and talent *disproportionately*, and build counter pressure to fight inertial forces. The team should share a common vision to build competitive advantage through innovations.

A late entrant has to have a clear innovative approach to build competitiveness as argued in the relevant literature on entrepreneurship and strategic management. The possibilities of sustaining competitive advantage are greater when the entrepreneurial initiatives are based on a strategic intent (Burgelman, 1983; Hamel and Prahalad, 1989; McGrath et al., 1995). The spirit of entrepreneurship should flow through the whole administrative system to become competitive. This is particularly so when the mission is to catch up with the development and overtake existing players. Hyderabad and Kolkata have very clear strategic intents that the chief ministers and other top level administrators have shared at every conceivable avenue.

State governments as organizations that are striving to build competitiveness need to possess qualities similar to those of firms competing in the marketplace to build and sustain competitive advantage. This is particularly so in fast changing liberalized and globalized economies where states are competing with each other. It is under such situations that we need high quality entrepreneurial leaders, who can envision the future growth trajectory and build resources not only in terms of infrastructure, but also in terms of creating an attractive, confidence boosting environment through brand (image) building. Time is one of the most crucial variables and is traded at a very high premium for a latecomer intending to develop an IT cluster. While the latecomer may benchmark and learn from the early movers and need not reinvent the wheel, everything has to be done at a rapid pace to out-compete locations. Moreover, unlike firms in many other industries, IT firms have a rapid decision cycle on most strategic decisions including the choice of investment destinations. Therefore fast response, short lead time, quick decisions and single window clearances by the government provide a decisive edge over other locations. It is here that the role of the entrepreneurial leader and the deftness of the team become all the more crucial.

The ex-Chief Minister of Andhra Pradesh, Chandrababu Naidu, and the current Chief Minister of West Bengal, Buddhadeb Bhattacharjee, are persons who exhibit many of the qualities of an entrepreneurial leader. Though no quantitative comparison is attempted here, a number of anecdotal evidence support this argument. This includes the list of new policy initiatives adopted, the strategy to create appropriate infrastructure, the creation of local demand through innovative e-governance projects, image building, creating capabilities and deftness in the team and so on. Continuity of entrepreneurial policies and leadership over a sufficiently long period is yet another dimension.

The detailed analysis of how an IT cluster emerged in Hyderabad and Kolkata throws up some additional insights. The most important factor is the quality of entrepreneurial leadership in the local administration. It synthesizes and synergizes the other three factors: industry attractiveness, factor supply and local pool of entrepreneurs. It acts as a binding force to make the cluster a winner. The framework presented in Figure 5.1 is accordingly modified to capture the confluence of factors shaping the competi-

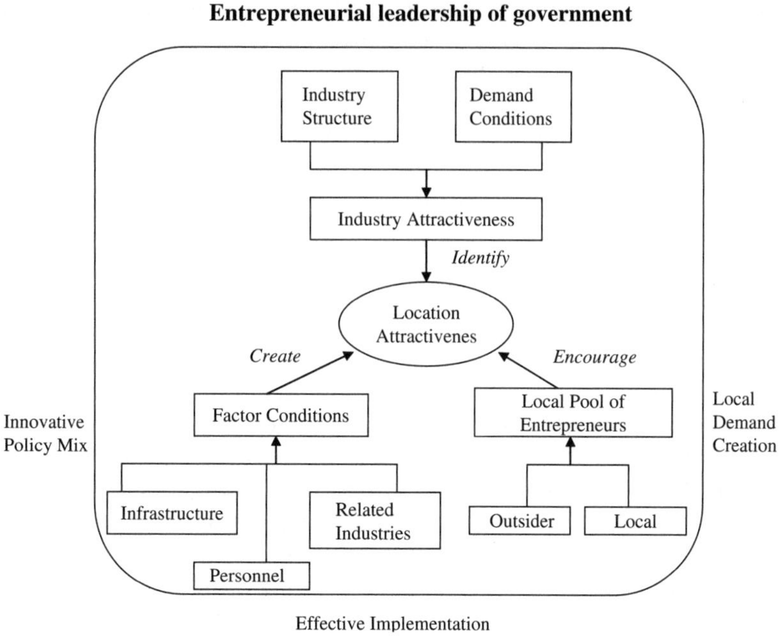

Entrepreneurial leadership of government

Industry Structure
Demand Conditions
Industry Attractiveness
Identify
Location Attractivenes
Create
Encourage
Factor Conditions
Local Pool of Entrepreneurs
Innovative Policy Mix
Local Demand Creation
Infrastructure
Related Industries
Outsider
Local
Personnel
Effective Implementation

Figure 5.2 Dynamics of formation of an industry cluster

tiveness of a location for the formation of an IT cluster as a late mover and shown in Figure 5.2.

IMPLICATIONS AND CONCLUSIONS

Drawing on the literature on industrial clusters, the key features of these three knowledge clusters are the following:

- Direct actions to create a positive image, supported by prompt delivery of services have attracted the key value adding ingredients into these locations. This led to the creation of a pool of skilled people who shared information across firms, though there did not exist much business linkages across firms.
- Firms worked together to represent their infrastructural needs that enabled the government to formulate or refine its policies.
- Many of the support services could take advantage of scale economies, and realized complementary value linkages among them, as is seen among the key players in a manufacturing concern.

- Information spillover across firms not only improved efficiency across firms but also resulted in higher levels of competitiveness.
- The presence of these firms, though without much to share across in terms of direct monetary benefits, created a new work culture for the cluster. Some of their practices led to greater levels of professionalism.

The evolution of such knowledge clusters in Hyderabad and Kolkata would not have happened, at a rapid pace, in the absence of the entrepreneurial leadership of the government.

We observe that at least three local level critical factors – relative factor conditions, entrepreneurial leadership of the state government and a pool of entrepreneurs – are needed for the successful evolution of a location into an industrial cluster as a late mover, provided the overall attractiveness of industry is high. For an early mover like Bangalore, these location factors evolved over a long period of time, on most occasions not by design, but by chance. Therefore the role that the local government there had in the past, and has even now, to maintain the supremacy of the Bangalore cluster is qualitatively different from that of the other state governments, which have to play a more direct role in shaping the formation of IT clusters as late movers.

The roles of industry attractiveness and factor conditions as two ingredients for cluster formation have been well documented. Also most clusters have grown because of a pool of local entrepreneurs, particularly in their early days of struggle. However the role of the state as an entrepreneurial leader, and the synergistic effect among these factors created by it have not been given due attention in the extant literature so far. We believe the major contribution of this chapter is that it highlights the importance of these factors among others in the formation of IT and other knowledge based clusters in an emerging market. This chapter also brings into sharp focus the differences that exist in the developmental models followed by the first movers as compared to the late entrants.

Many alternative models of development have been identified based on research in different country contexts. It is argued that the developmental approaches adopted by pioneering industrial nations differ substantially for nations that are in catch up mode (Abramovitz, 1986). Many studies (for example, Tyson, 1988; Vogel, 1988) on successful Newly Industrialized Countries (NICs), such as Japan, South Korea, Singapore and Taiwan, highlight the role of the government in either directly or indirectly influencing the quality and quantity of human resources, capital, technology and information to foster technological innovation and its diffusion. Though discounted in the purest capitalist sense, the role of government in leading the development process, particularly in poor countries is very high (Meir and Stiglitz, 2001). The purpose of government intervention is to move the economy from one level (lethargical equilibrium) to a higher level (prospects equilibrium). It is when the government is led by entrepreneurial leaders that such a paradigm shift takes place in line with the innovation arguments of Schumpeter (1949). Adelman (2001) has argued for a hyperactive government to accelerate the process of development in such a situation.

While industry attractiveness and factor conditions make a location attractive for cluster formation, it is the quality of entrepreneurial leadership of the state that determines the possibilities of a late entrant location becoming the leader. We believe that a detailed study of the strategic growth followed by Bangalore, Hyderabad and Kolkata provide valuable lessons for other locations trying to develop a number of knowledge intensive industries. For any other IT cluster to emerge in India as a late mover, the cluster formation process can be better planned and expedited. Any state government in India and similar emerging economies trying to emulate the IT cluster in Bangalore has to use the catch up and leap frog models by choosing any combination of vehicles of technology leverage. It has to play a crucial role in shaping the cluster by developing and promoting educational and research institutions, attracting investments in high technology areas by providing better factor conditions and creating local demand.

NOTE

1. We tend to equate the IT industry and the software industry in our current study and assume that both mean the same, as has been done by the majority of the authors while studying the industry.

REFERENCES

Abramovitz, M. (1986), 'Catching-up forging ahead and falling behind', *Journal of Economic History*, **46** (2), 385–406.

Adelman, I. (2001), 'Fallacies in development theory and their implications for policy', in G.M. Meier and J.E. Stiglitz (eds), *Frontiers of Development Economics: The Future in Perspective*, Washington, DC: World Bank.

Arora, A., V.S. Arunachalam, J.M. Asundi and R.J. Fernandes (2001), 'The Indian software services industry', *Research Policy*, **30** (8), 1267–87.

Barney, J.B. (2002), *Gaining and Sustaining Competitive Advantage*, Upper Saddle River, NJ: Pearson Education Inc.

Becattini, G. (1990), 'The Marshallian industrial district as a socio-economic notion', in F. Pyke and W. Sengenberger (eds), *Industrial Districts and Local Economic Regeneration*, Geneva, Switzerland: International Institute for Labour Studies, pp. 37–51.

Bossidy, L. and R. Charan (2002), *Execution – The Discipline of Getting Things Done*, New York: Crown Business.

Brenner, T. (2004), *Local Industrial Cluster – Existence, Emergence and Evolution*, London: Routledge.
Burgelman, R.A. (1983), 'A process model of internal corporate venturing in the diversified major firm', *Administrative Science Quarterly*, **28**, 223–44.
Business Standard (2003), 'Andhra Mouse catches up fast in Infotech', report, 26 November, 1.
Camagni, R.P. (1991), *Innovation Networks*, London: Belhaven Press.
Chakraborty, C. and D. Dutta (2002), 'Indian software industry: growth patterns, constraints and government initiatives', working paper, School of Economics and Political Science, University of Sydney, Australia, www.econ.usyd.edu.au/publication/2915 (ECON2002-1).
Choy, L.C. (1987), 'History and managerial culture in Singapore: "pragmatism", "openness" and "paternalism"', *Asia Pacific Journal of Management*, **4** (3), 133–43.
Chang, H.J. (1995), 'Role of the state in economic change: entrepreneurship and conflict management', in J.H. Chang and R. Rowthorn (eds), *The Role of the State in Economic Change*, Oxford, UK: Clarendon Press, pp. 31–50.
Devarajan, T.P, K. Ramachandran and S. Ramnarayan (2003), 'Entrepreneurial leadership and thriving innovation activity', 7th International Conference on Global Business and Economic Development, Bangkok, Thailand.
Van Dijk, M.P. (1999), 'Small enterprise clusters in transition – a proposed typology and possible policies per type of cluster', mimeo, Copenhagen Business School, Denmark.
Ghemwat, P. (2000), 'The Indian software industry at the millennium', HBS Case No. 9-700-036, Boston, MA: Harvard Business School.
Granovetter, M. (1985), 'Economic action and social structure: the problem of embeddedness', *The American Journal of Sociology*, **91** (3), 481–510.
Hamel, G. and C.K. Prahalad (1989), 'Strategic intent', *Harvard Business Review*, May–June, pp. 63–84.
Hanna, N. (1994), *Exploiting Information Technology for Development: A Case Study of India*, Washington, DC: World Bank.
Heeks, R. (1999), 'Software strategies in developing countries', *Communications of the ACM*, **42** (6), 15–20.
Johannison, B. (1998), 'Personal networks in emerging knowledge based firms: spatial and functional patterns', *Entrepreneurship and Regional Development*, **10** (4), 297–313.
Kapur, D. and R. Ramamurti (2001), 'India's emerging competitive advantage in services', *Academy of Management Executive*, **15** (2), 20–33.
Keeble, D. and F. Wilkinson (eds) (2000), *High Technology Clusters, Networking and Collective Learning in Europe*, Aldershot, UK: Ashgate.
Krugman, P. (1991), *Trade and Geography*, Cambridge, MA: MIT Press.
Lahiri, N. (2004), 'Knowledge spillovers: the role of geography, technology and intra firm linkages in the global semiconductor manufacturing industry', unpublished doctoral dissertation, University of Michigan, Ann Arbor, MI.
Marquis, D.G. (1969), 'The anatomy of successful innovations', Innovation, **1** (7), 29–37.
Marshall, A. (1920), *Principles of Economics*, 8th edn, London: MacMillan.
Maslow, A.H. (1954), *Motivation and Personality*, New York: Harper & Row.
McGrath, R.G. and I.C. MacMillan (2000), *The Entrepreneurial Mindset: Strategies for Continuously Creating Opportunity in an Age of Uncertainty*, Boston, MA: Harvard Business School Press.

McGrath, R.G., I.C. MacMillan and S. Venkataraman (1995), 'Defining and developing competence: a strategic process paradigm', *Strategic Management Journal*, **16** (3), 251–75.
Meir, G.M. and J.E. Stiglitz (eds) (2001), *Frontiers of Development Economics: The Future in Perspective*, Washington, DC: World Bank.
Miller, D. (1983), 'The correlates of entrepreneurship in three types of firms', *Management Science*, **29** (7), 770–91.
NASSCOM (2002), *The Software Industry in India: A Strategic Review*, New Delhi, India: National Association of Software and Service Companies.
NASSCOM (2005), *The ITES: BPO Super 9*, New Delhi, India: National Association of Software and Service Companies.
Pereira, A.A. (2004), 'State entrepreneurship and regional development: Singapore's industrial parks in Batam and Suzhou', *Entrepreneurship and Regional Development*, **16** (2), March, 129–44.
Polanyi, K. (1994), *The Great Transformatiion: The Political and Economic Origins of Our Time*, Boston, MA: Beacon Press.
Porter, M.E. (1990), *The Competitive Advantage of Nations*, New York: Free Press.
Porter, M.E. (1998a), 'Clusters and the new economics of competition', *Harvard Business Review*, **76** November–December, 77–90.
Porter, M.E. (1998b), 'The Adam Smith address: location, clusters, and the "new" microeconomics of competition', *Business Economics*, **33** (1), 7–13.
Porter, M.E. (2000), 'Location, competition and economic development: local clusters in global economy', *Economic Development Quarterly*, **14** (1), 15–34.
Ramachandran, K. (1986), 'Appropriateness of incentives for small scale enterprise location in less developed areas: the experiences', unpublished PhD thesis, Cranfield Institute of Technology, UK.
Ramachandran, K. (1989), 'Small enterprise promotion: role of incentives', *Entrepreneurship and Regional Development*, October–December.
Ramachandran, K. (2003), 'Customer dissatisfaction as a source of entrepreneurial opportunity', *Nanyang Business Review*, **2** (2), July–December, 21–38. This paper was reprinted by the *Singapore Accountant*, **20** (2), March–April, 2004.
Rastin T. (2003), 'Model for development: a case study of Singapore's economic growth', *Critique: A Worldwide Student Journal of Politics*, Fall, 1–10.
For Eastern Economic Review (2004), 'Cheerleader-in-Chief', **167** (17), 22 April, 38–9.
Roberts, N. (1984), 'Transforming leadership: sources, process, consequences', paper presented at the Academy of Management conference, Boston, MA, USA.
Rocha, H.O. (2004), 'Entrepreneurship and development: the role of Clusters', *Small Business Economics*, **23**, 363–400.
Salzman, H. (2000), 'The information technology industries and workforces: work organization and human resource issues', research report, Centre for Industrial Competitiveness, University of Massachusetts, November.
Saxenian, A. (1994), *Regional Advantage, Culture and Competition in Silicon Valley and Route 128*, Cambridge, MA: Harvard University Press.
Saxenian, A. (2000), 'The origins and dynamics of production networks in Silicon Valley', in M. Kenny (ed.), *Understanding Silicon Valley*, Stanford, CA: Stanford University Press, pp. 141–64.
Schein, E.H. (1996), *Strategic Pragmatism: The Culture of Singapore's Economic Development Board*, Cambridge, MA: MIT Press.
Schumpeter, J.A. (1949), *The Theory of Economic Development*, Cambridge, MA: Harvard University Press.

Scott, A.J. (1988), *New Industrial Spaces*, London: Pion.
Shane, Scott (2005), *A General Theory of Entrepreneurship: The Individual Opportunity*, Cheltenham, UK and Northampton, MA, USA: Edward Elgar.
Siegel, R., E. Siegel and I.C. MacMillan (1993), 'Characteristics distinguishing high-growth ventures', *Journal of Business Venturing*, **8** (2), 169–80.
Sonobe, T. and K. Otsuka (2003), 'Productivity effects of TVE privatization: the case study of government and metal casting enterprises in the Greater Yangtze river region', in T. Ito and A.O. Kruger (eds), *Governance, Regulation and Privatization*, Chicago, IL: University of Chicago Press.
Stevenson, H.H. and D.E. Gumpert (1985), 'The heart of entrepreneurship', *Harvard Business Review*, **63** (2), 85–94.
Stevenson, H.H and C.J. Jarillo (1990), 'A paradigm of entrepreneurship: entrepreneurial management', *Strategic Management Journal*, **11**, Summer 1990, 17–27.
Storper, M. (1997), *The Regional World: Territorial Development in a Global Economy*, New York: Guilford Press.
Stuart, T. and O. Sorenson (2003), 'The geography of opportunity: spatial heterogeneity in founding rates and the performance of biotechnology firms', *Research Policy*, **32** (2), 229–53.
Sturgeon, T.J. (2003), 'What really goes on in Silicon Valley? Spatial clustering and dispersal in production networks', *Journal of Economic Geography*, **3**, 199–225.
STPI (2007), 'Case study: Hyderabad', unpublished research report, Software Technology Parts of India, Ministry of Communications and Information Technology, Government of India.
Taeube, F.A. (2004), 'Culture, innovation and economic development: the case of the South Indian ICT clusters', in S. Muni and H. Romijn (eds), *Innovation, Learning and Technological Dynamism of Developing Countries*, Hong Kong: United Nations University, pp. 76–94.
Tyrini, I. (1994), 'The Japanese management structure as a competitive strategy', in Hellmut Schiitte (ed.), *The Global Competitiveness of the Asian Firm*, New York: St. Martin's Press, pp. 35–48.
Tyson, D. (1988), 'Making policy for national competitiveness in a changing world', in A. Furino (ed.), *Cooperation and Competition in the Global Economy*, Cambridge, MA: Ballinger Publishing Company.
Utterback, M. (1971), 'The process of technological innovation within the firm', *The Academy of Management Journal*, **14** (1), 75–88.
Vogel, E.F. (1988), 'Competition and cooperation: learning from Japan', in A. Furino (ed.), *Cooperation and Competition in the Global Economy*, Cambridge, MA: Ballinger Publishing Company.
Vyakarnam, S. and Y.M. Myint (2005), 'Entrepreneurs and entrepreneurship as key drivers of the Cambridge technopole', Second AGSE International Entrepreneurship Research Exchange, Melbourne, Australia.
Zwingle, E. (2002), 'Mega cities', *National Geographic*, **202** (5), 70–99.

PART 3

Emergence of venture capital in entrepreneurial economies in emerging regions

6. A comparative analysis of the development of venture capital in the Irish software cluster[1]

Frank Barry and Beata Topa

INTRODUCTION

The dramatic economic progress achieved by Ireland over the course of the 1990s and beyond had seen this period come to be dubbed the 'Celtic Tiger' era. Over the course of a little over a decade, Irish real national income per head rose from less than 65 per cent of the Western European EU average to achieve rough parity by the end of the 1990s. Unemployment tumbled from a high of 17 per cent in 1987 to less than 4 per cent in the early years of the new millennium, and employment expanded by a dramatic 70 per cent, driven by the decline in unemployment, an increase in the proportion of married women in the labour force and a shift from massive emigration to substantial immigration.

The series of beneficial shocks – policy induced and otherwise – to which the economy was subjected in the late 1980s have been much explored. These included a change in fiscal strategy in 1987. The reining in of government expenditures at that time made room for future tax reductions. Successive governments used the newly emerged 'social partnership approach' – which sees government, unions and employers come together every three years to agree a general path for wages and working conditions over the course of the coming years – to purchase wage moderation via the promise of future tax cuts. Such tax cuts have been estimated to account for about one-third of the rise in real take home pay since the partnership process began.

The doubling of EU regional aid funds in 1989 made it possible to implement the badly needed infrastructural projects that had been put on hold as part of the change in fiscal strategy. Between 1989 and 1999 aid flows to Ireland through the Structural and Cohesion Funds amounted to almost 3 per cent of GDP per annum, though careful analysis suggests that the direct effects on GDP of these EU regional aid programmes would have been moderate – adding about half of one percentage point per annum to the GDP growth rate of the 1990s (Barry et al., 2001).

The coming into being of the EU Single Market in the early 1990s led to a substantial increase in both intra-EU FDI flows and US FDI inflows to Western Europe, and Ireland captured a sharply increased share of these investments. Part of this increased share can be ascribed to the outlawing of restrictive public procurement policies as part of the Single Market initiative, as some of the larger EU member states had heretofore used such policies to draw transnational corporations away from Ireland's low corporation tax regime. With the outlawing of these practices, Ireland's attractiveness as an FDI location increased.

The downward trend in employment in Irish owned manufacturing companies reversed with the improvement in the investment climate in the late 1980s and early 1990s, and services sector employment expanded considerably.

The sector with which this chapter is concerned – the Irish owned (or 'indigenous') software sector – is of particular interest in that it is the sole high-tech sector in which domestic companies account for a substantial share of Irish employment. As shown in Table 6.1, domestic firms account for less than 20 per cent of employment in most high-tech sectors, whereas the equivalent figure in software is closer to 50 per cent.

The software sector, furthermore, is far more heavily clustered – around Dublin, the capital city – than are any other of the high-tech sectors, which are in turn more tightly clustered than are manufacturing or services more generally.

Table 6.1 Foreign firm employment as share of sectoral total; high-tech sectors

	Employment in foreign owned firms	Employment in domestic firms as % of sectoral total
Chemicals	17 874	23
. . . of which Pharmaceuticals	8 573	18.5
Office and data processing	18 303	11.7
Electrical machinery and apparatus	9 438	37.7
Radio, TV and communications	12 785	14.7
Medical and optical equipment	15 335	15.3
Software	15 300	45.2

Source: Central Statistics Office (2000); Enterprise Ireland (2002b).

The chapter analyses the emergence and evolution of the Irish indigenous software cluster. The next section sets the foundations by looking at the characteristics of the entire software sector, comprising both indigenous and foreign owned firms. The following section then considers the software/venture capital nexus that allows Dublin to be characterized as a high-tech cluster. It also briefly discusses the evolutionary two-stage model that has been advanced to explain the emergence of a similar nexus in Israel and that underlies our comparative perspective. The remaining sections detail the government support provided to the software sector, in line with phase 1 of the model, and the growth of government supported venture capital, in line with phase 2, and offer some conclusions.

ORIGINS, CHARACTERISTICS AND EVOLUTION OF THE SOFTWARE SECTOR IN IRELAND

Table 6.2 reports a measure of the importance of computer software employment in Western European EU countries. It details the share of software sector employment in total manufacturing and market services employment in a country, relative to the EU15 average. The highest shares in the Western European EU15 are exhibited by Sweden, the UK and Ireland.

Table 6.2 The relative importance of computer software employment in EU15 countries

EU15 Country	Share of software sector employment to total employment in country
Belgium	0.89
Denmark	1.25
Germany	0.61
Spain	0.62
France	1.05
Ireland	1.32
Italy	1.04
Netherlands	1.25
Austria	0.78
Portugal	0.27
Finland	1.25
Sweden	1.95
United Kingdom	1.47

Source: Barry and Curran (2004).

Within the software sector there is an important distinction between mass market packaged products, on the one hand, and other software activities, including custom and niche software and business solutions, on the other hand. The EU market is roughly equally divided between the two segments, with packaged software emerging as the most rapidly growing ICT subsector over the last decade (OECD, 2002).[2]

According to OECD (2002), Ireland and the USA are by far the largest software exporters in the OECD, accounting for shares of 29 per cent and 26 per cent, respectively. The vast bulk of Irish and US software exports are of packaged software, with Ireland accounting for around 50 per cent of all mass market packaged software sold in Europe.[3]

It is widely suspected that the figures for Irish output are inflated by the transfer pricing practices of corporations seeking to maximize the benefits of Ireland's low corporation tax environment. Nevertheless, even in employment terms, the packaged software sector is more important in Ireland than in other EU economies. Eurostat data register employment in this sector (which is classed as NACE 2233, 'reproduction of computer media', and included as part of manufacturing) in only eight EU countries, with employment numbers as shown in Table 6.3.

The mass market packaged software sector in Ireland is engaged in the manufacturing, localization and distribution (MLD) of software packages. Around 50 per cent of employees in these operations are typically engaged in the manufacturing stage, which does not require highly skilled labour, while around 30 per cent are involved in localization. In the case of Microsoft's Irish operations some 90 per cent of staff involved in localization had third-level qualifications in information technology or linguistics, while 35 per cent were nationals of mainland European countries (Coe, 1997).

Table 6.3 EU15 employment in mass market packaged software (2000)

NACE 2233 Reproduction of computer media	
Spain	663
France	875
Ireland	5591
Italy	342
Netherlands	168
Finland	16
Sweden	194
United Kingdom	3576

Source: Eurostat New Cronos database.

Some of the subsidiaries of packaged software MNCs outsource activities in Ireland, leading to the development of a software supporting subcontracting sector in activities, such as localization and translation, printing, disk manufacturing and logistics.

Even though these activities are not very high tech in nature, the sector has nevertheless moved up the value chain over time. The key players in the MLD sector (including Microsoft, Lotus, Oracle, Symantec, Informix and Corel) first established software manufacturing facilities in Ireland around the mid 1980s, duplicating and shrink-wrapping disk copies of the software programs developed by the parent company, and arranging for the printing and assembly of manuals. The second phase, again beginning with Lotus and Microsoft, saw these companies adding localization to the process. This involves translating the original products into other languages and cultural and technical formats appropriate to the destination markets. Besides translation there is some programming involved in preparing the text to be translated and then reincorporating it back into the program. The third phase of the sector's development saw the transfer of the responsibility for distribution, which had previously been handled by local distributors, to the Irish operations themselves. Thus Ireland became an operations hub (Crone, 2002; Coe, 1997; Ó Riain, 2004).

Crone (2002) finds that MLD activities account for about half the jobs in the foreign owned software sector in Ireland. The other half are accounted for by the software development sector, which is substantially more highly skilled. One segment consists of branches of major computing services or IT consulting companies (including EDS, IBM, ICL and Accenture), while the other is an adjunct to non-software electronics corporations, such as Motorola and Ericsson, with operations focused on the production of embedded software and applications for products such as mobile phones.

The remaining segments of the software sector are classified in the international production data as NACE 72, which comprises computer services and related activities. This segment includes custom software (provided for individual companies), niche software (which is written for specific business sectors) and other software services which are provided both for organizations and for consumers.

Within this segment countries like Sweden, Denmark, Finland, the Netherlands and the UK have higher weights than the rest of Europe. These are all countries with high computer penetration rates and other 'information society' attributes. Their relatively strong showing in this sector reflects the fact that many computer services are essentially non-tradable.[4]

The indigenous Irish software segment stands out however in being highly export oriented, suggesting that it exhibits different properties from these non-tradable elements. While UK software and computer services companies are found to obtain only around one-third of their revenues from exports, for example, and French and German companies from 25 to 30 per cent, exports accounted for 85 per cent of the revenues of Irish indigenous firms in 2002 (up from 41 per cent in 1991).

The strong export orientation of the Irish indigenous software sector shall be discussed more fully below, when we consider the types of government support offered to indigenous software firms in Ireland. We turn our attention first, however, to the characteristics of the Irish economy that led to the initial emergence of the indigenous software sector. The combined work of geographer Mike Crone (2002), IT journalist John Sterne (2004), sociologist Séan Ó Riain (2004) and public policy academic Anita Sands (2005) points to the conjunction of the following factors:

- Government infrastructural and human capital development programmes that were directed primarily towards attracting FDI.
- The role of foreign owned software and electronics sector MNCs as managerial and entrepreneurial incubators.
- The diversity of the economy's inward FDI sectors as sources of early stage demand.

Government actions to improve the investment climate with respect to telecommunications and education have been described by MacSharry and White (2000) – the former an erstwhile finance minister in the Irish government and the latter a former managing director of Ireland's Industrial Development Agency (IDA) – as largely concerned with enhancing the attractiveness of the country to FDI, though not specifically with foreign software firms in mind.[5]

They describe, for example, how the concerns expressed by the IDA's foreign manufacturing sector client companies over the poor state of the telecommunications system in the 1970s led to control of the system being wrested from the hands of a moribund government department. The telephone service was commercialized and one of the most advanced digital based networks in Europe put in place shortly thereafter (Burnham, 1998). Several of the most important domestic software entrepreneurs emerged through involvement with this massive telecommunications development programme, while the improved infrastructure itself of course also enhanced the environment in which Irish entrepreneurs operated.

A similar conclusion arises with respect to the expansion in tertiary science based education that occurred over this period. A Manpower Consultative Committee had been established in 1978 to provide a forum for dialogue between the IDA and the third-level education system. The state agency, concerned by the looming disparity between electronics graduate outflows and its own demand projections, convinced the government to fund a massive expansion in educational capacity in these areas. The output of engineering graduates, as a result, increased by 40 per cent between 1978 and 1983, while the output from computer science increased ten-fold over this same period. Ireland has since then exhibited one of the highest proportions of science and engineering graduates in the 20–34 age range in the world (Barry, 2007). Many of the most innovative Irish software companies were in turn spun off from the expanded university computer science and engineering departments.

There were some spillovers also from the substantial foreign MNC presence in the economy.[6] These arose through the role that foreign MNCs located in Ireland played as a source of sophisticated early stage demand and as incubators in supplying future entrepreneurs with some of the managerial and sectoral experience necessary for future success (documented by O'Malley and O'Gorman, 2001).

By the 1990s indigenous software had become one of the highest growth sectors in the rapidly growing Celtic Tiger economy. Over the second half of the 1990s employment in the sector grew at 24 per cent per annum, revenues at 30 per cent and exports at 37 per cent (Crone, 2004). It furthermore accounted for the bulk of venture capital (VC) investments in Ireland.[7] Comparative data on computer-related VC investments (which in Ireland equates largely to software) in 2003 are reported in Table 6.4.

Table 6.4 Sectoral distribution of VC investments (%) in the USA, Europe and Ireland, 2003

Sector	USA	Europe	Ireland
Communications	23.1	16.9	11.1
Computer related	**36.7**	**6.00**	**83.3**
Electronics	1.30	1.90	1.10
Biotech	18.8	2.30	0.30
Medical	9.20	6.00	2.20
Industrial/energy	3.90	11.9	0.00
Consumer related	1.30	19.4	0.00
Financial services	2.10	2.30	0.40
Other	3.50	33.3	1.50

Source: Authors calculations from EVCA Yearbooks for Ireland and Europe and PricewaterhouseCoopers, Thomson Venture Economics, NVCA Money Tree Surveys for the USA.

RESEARCH STRATEGY AND COMPARATIVE PERSPECTIVE

Kenney (2004), in discussing the emergence of high-tech clusters, characterizes the best-known model – that of Silicon Valley – as an 'ecosystem' comprising two intertwined but analytically separable economies. The first is of a conventional nature, consisting of established firms, universities, research laboratories and so on, which produces both output and innovations. The second, which distinguishes Silicon Valley from most other industrial clusters, consists of the institutional infrastructure that has evolved, with venture capital at its core, to enable the creation and growth of new start-up firms.

This ties in with the literature on why government attempts to establish venture capital industries have frequently ended in failure. Such failures are documented in the case of peripheral regions in the USA by Florida et al. (1990) and Florida and Smith (1994), who note that investments made in regions without the appropriate background conditions perform poorly, while investment of local funds outside the region entails capital transfer from the target area.[8] In surveying the litany of failures globally, Avnimelech et al. (2005) suggest that the problem has been one of conceptualization, where the absence of VC is seen as a supply side deficiency – a lack of available funds – rather than as a more deep-rooted condition encompassing also the absence of the demand side factors necessary to sustain a VC industry. As they point out: 'a vibrant VC industry is dependent upon a flow of investment opportunities capable of growing in value quickly enough to provide capital gains justifying the investment risks'. (p. 212)

These considerations have led Avnimelech and Teubal (2006a) to develop a temporal Innovation and Technology Policy Cycle model, drawn upon the Israeli experience, to outline a medium- to long-term policy effort to stimulate innovation and innovation intensive clusters in industrializing economies.[9] Their perspective suggests that the successful emergence of Silicon Valley type high-tech clusters is driven by the co-evolution of venture capital and high-tech start-ups.

The early phase they identify refers to the emergence of innovation capabilities and the diffusion of R&D within the economy, and entails direct government support to business sector R&D and innovative SMEs or start-ups. The background conditions necessary to produce an environment conducive to the successful establishment of a VC sector arose in Israel through the sharp increase in military R&D spending following the Six Day War of 1967, significant investments by foreign multinationals in R&D laboratories in the country and a consciously orchestrated process to expand R&D and innovation in the business sector following the establishment of the Office of the Chief Scientist (OCS) at the Ministry of Industry and Trade in 1969.

High quality start-ups are known to be crucial in innovative industries because of the limitations of incumbent companies in undertaking major or radical innovations. Path dependence arises when one considers the source of start-ups, many of which – in both the USA and Israel – are spun off from incumbent high-tech companies, which also represent an important source of founders and managers of new VC firms.

A later phase – in the wake of the emergence of high-tech sectors which create the appropriate demand conditions for VC – sees a venture capital sector emerge through the implementation of targeted VC policies, alongside an acceleration in the growth of high-tech start-ups and IPO and mergers and acquisitions (M&A) activity.

The 1992 Inbal programme represented the first significant government effort to create a VC industry in Israel. It was hidebound by bureaucratic oversight procedures, however, and, though the programme insured against downside risk, the funds were nevertheless exposed to the vagaries of the stock market. The Yozma programme that began operations in 1993 proved to be far more successful. Yozma was a $100 million government owned VC fund with two functions: the first to operate as a fund of funds ($80 million was invested in ten private VC funds, which had to be matched by a total of $120 million in private funding from 'significant foreign partners'), while the second saw $20 million retained in the government owned Yozma Venture Fund to be invested directly in early stage activities. This represented the backbone of an industry that invested in excess of $1 billion in Israel in 2001.

Avnimelech et al. (2005) compare the emergence of VC financing in Israel to that in the USA. The USA is of course the birthplace of venture capital. By the late 1970s the industry had consolidated to become a part of the US national innovation system; the flow of pension fund monies into Silicon Valley VC funds had freed the region from dependence on New York and Chicago investors, and, shortly thereafter, the exit process for VC funded firms on the NASDAQ had been routinized and VC and high-tech start-ups had co-evolved to become self-reinforcing. Unlike in the case of Israel, however, there was little targeted VC policy in the USA. Avnimelech and Teubal (2006b) argue that the background conditions in Israel could probably not by themselves have triggered the supply of VC without the assistance of a government programme such as Yozma because of market or system failures, including the lack of pre-established VC/SU reputations and critical mass to enable the establishment of partnerships with foreign VCs, and coordination problems between start-ups, VC organizations and risk capital. Having succeeded in getting the VC sector up and running,

Israeli government support had become much less significant by the late 1990s and the authorities were able to take a back seat in the process.

The experience of both the USA and Israel suggests then that a certain level of high-tech activity and sophistication, a continued stream of new technological and business opportunities and the creation of a critical mass of start-ups are required as preconditions for the emergence of a successful VC sector. Israel represents a particularly valuable case study for present purposes because its VC industry was established more recently and more deliberately than that of the USA.

The Innovation and Technology Policy Cycle model of Avnimelech and Teubal (2006a) is generic in that it allows for different variants reflecting different country contexts, although it emphasizes throughout the importance of direct business sector support, at least in the early stages.[10] Possible country variants, they suggest, can include differences in programmes in support of scientific research and university training; horizontal versus targeted programmes; the function or functions supported (for example, whether technology transfer, learning or R&D), and the instruments applied (whether subsidies, loans, tax benefits or other alternatives).

This chapter adds to the literature on the successful emergence of domestic high-tech/venture capital clusters in late developing regions by looking at the experience of Ireland over the years leading up to and including the Celtic Tiger era. Our research question asks whether the Irish experience offers support for the Innovation and Technology Policy model that has been constructed on the basis of Israel's experience in similar sectors. We have already seen how some of the background conditions necessary to proceed to the second stage had been satisfied in Ireland by the 1990s. We have yet to chart the policies followed by the Irish authorities to nurture the indigenous software sector however, as Avnimelech and Teubal (2006b) argue to be necessary in late developing economies. This is the topic of the next section of the chapter, which addresses the strategy of the Irish authorities as the cluster entered the second stage, where intervention by the authorities, they suggest, may be required to get the VC element up and running.

GOVERNMENT SUPPORT FOR THE INDIGENOUS SOFTWARE SECTOR IN IRELAND

What role did the state agency, Enterprise Ireland (which had been spun off from the IDA in the 1990s to focus on the needs of indigenous industry), play in response to the emergence of high-tech software start-ups? Before addressing this question, some background details on Irish industrial support are provided more generally.

Irish state financial support to industry, in terms of euro per person employed, was some 15 per cent above the Western EU average in the mid to late 1990s, but the gradual tightening of EU restrictions on state aids forced them to become more horizontal in nature. In line with this Enterprise Ireland shifted its focus from 'capacity' support for employment creation and fixed asset investment to 'capability' support in areas such as human resource development, R&D, marketing and market development. In response to the recommendations of the Industrial Policy Review Group that the agency should shift from grants to equity 'to meet gaps in financial markets for venture capital and seed capital' (IPRG, 1992, p. 12), it progressively increased the proportion of support provided through equity, in the form of both ordinary shareholdings and preference shares.[11] Equity participation increased from 5 per cent of total financial supports in 1989 to 28 per cent in 1998 (Forfás, 2000, Table 5.2).[12]

The shift in emphasis towards capability development has required firms to reach agreement with the state agency on an integrated development plan before aid can be accessed. While this may sound overly interventionist, it has the effect of forcing emerging firms to assess as objectively as possible their own strengths and weaknesses, and the agency will then customise a support package that may include (1) helping companies to monitor markets and exploit new market opportunities, (2) encouraging process and operations improvement and the development of better products and services through improved access to appropriate research, and (3) the promotion of increased management and employee training levels.

The agency is particularly supportive of what it terms high potential business start-ups (HPSU). These are export oriented firms that, in the case of international services, (1) are located in a product market that has grown by at least 20 per cent in the previous year; (2) are based on technological innovation or the exploitation of a rapidly developing market niche; (3) are founded and promoted by experienced managers, entrepreneurs, academics or highly skilled technical graduates, either from within Ireland or returning from abroad; (4) are deemed to have the potential to grow within two years to have annual sales of 1.3 million euro and employ ten or more people; and (5) show clear evidence of being able to continue to grow substantially and of being in a position to fund such growth.

The agency works intensively with such firms to ensure access to the best external management advice; it helps them to attract expertise to their boards – for example, through appointment of experienced non-executive directors – and to build an appropriate management team; it provides support for in-company training and for product and process development through direct support for in-company R&D and through establishing technology innovation networks, and it helps them develop contacts with private sector financiers. It may also, where necessary, offer direct financial support (Forfás, 2000).

The firm assessment process operates to rigorous standards. To be deemed eligible for funding (though with no automatic entitlement), projects seeking support must first successfully pass a formal cost-benefit analysis. Qualitative and other factors that are difficult to quantify are then taken into account in a Quality Ranking Matrix which focuses in particular on projects from well-managed innovating companies in high growth, high productivity export oriented sectors.[13]

Ó Riain (2004, pp. 98–105) provides further details of the hands-on approach that has operated in the case of indigenous software firms, suggesting that (1) mentoring programmes that pair small companies with experienced industry figures and (2) the Enterprise Development Programme that provides one-on-one support and advice in terms of business plan development have been of particular importance. Indeed, according to Walsh (1985), the latter had been instituted in 1978 partly in response to the lack of venture capital finance available at that time. The state has also provided a substantial proportion of the R&D funding for indigenous software companies and, more recently, as discussed in the next section, of venture capital funding as well.[14]

Bearing this in mind, Crone (2004) notes that the significant number of indigenous software companies that have attracted VC funding suggests that the Irish sector can be viewed as following the 'Silicon Valley' model of high-tech development. Seven such firms were floated on international stock markets including the NASDAQ in the mid to late 1990s.

GOVERNMENT SUPPORT AND THE GROWTH OF VENTURE CAPITAL IN IRELAND

In 1984 the Irish government introduced a special tax incentive programme – the Business Expansion Scheme (BES) – to encourage long-term equity capital investments in new and small companies operating in particular sectors of the economy that 'would otherwise find it difficult to raise such funding and would instead have to rely on expensive loan finance' (Enterprise Ireland, 2007). EVCA data are available only from 1984 so it is not possible to evaluate the changes induced by the introduction of the BES. It is noteworthy, however, that in 1985 the share of private individuals in the accumulated amount available for venture capital funds was almost 25 per cent.

The real breakthrough in Irish VC took place in 1994, however, when two major government initiatives were announced. The first was the issuing of guidelines, supported by the Irish Association of Pension Funds (IAPF), that pension funds should place 0.08 per cent of their assets annually into venture capital funds over the next five years. The suggestion that pension funds 'should support the venture capital industry by becoming a recognized form of finance for entrepreneurial companies' had its roots in a report commissioned by the Irish Association of Pension Funds (IAPF, 1994), the Irish Insurance Federation and the Department of Finance, at the request of the Minster of Finance, which found that pension fund investments in the domestic market were negligible in comparison with the situation in the USA and the UK.[15]

Since then, as seen in Figure 6.1, the proportion of new funds accounted for by the pension funds sector has hovered around the European average.

The second major step taken in 1994 saw the establishment by the state agency Enterprise Ireland of a five-year plan – the Seed and Venture Capital Measure (1994–9) – co-financed by EU regional aid. The programme was targeted at establishing venture/seed capital funds. Financing was provided on condition that a minimum of 50 per cent of the capital would be privately funded. The EU and national funding amounted to a total of 44 million euro and this was matched at the beginning by 40 million euro in private investments. Returns were fed back into further investments. Although at the inception there were difficulties in getting the private sector involved, ultimately a sum of 119 billion euro had been invested in 130 companies by the 15 operational funds by 2003 (Enterprise Ireland, 2002a). Crucially, from a governance point of view, these VC funds are run on a purely commercial basis, with investment decisions taken solely by private sector VC fund managers.

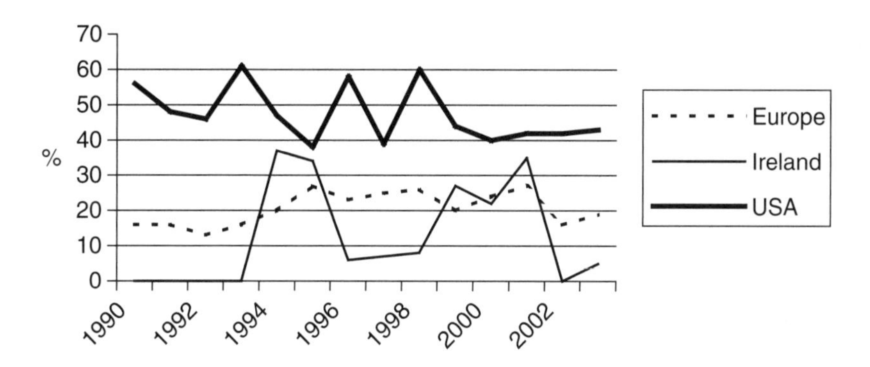

Source: Enterprise Ireland (2005).

Figure 6.1 Proportion of new funds accounted for by pension funds: Europe, Ireland and the USA

In 2001 the Seed and Venture Capital Fund Scheme was recommenced under the National Development Plan 2001–2006 with funds amounting to 95 million euro. The objective of the programme was to leverage 400 million euro in private funding. This had already been achieved by 2002, and by 2004 the 15 funds (with about 500 million euro in capital raised) established under the programme had made investments in 75 companies totalling 133 million euro (Enterprise Ireland, 2005).

Over the entire period since these government initiatives were taken in the mid 1990s, government funds as a share of total new funds in Ireland have typically been greater than is the case for Europe on average (Table 6.5).

The classic case for government intervention of this type is provided by Stiglitz and Weiss (1981), who note that the informational asymmetries that can preclude access to external capital for new small firms may be counteracted by the signalling effect of public funds. Figure 6.2, which plots privately raised new funds on the left-hand scale and the government percentage share of new funds on the right, suggests that this indeed appears to have been the case in Ireland. Michael Murphy, chairman of the Irish Venture Capital Association, concurs, noting that 'Enterprise Ireland acted as a catalyst; it helped draw in matching funds faster and accelerated the on-going development of the market'.[10]

The Irish authorities, furthermore, have been commended on establishing a fiscal and legal framework conducive to the development of venture capital. The 2003 report of the European Venture Capital Association published an evaluation of the extent to which member countries maintained an environment which was favourable both for the demand side (venture capital investors) and the supply side (entrepreneurs) of the industry. On a scale running from 1 (most favourable) to 3 (least favourable), the average composite score for the Western European EU (the EU15) was 2.04. Ireland achieved a score of 1.58, placing it second to the UK, which scored 1.2. The report highlighted as beneficial aspects of the Irish environment: (1) overall tax policy, with a low corporate tax rate and tax incentives for private individuals – including the Business Expansion and Seed Capital schemes discussed earlier – as well as R&D incentives; (2) the most favourable entrepreneurial environment in the EU, with the lowest time, cost and capital requirements for setting up private or public limited companies; (3) the lack of restriction on pension funds investments in private equity; and (4) the availability and optimal regulation of limited partnership funds which provide a suitable legal structure for venture capital funds.

Ireland's achievements in establishing a VC industry are reflected in the fact that, even though indigenous industry is less concentrated in high-tech sectors than is the case for the average Western European EU economy,

Table 6.5 Sources of new funds: Europe, Ireland and the USA (percentage by type of investor)

		1990	1991	1992	1993	1994	1995	1996	1997	1998	1999	2000	2001	2002	2003
Corporations	EUR	4	5	5	5	10	5	4	11	11	10	11	6	7	5
	IRE	0	0	0	3	2	12	7	5	1	3	11	2	0	4
	USA	7	5	4	8	9	5	20	25	12	14	4	3	2	2
Private	EUR	4	5	3	3	3	3	7	4	8	7	7	7	6	3
	IRE	20	35	19	40	9	0	0	14	26	22	2	2	0	10
	USA	13	14	12	7	12	17	7	12	11	10	12	9	9	10
Financial institutions	EUR	55	48	44	40	41	36	41	42	40	46	35	36	39	22
	IRE	65	21	27	23	35	30	45	17	13	32	39	31	11	12
	USA	10	6	17	12	10	20	3	6	10	16	23	25	26	25
Pension Funds	EUR	16	16	13	16	20	27	23	25	26	20	24	27	16	19
	IRE	0	0	0	0	37	34	6	7	8	27	22	35	0	5
	USA	56	48	46	61	47	38	58	39	60	44	40	42	42	43
Government	EU	3	2	9	7	3	3	2	2	6	5	6	6	11	7
	IRE	0	1	0	0	0	0	16	36	13	4	11	27	9	17

Source: Authors calculations on the basis of EVCA and NVCA data.[17]

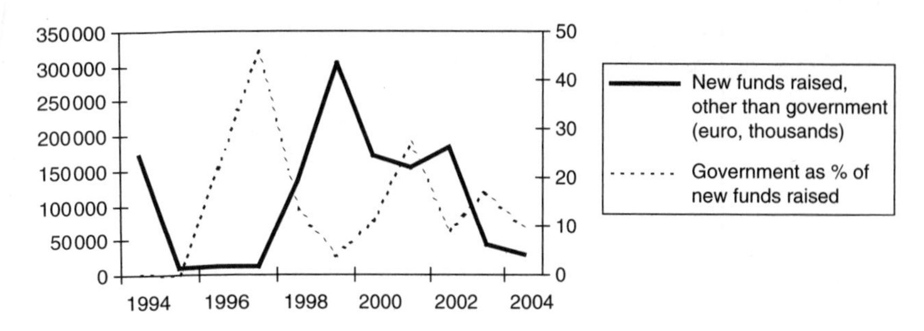

Source: Authors calculations on the basis of EVCA data.

Figure 6.2 Government intervention and privately raised new funds

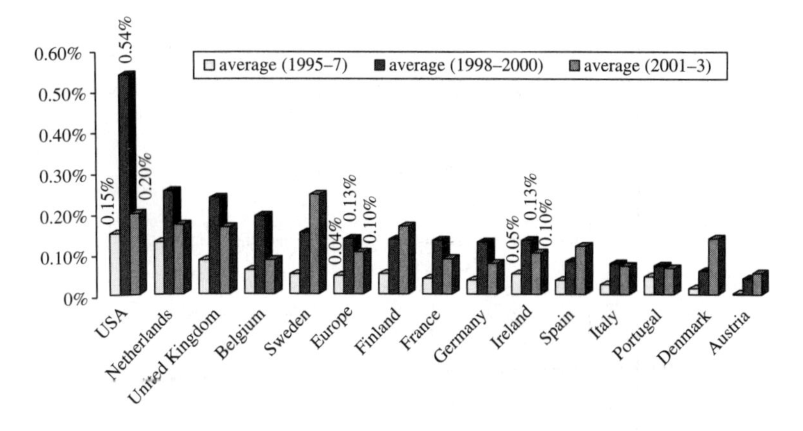

Source: Authors calculations on the basis of EVCA Yearbooks and NVCA Yearbook (2004),[18] OECD data (GDP and GNP measures at current prices); exchange rates: Eurostat.

Figure 6.3 Venture capital investments as a percentage of national income; various countries and regions; average values for 1995–7, 1998–2000 and 2001–3

venture capital investments as a share of national income have matched the Western European average (Figure 6.3).[19]

CONCLUDING COMMENTS

We noted, at the outset, the emerging consensus that government attempts to establish a VC industry can only be expected to be successful if appropriate investment opportunities are available in the region. This is formalized in

the Innovation and Technology Policy Cycle model of Avnimelech and Teubal (2006a), which is concerned with the emergence of innovation intensive clusters in industrializing economies. The early phase identified in the model relates to the emergence of innovation capabilities and the diffusion of R&D across the economy. Their model draws specifically on the Israeli experience where the background conditions were established in the wake of the Six Day War of 1967 through a sharp increase in military R&D spending, significant investments by foreign multinationals in R&D laboratories and a state orchestrated process to expand innovation in the business sector.

In Ireland the process by which the background conditions were established was different. As detailed earlier, the constellation of factors that led to the emergence of a dynamic indigenous software sector included government infrastructural and human capital development programmes (that were directed primarily towards attracting FDI), the diversity of the economy's inward FDI sectors as sources of early stage demand, and the role that foreign owned software and electronics sector MNCs played as managerial and entrepreneurial incubators. Path dependence via spin-offs from incumbent high-tech companies then took hold, as discussed in Chapter 1. As in Israel, there was substantial government support of business sector R&D and innovative SMEs and start-ups.

The second phase of the Avnimelech and Teubal model, in the case of successful high-tech clusters, sees venture capital emerge, alongside an acceleration in the growth of high-tech start-ups, and IPO and M&A activity. They describe how VC funding was purposefully developed by the Israeli state, after a number of false starts, through the enduring Yozma programme of 1993. This government owned VC fund had two functions: the first to operate as a fund of funds invested through private VC companies, and the second to invest directly in early stage activities.

Enterprise Ireland followed a similar policy beginning in 1994 when it established venture and seed capital funds co-financed by EU regional aid. Details of the role played by government as the cluster entered the second stage have been provided. Overall, then, the emergence of a domestic high-tech cluster in Ireland has clear parallels with the Israeli experience.[20]

Avnimelech and Teubal (2006b) argue that the background conditions in Israel would have been unlikely to trigger the necessary supply of VC without government involvement because of market or system failures. The evidence presented here supports this proposition in the Irish case also. How, then, are we to understand how VC emerged in the USA with no direct government involvement in the process? One likely explanation is provided by the Coase theorem (named for Ronald Coase, winner of the 1991 Nobel Prize in Economics), which holds that, in the absence of transaction costs, interested parties will bargain privately to correct any externality. It would have taken a long time for interested parties in the USA to achieve this outcome, the possibilities for which are clearly much greater in the case of a first-mover economy like the USA than for later developing regions.

Finally, in attempting to draw lessons for other emerging regions, it is important to bear in mind that the successful application of the kinds of interventionist strategies followed in the Irish and Israeli cases is heavily dependent on strong standards of public sector governance. Otherwise, the possibilities of abuse that arise in state investments or subsidies of any kind may mean that fear of 'government failures' should take precedence over the desire to correct market failures.

NOTES

1. We are grateful to Morris Teubal and Martin Kenney for stimulating our interest in venture capital over the course of a series of workshops on high-tech clustering held in Sweden, the proceedings of which have been published in Braunerhjelm and Feldman (2006).
2. Microsoft is by far the largest packaged software firm in the world, followed by IBM which has about one-half of Microsoft's level of packaged software sales (OECD, 2002). Local firms, on the other hand, tend to dominate the non-packaged segment.
3. Over two-thirds of Irish software exports go to the EU while one-quarter goes to the rest of Europe, Middle East and Africa (EMEA) triad – the Middle East and Africa (Crone, 2002).
4. The Commission of the European Communities (2003) reports that the overall EU market for (largely non-tradable) IT services is about twice the size of the market for (tradable) software products.
5. The fact that the state's development agencies retained a strong focus on global marketplace trends, however, increased the probability that even unanticipated outcomes would have been beneficial rather than detrimental.
6. Ireland is the most FDI intensive economy in Europe, with foreign owned firms accounting for almost 50 per cent of Irish manufacturing employment, compared to an average figure of 23 per cent for the Western European EU member states and 33 per cent for the three largest Central and Eastern European economies. Ireland also records the highest share of services sector employment in foreign owned firms out of the 17 EU countries plus the USA and Norway for which data are available.
7. This perhaps relates to the higher entry barriers found for manufacturing sectors by Oakey (1995), for example.
8. Mason and Harrison (2002) critiqued British government proposals to increase the amount of venture capital available in less well-endowed areas on similar grounds, because of the lack of capability on the part of start-ups in these regions to usefully absorb venture capital.
9. Avnimelech and Teubal (2006b) present a lengthy discussion of the methodological issues that arise in this type of research. They describe their approach as being based on grounded theory, one of the main purposes of which is to transform tacit knowledge into codified knowledge and which is appropriate to newly emerging research areas (Partington, 2000).
10. Direct measures in support of innovation and innovative SMEs contrast with indirect measures, such as promotion of institutions supporting the business sector (for example, universities, technology centres and government laboratories) and promotion of venture capital itself.

11. The Industrial Policy Review Group – one of a number of periodic external assessments of the state's development agencies – recommended that the agency should become much more an 'aggressive venture capitalist' and should be prepared to take stakes as high as 50–60 per cent (IPRG, 1992, p. 72).
12. Preference shares with a low coupon rate are used to provide a form of long-term finance at low cost to SMEs that are unable to raise development finance from the market on similar terms. Evidence on the significant returns earned by Enterprise Ireland from dividend income, the redemption of preference shares and the sale of ordinary shareholdings in client companies is provided in Forfás, 2000, Table 5.3.
13. The focus of the development agencies on export development has been criticized in some quarters as overly mercantilist. It has been pointed out in defence however that non-traded sector firms are likely to be competing largely with each other, which would put the state in a vulnerable position were it to support some and not others. In the case of software the strong focus of the relevant agency has always been on software product firms, which tend to be much more export oriented than software services.
14. State expenditure on capacity and capability support is a multiple of state investments in the privately managed venture capital funds discussed in the next section.
15. The guidelines were issued as an alternative to legislation which would have required pension funds to make certain commitments to venture capital.
16. Quoted by Cowley (2003, p. 70).
17. Figures for particular countries and regions do not sum to 100 as only certain categories of investors are included here. Categories excluded for Ireland and Europe are funds of funds, academic institutions, capital markets, realized capital gains and the class of 'not available' data, while those excluded for the USA are endowments and foundations.
18. Europe and Ireland: early stage investments include 'seed' and 'start-up', the US early includes 'start-up/seed' and 'early'. To make a comparison between European and American data possible, the category 'later' was excluded from the US VC activity.
19. GNP is conventionally used as the national income denominator for Ireland in order to exclude the vast profits recorded by foreign MNCs in the country (because of its low corporation-tax regime), which are included in GDP. In none of the other countries shown is there a substantial difference between the two measures.
20. It appears too early to be able to say in the Irish case however – as has been said of Israel – that the authorities can now take a back seat in the venture capital process.

REFERENCES

Avnimelech, G. and M. Teubal (2006a), 'From direct support of business sector R&D/innovation to targeting venture capital/private equity: a catching-up innovation and technology policy life cycle perspective', *Economics of Innovation and New Technology*, special issue on The Governance of Technological Change, **17** (1), 18–42.

Avnimelech, G. and M. Teubal (2006b), 'Creating venture capital industries that co-evolve with high-tech: insight from an extended industry life cycle perspective on the Israeli experience', *Research Policy*, **35**, 1477–98.

Avnimelech, G., M. Kenney and M. Teubal (2005), 'The life cycle model for the creation of national venture capital industries: The US and Israeli experiences', in E. Giulani, R. Rabellotti and M.P. van Dijk (eds), *Clusters Facing Competition: The Importance of External Linkages*, Aldershot, UK: Ashgate, pp. 195–214.

Barry, F. (2007), 'Third-level education, foreign direct investment and economic boom in Ireland', *International Journal of Technology Management*, **38** (3), 198–219.

Barry, F. and D. Curran (2004), 'Enlargement and the European geography of the information technology sector', *World Economy*, **27** (6), 901–22.
Barry, F., J. Bradley and A. Hannan (2001), 'The single market, the structural funds and Ireland's recent Economic Growth', *Journal of Common Market Studies*, **39** (3), 537–52.
Braunerhjelm, P. and M. Feldman (eds) (2006), *Cluster Genesis: Technology-Based Industrial Development*, Oxford, UK: Oxford University Press.
Burnham, J. (1998), 'Global telecommunications: a revolutionary challenge', *Business and the Contemporary World*, **2**, 231–48.
Central Statistics Office (2000), *Census of Industrial Production*, Dublin, Ireland: Stationery Office.
Coe, N.M. (1997), 'US transnationals and the Irish software industry: assessing the nature, quality and stability of a new wave of foreign direct investment', *European Urban and Regional Studies*, **4** (3), 211–30.
Commission of the European Communities (2003), *European Business: Facts and Figures, Part 6: Business Services, 1991–2001*, Luxembourg: Office for Official Publications of the European Communities.
Cowley, L. (2003), 'Irish venture: a winning formula?', *European Venture Capital Journal*, September, 70–71.
Crone, M. (2002), 'A profile of the Irish software industry', consultation paper, www.qub.ac.uk/nierc, 13 February 2007.
Crone, M. (2004), 'Celtic Tiger cubs: Ireland's VC-Funded software start-ups', paper presented at the Institute of Small Business Affairs National Entrepreneurship and SME Development Conference, Newcastle-Gateshead, UK November.
Enterprise Ireland (2002a), 'Seed and venture capital report', www.enterprise-ireland.com, 13 February 2007.
Enterprise Ireland (2002b), 'National software directorate', www.nsd.ie/htm/ssii/stat.htm, 13 February 2007.
Enterprise Ireland (2005), 'Seed and venture capital programme 2000–2006', www.enterprise-ireland.com, 13 February 2007.
Enterprise Ireland (2007), 'Business expansion scheme', www.enterprise-ireland.com/Grow/Finance/Business_Expansion_Scheme.htm, 13 February 2007.
Florida, R. and D.F. Smith (1994), 'Venture capital and industrial competitiveness', report to the US Department of Commerce, Economic Development Administration, Washington, DC, May.
Florida, R., M. Kenney and D.F Smith (1990), 'Venture capital, innovation and economic development', report to the US Department of Commerce, Economic Development Administration, Washington, DC, June.
Forfás (2000), *Enterprise 2010: New Strategy for the Promotion of Enterprise in Ireland in the 21st Century*, Dublin: Forfás.
Industrial Policy Review Group (IPRG) (1992), *A Time for Change: Industrial Policy for the 1990s*, Dublin: Stationery Office.
Irish Association of Pension Funds (1994), *Asset Governance: Guidance for Irish Pension Schemes*, Dublin, Ireland: Irish Association of Pension Funds.
Kenney, M. (2004), 'Supportive economic institutions: Silicon Valley's lessons for developing countries', in A. Bartzokas and S. Mani (eds), *Financial Systems, Corporate Investments in Innovation and Venture Capital*, Cheltenham, UK and Northampton, MA, USA: Edward Elgar.
MacSharry, R. and P. White (2000), *The Making of the Celtic Tiger: The Inside Story of Ireland's Booming Economy*, Dublin, Ireland: Mercier Press.

Mason, C. and R. Harrison (2002), 'Is it worth it? The rates of return from informal venture capital investments', *Journal of Business Venturing*, **7**, 211–36.
Oakey, R. (1995), *High Technology New Firms: Variable Barriers to Growth*, London: Paul Chapman Publishing.
OECD (2002), *Information Technology Outlook*, Paris: OECD.
O'Malley, E., and C. O'Gorman (2001), 'Competitive advantage in the Irish indigenous software industry and the role of inward foreign direct investment', *European Planning Studies*, **9** (3), 303–21.
Ó Riain, S. (2004), *The Politics of High-Tech Growth: Developmental Network States in the Global Economy*, Cambridge, UK: Cambridge University Press.
Partington, D. (2000), 'Building grounded theories of management action', *British Journal of Management*, **11**, 91–102.
Sands, A. (2005), 'The Irish software industry', in A. Arora and A. Gambardella (eds), *From Underdogs to Tigers: The Rise and Growth of the Software Industry in Brazil, China, India, Ireland and Israel*, Oxford, UK: Oxford University Press.
Sterne, J. (2004), *Adventures in Code: The Story of the Irish Software Industry*, Dublin: Liffey Press.
Stiglitz, J. and A. Weiss (1981), 'Credit rationing in markets with imperfect information', *American Economic Review*, **71**, 393–410.
Walsh, M. (1985), 'Venture capital in Ireland', *Irish Journal of Irish Business and Administrative Research*, **7** (1), 33–45.

7. Policy intervention in the development of the Korean venture capital industry[1]

Seungwha (Andy) Chung, Young Keun Choi, Jiman Lee, Sunju Park and Hyun-Han Shin

INTRODUCTION

Many governments around the world actively design and implement policy initiatives to promote small businesses as they are an important source of the national income and employment. The public policy for entrepreneurial companies with technological orientation is a differentiated part of those policy initiatives. The success of entrepreneurial companies is defined more by uncertain market forces once they start up with unproven technological ideas. Generally speaking, the serious application of competitive market mechanisms to start-up companies is the best way to promote innovative activities among private companies. However, governments of developing countries trying to catch up with technological advancement have a legitimate incentive to seriously consider socioeconomic externalities of sponsoring entrepreneurial companies. Thus they often intervene in the market for corporate creation and development.

From the late 1990s the Korean economy in general shifted its focus from traditional heavy industries to the information and telecommunications industry as a strategic policy sector. Faced with the unexpected financial crisis that started in late 1997, the economy accelerated industrial restructuring processes toward the new economy, supporting the rise of high technology ventures. The purpose of this study is to review the Korean government's unique policy drivers for promoting venture related industries under changing economic environments, to evaluate their effects on the development of entrepreneurial capabilities at a societal level, and also to draw out new propositions by comparing the pattern the Korean venture industry has shown for the last decade with ones presented in existing studies. We pay attention to the inter-related development of three key elements of this industry, that is, entrepreneurial companies, venture capital and the exit market. Needless to say, government policy drivers are designed to promote their synergetic interactions. It is easy to expect their short-term positive influences, but hard to evaluate their long-term sustainability.

Even though our overall focus is on the development pattern of the Korean venture industry in general, we would like to give special attention to the information technology sector, one of the most competitive industrial sectors in Korea. During the period from 1995 to 2004 the gross production in the manufacturing part of the information technology sector grew 33 per cent per year, reaching around $184 billion in total (Korea Institute for Industrial Economics and Trade, 2006). Its contribution to the GDP and to the manufacturing industry overall increased from 4.9 per cent in 1995 to 7.2 per cent in 2004, and from 17.7 per cent in 1995 to 25.1 per cent, respectively. In a sense, venture promotion policies in Korea substantially started with this growth of the information technology sector during the 1990s (Choi et al., 2002). The government wanted to accelerate the development of information and telecommunications technologies by sponsoring technology-based entrepreneurial start-ups.

However, right after the burst of the Internet bubble in 2000, the societal expectation of the new economy and venture industry in general went dramatically sour along with the stock market collapse around the world. This event turned the clock back, and shrank not only the market for IPOs, but also overall venture related industries. The government reaction was to continue the support for the industry and to establish a more regulative system. It would be interesting to see if the entrepreneurial activities and capabilities are sustained as the government intended half a decade after the bubble economy. In the next section we shall discuss economic meanings of government policies promoting entrepreneurship.

PUBLIC POLICY FOR PROMOTING ENTREPRENEURSHIP

The concept of industrial policy was actively utilized by the Japanese government, especially the Ministry of International Trade and Industry (MITI) around the 1970s. Such terms as industrial rationalization, advanced industrial organization and new industrial systems were used as variations of the concept (Cheon, 2000). Johnson (1982) defines industrial policy as government action which leads and coordinates industrial activities in order to increase the productivity and competitiveness of either a national economy or a certain industry by aiding or retrenching various industries. It aims to improve efficiency in a certain field of industries, but not to improve general economical conditions at the macro-economic level (Zysman and Tyson, 1983).

Following Japan's lead, the Korean government has actively engaged in industrial policy to directly enhance the national competitiveness of selected industries, and to rationalize the allocations of economic resources across various industrial sectors (Cho, 1994; Chung, 2003; Kim et al., 2000). As early as the 1980s, the Korean government selected the information technology sector as a strategic area for public support. The government support for venture related industries can also be understood in this continuum. However the policy to promote overall venture industry is unique compared with traditional industrial policies for heavy industries, such as the automobile or shipbuilding industry, due to the knowledge intensive nature of the industry.

Entrepreneurial ventures are essentially a business organization run on the basis of new technology and ideas (Bollinger et al., 1983). They are strongly market oriented enterprises pursuing profitability. For this reason, many venture capitalists put more weight on the growth and profit potential of their investment targets rather than technology itself (Lee, 2003). In contrast, the government tends not to consider profitability as the only factor in its support for venture industry (Brown and Jackson, 1990; I. Lee, 2002). It has to keep the balance between the promotion of technical advancement, job creation and business start-ups.

Economic Significance of Entrepreneurial Policy

Most of all, entrepreneurial companies develop a groundbreaking technology, commercialize that technology and disseminate the technology throughout society (Bollinger et al., 1983). Research studies find that they are actually successful in developing innovative technologies. Also products and services based on the innovative technologies have been developed by small-sized entrepreneurial ventures. Acs and Audretsch (1990) show that the frequency of technological innovation of small- and medium-sized ventures is relatively higher than existing large companies in high technology industries. In the 1980s, according to Scherer (1991), 225 innovative technologies per million persons were developed by the large companies with more than 500 employees, but 322 technologies per million persons by the companies with less than 500 employees. Similarly, Rosen (1991) suggests that small firms rather than large ones develop more groundbreaking technologies with a higher level of socioeconomic impact. The government pays attention to this possibility of developing innovative technologies by supporting entrepreneurial ventures.

Second, with their innovative capabilities, entrepreneurial ventures contribute to the restructuring of national industries by providing new goods and services, and sometimes by creating a new field of industry. According to the Korean Small and Medium Business Administration (2001), the entrepreneurial ventures play a major role in developing the groundbreaking technologies and R&D expenditure of the entrepreneurial ventures reached 7.1 per cent and 8.0 per cent in 1999 and 2000, respectively, but only 1.89 per cent and 1.81 per cent in the case of large enterprises. Even though the results of R&D could not be indexed as in the study carried out, it clearly shows that the venture firms in Korea have invested more resources in R&D than the large ones.

Entrepreneurial investments reallocate national resources toward high technology industries. In Korea, for example, the national resources have been rapidly reallocated to high technology industry during 1999 and 2000. This economic restructuring helped to overcome the financial crisis in a short period in time. As a result, the gross output of information and telecommunications industries in Korea had grown about ten times from $15.2 billion in 1990 to $141.7 billion in 2000. The value added of that industrial sector exceeded the GDP growth rate by far during the same period, growing 18.9 per cent annually from $33.7 billion in 1996 to $67.2 billion in 2000 (Lee, 2003). The proportion of the information technology sector in the GDP had grown from 8.1 per cent in 1996 to 13.0 per cent in 2000.

Last but not least, entrepreneurial ventures create new employment. In Germany, for instance, new ventures contributed 33 per cent of new employment, and 13.7 per cent of the total employment were created by new ventures that were less than 17 months old in 1987 (Hamermesh, 1993). During the 1980s, the Fortune 500 companies lost 4 million jobs while entrepreneurial ventures with less than 100 employees created 16 million jobs (Birch, 1990). In Korea, similarly, the employment growth rate of entrepreneurial ventures reached 18.8 per cent in 1999 and 24.3 per cent in 2000, respectively, but only 1.4 per cent and 3.8 per cent for large enterprises (Small and Medium Business Administration, 2001). The fact that entrepreneurial ventures contribute to job creation became one of the major reasons why policy makers are putting intensive efforts to foster them. Understanding this positive impact on employment, the Korean government had actively promoted entrepreneurial start-ups since the late 1990s. However, if the goal of this policy is solely to increase the number of jobs created by entrepreneurial ventures, the mass production of entrepreneurial ventures would be nothing but a short-term goal. As a side effect, such efforts of government might generate an increase in the number of faltering enterprises. Also the government policy would focus more on providing the entrepreneurial ventures with direct investment generating an increase in employment in short run rather than building up a long-term market system such as establishing infrastructure.

POLICY INITIATIVES AND CHANGES IN THE VENTURE INDUSTRY

We define venture industry as being comprised mainly of entrepreneurial companies, venture capital and the exit market (Chung, 2003). The ultimate goal of public policy for this venture industry would be to establish the efficient economic infrastructure and financial systems that support the dynamic creation and growth of entrepreneurial companies (Robbins-Roth, 2001). The industrial policy for new technology-based venture firms in Korea needs to be understood in the framework of changing industrial policies. Industrial policy, in general, includes the policy for industrial restructuring, the policy for industrial organization and the policy for technical advancement (Kim et al., 2000). Up until the early 1980s, the Korean government had not achieved any satisfactory outcome in venture promotion because they pursued it only as part of the policy for technical advancement. As the government tried to restructure strategic industries, favoring new information technology from the mid 1990s, the Korean venture industry entered into the growth stage. Figure 7.1 summarizes policy actions and institutional changes in the Korean venture industry from the 1980s until the early 2000s.

The Start-up Period: Up to 1995

The Korean government became aware of the importance of technical advancement for industrial competitiveness and the necessity of public policy for technical innovation in the private sector. In parallel, they emphasized the necessity of developing new technology-based firms as a way of restructuring and improving manufacturing industries. Private equity markets were essential in supporting new technology-based firms.

Based on this awareness, the government established three government run venture capitals in the early 1980s, Korea Technology Development Corporation (KTDC, later renamed as KTB), Korea Development Investment Corporation (KDIC) and Korea Technology Finance Corporation (KTFC). However investment activities of these venture capital firms were miniscule due to the lack of societal understanding about the function of venture capital in the early 1980s. Without an exit mechanism for investment returns, there was no incentive to put capital into venture capital markets.

At this stalemate the government passed an 'Act for Supporting Small and Medium Businesses' and an 'Act for Financing New Technology-Based Businesses' in 1986 in order to increase the venture capital supply. These laws were helpful in establishing new venture capital firms. At that period

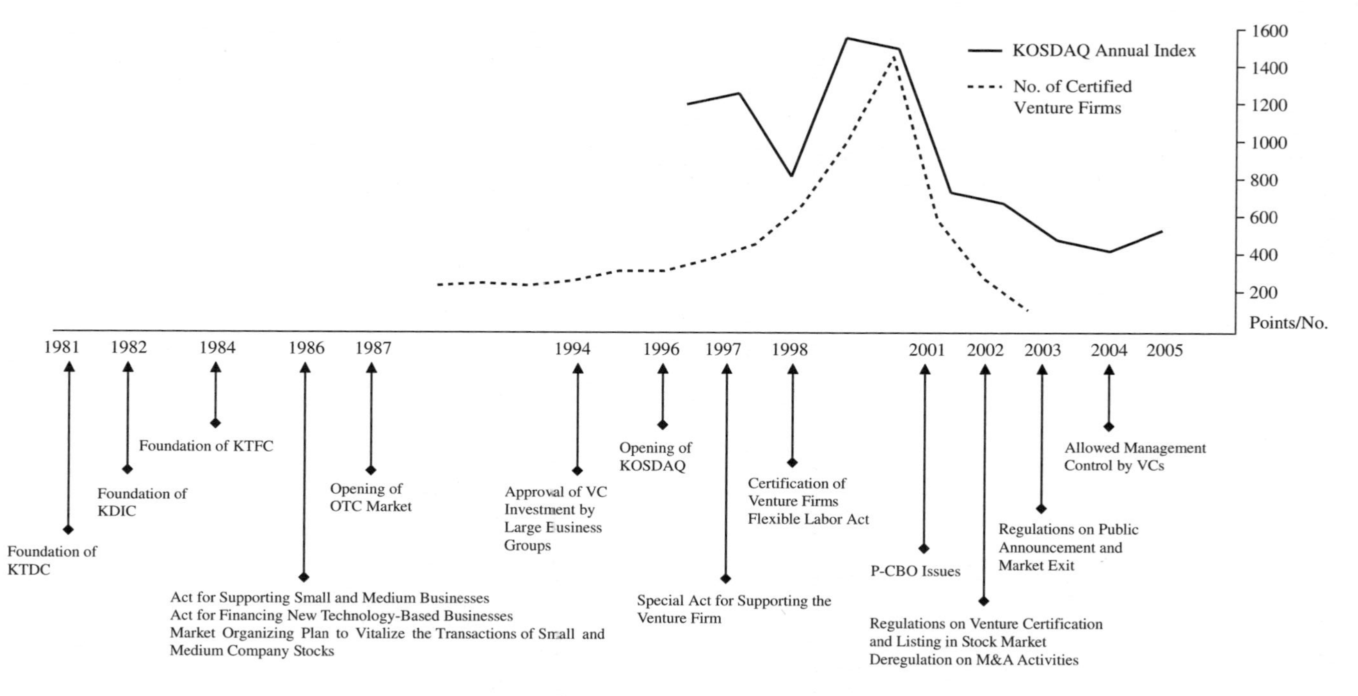

Figure 7.1 *Policy initiatives over time and changes in entrepreneurial markets*

of time, establishment of the Korean venture capital firms was not determined by supply and demand in the market, but by government support and regulations. The government regulated qualifications of the companies that venture capitalists consider as investment candidates. For example, they could invest in small firms at the initial and growth stages, aged no more than 14 years, and put in money as equity investment, but not as a loan.

The government tried to establish venture capital markets by regulating venture capital industry to invest more than 40 per cent of the total capital as equity for venture capital firms, and more than 50 per cent of the capital as equity for venture investment funds, respectively. In order to compensate for these strict regulations, they gave venture capital firms some tax benefits and other financial incentives. By restricting investment target companies and investment methods, the government tried to form venture capital markets focusing only on companies at the early and growth stages. The government induced venture capital firms to focus on technology-based industries by listing industry categories that venture capital firms could invest in. Despite the government efforts, venture industry was not developed as expected as there were not many technology-based companies lacking a specialized workforce.

Until the mid 1980s there was no way to convert invested capital into cash. So, the government opened the over the counter trading market in April 1987 after their announcement of 'the Market Organizing Plan to Vitalize the Transactions of Small and Medium Company Stocks' in December 1986. This did not, however, function properly as a market with the annual trading volume of the market amounting to only a daily trading volume of the regular stock exchange. It also failed to work as a primary financing market for venture companies going public. The government could not consistently keep the regulations due to this weak market condition. As the venture capital markets went from bad to worse until the mid 1990s, the government allowed conglomerates (*chaebols*) to establish venture capital firms in 1994. Until then, conglomerates entry into venture capital markets had been banned. The government expected conglomerates to put more capital in venture capital markets by allowing them to engage in venture businesses.

The Growth Period: From 1996 to 2000

Since the mid 1990s, entrepreneurial companies had been founded and actively grown in the fields of communications equipment and computer hardware and software so that venture capitalists increased their investments in these industries. Technological development in the information technology sector stimulated the founding and growth of new technology-based companies.

In 1996 the Korean government implemented two key policy initiatives to support venture industry. The first was the opening of the KOSDAQ stock market for growing companies in July 1996. It was contributory to establishing a new stock market where the stocks of young companies are traded. Thus the KOSDAQ market was differentiated from the existing Korea Stock Exchange. With the opening of the KOSDAQ market, the government expected the venture industry to be vitalized through efficient capital flows and investment recycling. Consequently, going public in the KOSDAQ for new technology-based companies has become easier. In addition, the government permitted institutional investors and pension funds to invest in the over the counter market. Foreign investors could also make direct investments in the KOSDAQ market whereas mutual funds and investment funds could invest indirectly in the KOSDAQ market. The KOSDAQ market grew rapidly as a center of the direct financing market. Given this institutionalization of the KOSDAQ market, superior entrepreneurial companies in the information technology sector started to go public in that market. After the opening of the KOSDAQ market in 1996, it had taken only four years for the gross market value of the KOSDAQ listed companies to surpass that of the regular Korean Stock Exchange, though with the Internet bubble.

The second key policy initiative, which is quite unique to the Korean government, was the introduction of the certification system for venture firms in 1997. The government began to plan a support policy for new technology-based companies since the mid 1990s. The traditional economic structure, focused on large companies, was useful for the quantitative growth of the Korean economy until the early 1990s. However it became a barrier against continued economic development in the era of new economy and information technology. The government selected new technology and knowledge-based industry as strategic targets for further development of promising companies. They came up with 'the Special Act for Supporting the Venture Firm' (the Venture Special Act hereafter). According to this Venture Special Act, a 'venture firm' is defined as a small and medium enterprise aiming to develop, apply and commercialize a new technology.

Any entrepreneurial company can be certified as a 'venture firm' if it meets one of the four requirements. First, a venture firm is an enterprise in which the 'total investment (including debt)' of venture capital is more than 20 per cent of equity; or in which venture capital holds more than 10 per cent of 'equity (stocks)'. Second, a venture firm is an enterprise in which the ratio of R&D investments to annual sales is more than 5 per cent.

Third, a venture firm is an enterprise in which the sales or exports based on its patents are more than 50 per cent or 25 per cent of the gross sales, respectively; or in which the gross sales or exports based on a new technology and/or knowledge-based business are more than 50 per cent or 25 per cent of the gross sales, respectively. Lastly, a venture firm is an enterprise which receives excellent ratings by a technology evaluation agency licensed by the government.

The growing Korean venture industry faced a dramatically new economic environment right after the 1997 national financial crisis. Extreme economic restructuring processes had provided an unexpected opportunity to venture firms in accessing new businesses, workforce and capital markets. Venture firms could grow fast while large companies collapsed and went through restructuring processes. About half of the top 30 largest companies went through legal reorganization. Other large companies also engaged in strategic restructuring. These large companies spun off underperforming divisions and started to outsource non-core businesses. These changing economic conditions provided venture firms with a wide range of new business opportunities.

Labor markets also became quite pliable with the help of 'the Flexible Labor Act' instituted in 1998. This law was originally intended to make organizational restructuring easier for large companies. The change of traditional lifetime employment into new flexible employment was an opportunity for venture firms to recruit high quality employees.

Finally, the growth of capital markets was explosive from 1999 to 2000. The year 1999 is often called the take-off year for the Korean venture industry. The rapid growth of the venture capital market through KOSDAQ had resulted from the restructuring of the overall banking sector, and later from the low interest financial environment getting over the economic crisis. The growth of the KOSDAQ market accelerated the capital inflow to the venture industry and thus the development of venture firms in general.

The Shakeout Period: From 2001 to the Present

The shakeout of the Korean venture industry was precipitated by a crash in the NASDAQ market in spring 2000. After the NASDAQ index hit 5000 points, which was its highest at the time, it continued to plummet down to 1500 points. With rapid market readjustments during the second half of 2000, high technology venture firms faced a dramatic drop in stock prices. Internet companies were hit hardest as elsewhere. Moreover the domestic factors got aggravated due to insufficient restructuring, misdeeds of venture managers and unfair trading in the KOSDAQ market. So, in June 2001, the price index of the KOSDAQ market had dropped by more than 70 per cent compared with the highest level previously reached.

The stock market crash led to a slump in the IPO market, which in turn made it harder for venture firms to raise capital through the primary stock market. Moreover this also froze the venture capital market, a major source of capital for entrepreneurial companies. Venture capitalists began to take stricter measures in evaluating venture firms. Thus, many start-up companies found it even harder to raise sufficient capital. Projects that would have received enough capital before either got financing at a higher rate or did not get any financing at all. In hindsight, a tremendous drop in the market value of venture firms actually resulted from short-term oriented investment practices and overvaluation. Between 1999 and 2000 many people started to have their own business plan and competed with one another to invest in good looking deals.

With the overall venture industry experiencing a dramatic shakeout, the government implemented four major policy initiatives. First, the certification system for venture firms and the KOSDAQ registration standards were monitored more strictly. So far, unqualified venture firms tended to attract private equity investments. They were easier targets of misbehaving government officials and of stock price manipulations in the KOSDAQ market. Accordingly, in 2002, the government made the venture evaluation agencies guarantee the truthfulness of their certification. Venture firms themselves declared an ethical business statement through the Korea Venture Business Association. The government also raised the registration standards for the KOSDAQ market (B. Lee, 2002).

Second, the government began to encourage mergers and acquisitions (M&A) markets as well as to strengthen regulations on the KOSDAQ market for poor performing companies. The Venture Special Act was amended in April 2002 to permit stock exchanges ('swaps') between venture firms and to simplify legal procedures for M&As. Since 2003 the government strictly applied registration standards for the KOSDAQ listed companies such as public disclosures and registration cancellations, in consultation with the Financial Supervisory Commission. As a result, some companies used a so-called 'backdoor listing' technique of buying existing listed companies to go public. This was helpful for both venture capitalists in harvesting and public investors in safe investment.

Third, the government directly supported venture firms through two different ways of funding, that is, providing public funds to venture capitalists and contributing to its own investment fund. The main government authority in charge of supporting venture capital markets was thus far the

Small Business Corporation, which was directed by the Small and Medium Business Administration. With the prospering KOSDAQ market and information and telecommunications industry in 1999, the Ministry of Information and Communication also participated in venture capital funds. In addition, the National Pension Service's participation in venture capital funds played an important role in recovering venture capital markets lacking private funds.

In parallel, the government began to directly contribute capital to venture firms in the form of primary collateralized bond obligation (P-CBO) in 2001. A venture firm requesting an investment was evaluated by a securities company for funding support. Later, the P-CBOs became bad debts that the government was liable for. This was caused partly by the system of evaluation and investment decision by securities companies, which were not specialists in venture capital investment. So, the P-CBO funding system, in a sense, delayed the restructuring of the Korean venture industry with the indiscreet funding mechanism (I. Lee, 2002).

Lastly, venture capitalists were allowed to actively participate in the management of invested companies. The Korean venture capitalists had not been active investors so far. This made them avoid early stage companies that desperately needed more capital investment. From December 2004 investment in temporary management control was permitted. In June 2005 venture capitalists were even allowed to take the management control of companies less than seven years old. With this change, venture capitalists could monitor and control venture firms tightly, and aid them in acquiring management skills from venture capitalists.

PERFORMANCE ASPECTS OF THE VENTURE INDUSTRY

The main objective of venture promotion policy would be to create the socioeconomic conditions, institutions and infrastructure that cause a steady development of venture industry. The venture industry, as previously discussed, is comprised of entrepreneurial companies, venture capital and the exit market. Here we address the issue of how effectively the Korean venture promotion policies have achieved the goal by asking the following questions. (1) Have these policies facilitated the creation of venture firms? (2) Have venture capitalists distributed their resources according to the Korean government's policy objectives of supporting technology-based businesses? (3) Has the KOSDAQ stock market performed its role as an exit market effectively? By answering these questions, we hope to gain an insight on whether or not the socioeconomic infrastructure suitable for venture industry has been developed and sustains in Korea as the government expected.

Effects on Start-ups

Among the three major constituencies of venture industry, venture firms are both the core of the industry and the engine of wealth creation. So, the vitalization of entrepreneurial start-ups is an important outcome variable when measuring the effectiveness of the government's venture industry policy. By examining the number of government certified venture firms per year and the distribution of founding years of venture firms listed in the KOSDAQ market, we can evaluate the policy performance in vitalizing start-up activities.

The number of new corporations, including venture firms, decreased 19.2 per cent from 41 460 in 2000 to 33 497 in 2003. As shown in Figure 7.2, the number of certified venture firms grew rapidly, peaked in 2001 and declined since. As the market underwent an adjustment period, a small upward tendency has recently been noticed.

We can also examine the distribution of the years that certified venture firms were established. Among the certified venture firms still in operation at the end of 2003, many were founded in 2000 at the peak of the KOSDAQ market (Figure 7.3). The number of start-ups has declined dramatically after the market crash.

Figure 7.4 depicts the distribution of industrial sectors to which certified venture firms belong. The certification of venture firms in the information technology sector soared during 2000 and 2001. Interestingly, the trend of more than half of venture firms belonging to the information technology industry has been maintained throughout.

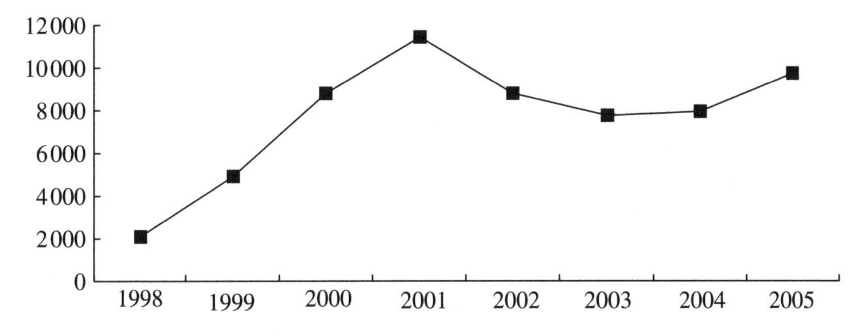

Source: VentureNet (2006), www.venturenet.or.kr.

Figure 7.2 Total number of certificated venture firms by year

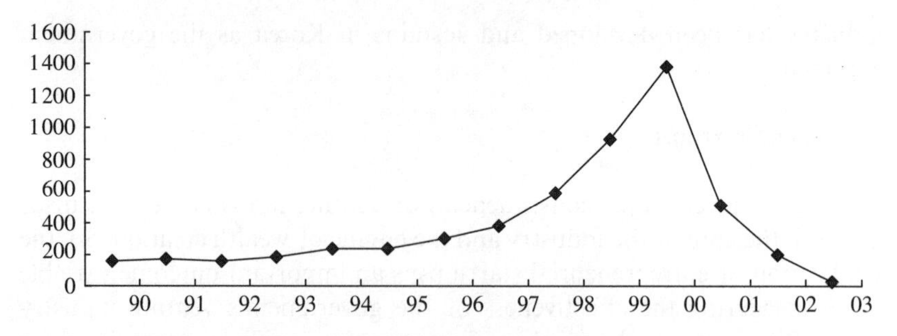

Note: Excluded are the companies founded before 1990 (total of 726 venture firms).

Source: Adapted from Small and Medium Business Administration (2001), 'Close investigation of the actual conditions of venture firms', and the KIS-Value database.

Figure 7.3 Number of certified venture firms by the founding year

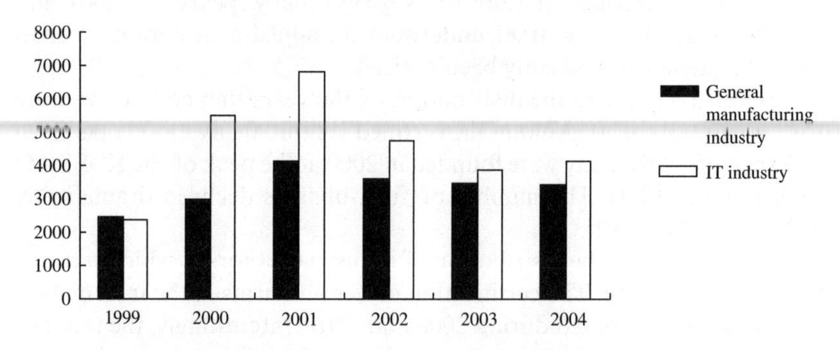

Source: Small and Medium Business Administration (2001, 2005).

Figure 7.4 Information technology (IT) venture firms as a proportion of total

Effects on Venture Capital

The venture capital firm is a profit organization that invests in entrepreneurial companies with the funds raised from professional investors and with the goal of creating and growing them. Ideally, venture capital firms contribute to the government's goals of creating jobs, driving technology development, establishing high technology industry and enabling community development as well as economic growth. However it can happen and has happened that venture capital firms ignore these goals while focusing soley on maximizing their private returns. Therefore there exists the need for the government to oversee the operation of venture capital firms. To this

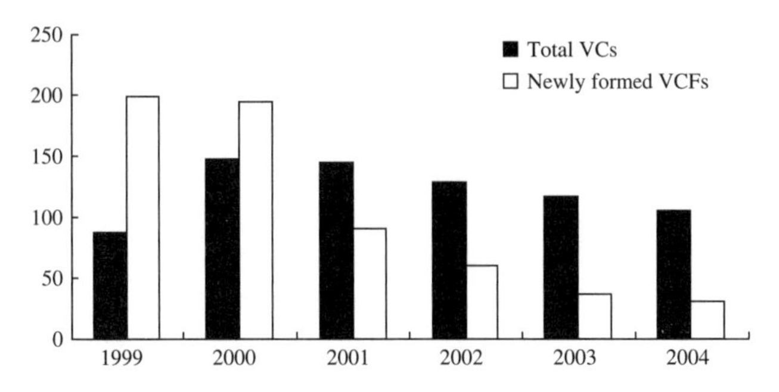

Source: Small and Medium Business Administration (2001, 2005).

Figure 7.5 Number of venture capital firms (VCs) and venture capital funds (VCFs)

end, the Korean government has instituted special laws and regulations to control and support venture capital industry as we discussed in the previous section. The effectiveness of the government's regulatory efforts in this regard can be measured by studying the annual changes in the number of venture capital firms, the investment pattern during the life cycle of venture firms and investment trends across various industry sectors.

Venture capital firms (VCs) in Korea are a limited company established by the fund contributing shareholders. In addition to the original fund, venture capitalists may recruit external funds and form a 'venture capital fund' (VCF) that typically has a five year partnership agreement. As shown in Figure 7.5, the number of VCs has declined since 2000. Going through market adjustments after the stock market crash, only the top 30 VCs have been investing in their normal capacities whereas the remaining 80 or so VCs have, in fact, stopped their investment activities, only harvesting their existing equity investments. The number of VCFs made a sharp downturn from 2001 as well. Considering the fact that a significant portion of the funds were formed in 2000 and thus will dissolve in 2006, the asset values of VCFs are expected to decrease even further in 2006.

The amount of venture capital investment increased dramatically during the period from 1999 to 2000 as shown in Figure 7.6. But it has declined sharply after the KOSDAQ market crash in 2000. The pace of decline, although slowed, has continued ever since. As shown in Figure 7.7, after 2000 the investment in early stage companies up to three years old has declined to 32 per cent in 2004, compared to 72 per cent in 2001 while the investment in later stage companies more than three years old has increased steadily. The trend of avoiding early stage ventures has continued and accelerated to some degree.

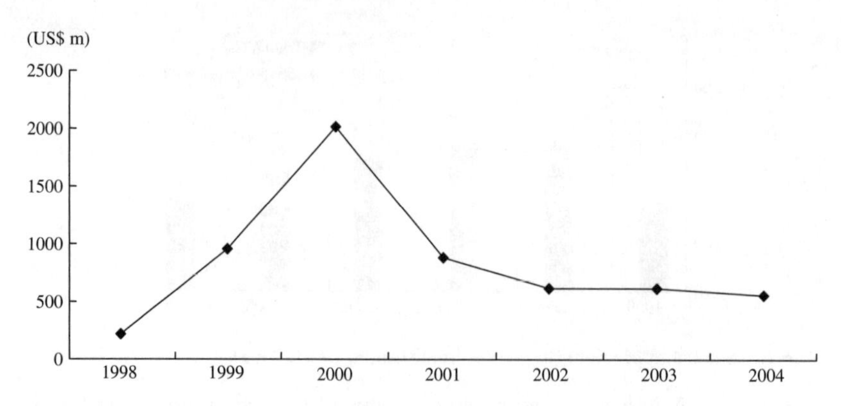

Source: Small and Medium Business Administration (2001, 2005).

Figure 7.6 Amount of venture capital investment per year

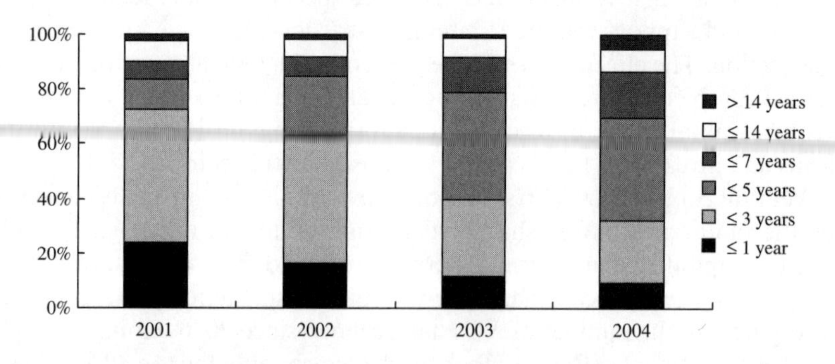

Source: Small and Medium Business Administration (2001, 2005).

Figure 7.7 Proportion of investment amount based on the age of venture firms

The venture capitalists have dramatically increased their investments in the information technology industry since 1999. Although the absolute amount of investment funding has decreased since the KOSDAQ crash, the investment in information technology companies still accounts for more than 50 per cent of the total investment (Figure 7.8). This shows that the government efforts have been successful in fostering the development of new technology-based firms in the information technology sector.

In order to establish the socioeconomic infrastructure for entrepreneurial activities, the government ought to be capable of enticing private funds into the venture capital market. As shown in Figure 7.5, however, the

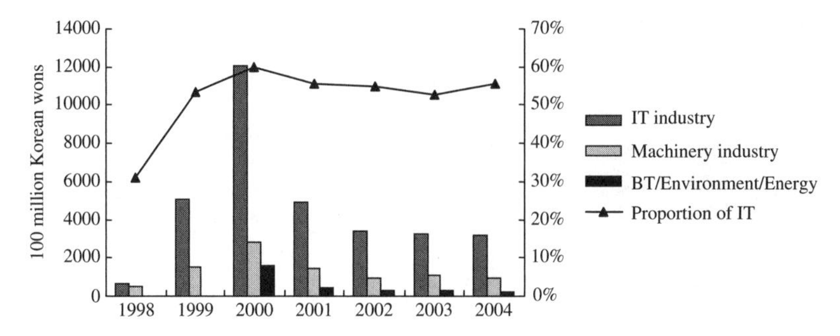

Source: Small and Medium Business Administration (2001, 2005).

Figure 7.8 Amount of venture capital investment based on industrial sectors

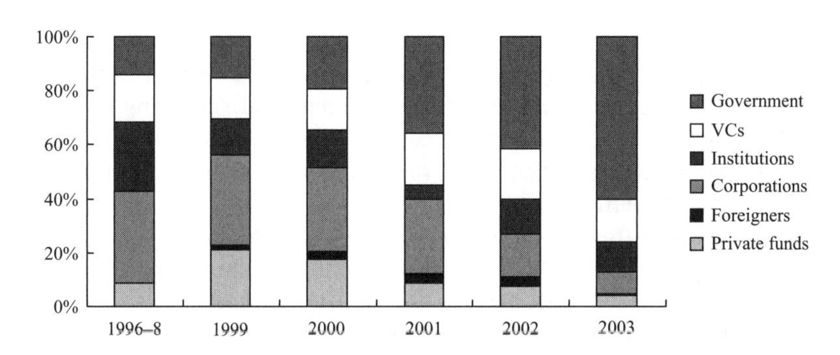

Source: Small and Medium Business Administration (2001, 2005).

Figure 7.9 Sources of venture capital funds

number of venture capital funds plunged after the peak in 1999 and the trend has persisted ever since. Moreover, when excluding the government related funds and the funds from venture capital firms, the ratio of private funds contributing to the overall venture capital funds has decreased steadily (Figure 7.9). This indicates the slowdown of influx of private funds to the venture capital market.

Effects on the KOSDAQ Stock Market

Although the purpose of the exit market is to fund corporate growth in the public market, it gives venture capitalists an opportunity to cash in their investments. Since the KOSDAQ market was established in 1996, it had experienced a fast-track growth especially from 1999 to 2000. Once, in

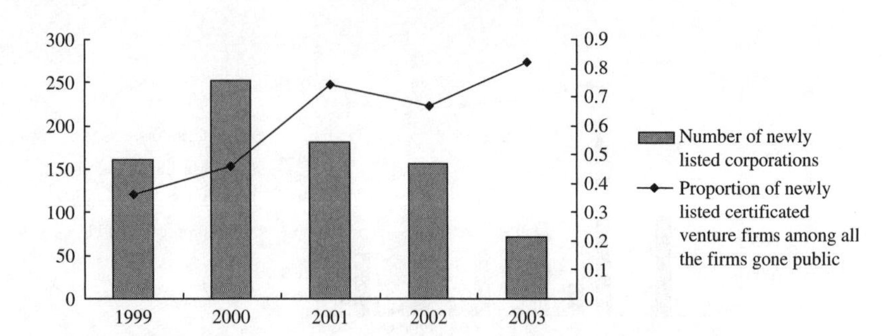

Note: Included are so-called 'back-door' listings through M&As.

Source: Kang et al. (2004).

Figure 7.10 Proportion of certificated venture firms in the KOSDAQ stock market

2000, it surpassed the regular Korea Stock Exchange market in terms of market value. It continues to perform the function of supplying capital to the listed companies even after a sharp decline. This outcome in the KOSDAQ market was driven by the interplay between market fluctuations and government actions.

We can review the current status of the KOSDAQ listings by venture firms, and by venture capital invested companies. As shown in Figure 7.10, the number of newly listed firms in the KOSDAQ market was on a steep increase up to 2000, but has rapidly decreased ever since. Even though the number of new KOSDAQ listings has decreased, the proportion of venture firms relative to the total new listings has risen to 80 per cent in 2003. This indicates that the KOSDAQ market continues to serve as a major capital market for venture firms. Also, the proportion of venture capital funded venture firms among the listed corporations has increased to 70 per cent in 2005 as seen in Figure 7.11. Overall, we can state that the KOSDAQ market continues to function as a primary stock market for new technology-based firms going public as well as an exit market for venture capitalists.

Effects on Information Technology Industry

The number of information technology companies listed in the KOSDAQ market surged dramatically in 1999. From that time on, the proportion of information technology companies of all the new listings has been a little over 50 per cent until now (Figure 7.12). So, information technology has become a major sector in the KOSDAQ market. This information technology sector

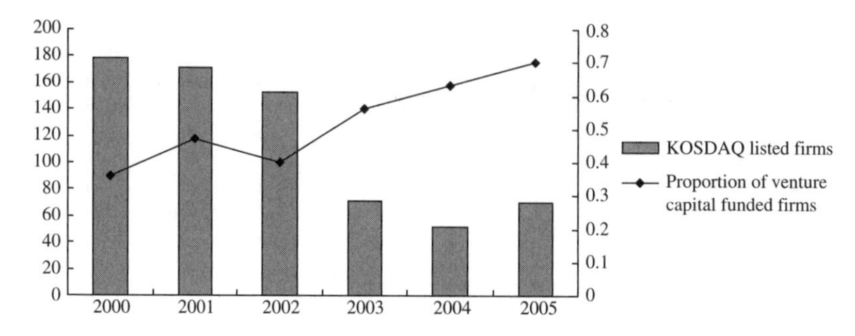

Note: Excluded are 'back-door' listings through M&As.

Source: Internal Document, Korea Venture Capital Association (2006).

Figure 7.11 Proportion of venture capital funded corporations in the KOSDAQ stock market

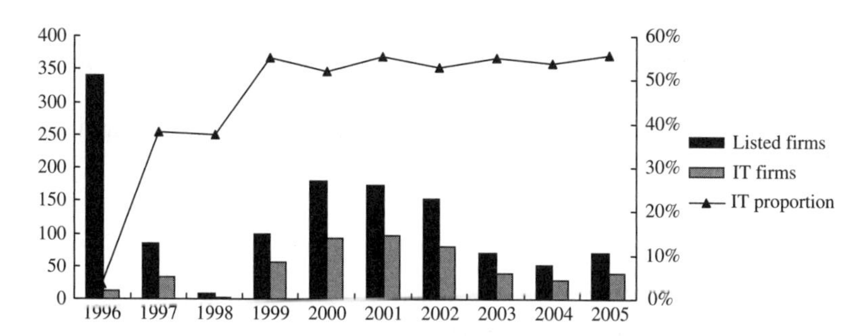

Note: When the KOSDAQ Stock Market opened on 1 July 1996, the number of listed corporations transferred from the existing over the counter market was 341 whereas the number of newly listed corporations in the opening year was only 25.

Source: Korea Stock Exchange (2006).

Figure 7.12 Proportion of information technology companies in the KOSDAQ stock market

includes information technology components, information technology products, semiconductors, software, computer services, digital contents, broadcasting service and Internet communication service areas.

The information technology sector in Korea made up 7.2 per cent of the GDP in 2004, an increase from 4.9 per cent in 1995. Compared with other industries, however, the information technology sector is the most active in entrepreneurial activities and performance by statistics. The proportion of information technology venture firms among all venture firms, the

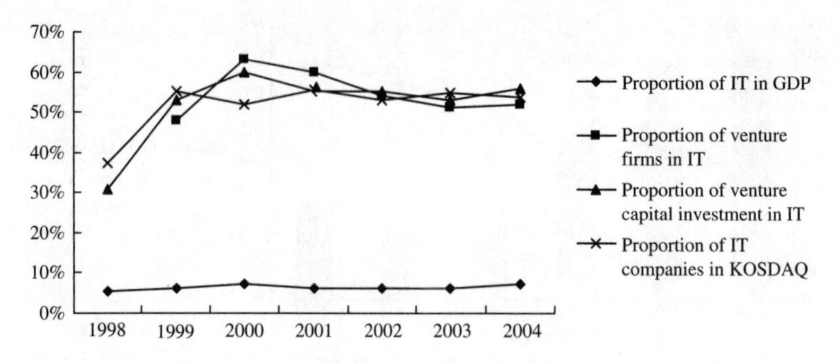

Source: Korea Stock Exchange (2006).

Figure 7.13 Economic impact of the information technology sector

proportion of information technology companies among venture capital invested firms, and the percentage of information technology companies among all the KOSDAQ listings are significantly higher than its contribution to the GDP (Figure 7.13). In fact all the three measures have been above 50 per cent ever since 2000. This indicates the importance of the information technology sector for sustaining entrepreneurial capabilities in Korea.

DISCUSSION

In this section we want to discuss several implications from our findings in the Korean entrepreneurial setting. Our discussion will lead to some propositions on entrepreneurial ventures, venture capital and industrial development. Regardless of the practical implications of this research, we are also well aware of our limitations that could have arisen from our research setting, which was only one case at a societal level. Therefore future studies might extend our findings in understanding the entrepreneurship of other countries and make regional comparisons.

Entrepreneurial Ventures

The early studies on the growth of entrepreneurial ventures primarily discuss the role of individuals such as entrepreneurs and founders (Cooper, 1993; Covin and Slevin, 1991; Gartner, 1985). But, as the importance of other factors besides founders in the establishment and growth of the ventures became appreciated, more serious studies on the enterprise concept and decision factors in the sphere of strategic management began to be carried out (Baum et al., 2001). There are only a few existing studies on the venture support systems besides the ones on environment as external factors of a business enterprise.

The literature shows that a change in social structure and cultural peculiarity are the initial conditions stimulating the start-up of entrepreneurial ventures and thus influencing the start-up rate (Cole, 1959; Lynn, 1991). Other conditions such as advances in technology (Casson, 1995; Klepper and Simmons, 1999; Kyläheiko and Miettinen, 1995), globalization (Carree, 1997) and economical growth (Reynolds, 1997) may stimulate the start-up rate. However the government also plays a major role in drawing potential entrepreneurs' attention (Verheul et al., 2001), affecting the start-up rate with policy drivers (Carree et al., 2002).

In the Korean setting, Lee et al. (1998) reviewed venture capital as one of the key venture support systems. The results of the study show that the success of a venture firm depends greatly on venture capital. Bae (2002) points out that the support in venture capital for Korean domestic venture firms is inadequate compared with those in the developed countries. He also argues that excessive funds that flowed in during the booming period of venture investment since 1999 have worsened the screening capabilities of venture capital. Koh et al. (2003) examines a model to measure the venture success factors. In addition to founders, infrastructure, resources, strategy and start-up processes, the external support systems and industrial environment, including government policy, were used as independent variables and business success was used as a dependent variable. They show that external support systems as well as industrial environments, such as market size and government policy, function as positive factors. Taken as a whole, the external factors have a significant effect on the success of venture firms.

In Korea, for instance, the cumulative number of venture firms in the last decade, as depicted in Figure 7.2, had reached its peak in 2001 with the sudden rise in the KOSDAQ stock market. Also this was the time Korean venture capitalists made the highest level of early stage investment. However the number has been falling continuously. As shown in the growth of venture firms in Figure 7.10, the number of firms matured enough for the IPO has also been declining since 2000. The Korean government has been trying to promote diverse policies to promote venture firms since the establishment of the KOSDAQ stock market. As previously discussed, existing studies show that external factors, such as government policy and support systems, extensively affect start-ups and growth of venture firms and yet the result was quite different in Korea. This unexpected phenomenon can be explained by two possibilities. First, the start-ups of venture firms increased when the venture capitalists most actively made their investment in the early stage. Second, development of new businesses was more active when expectations for socioeconomic compensation were high. In other words, more venture firms will be created and developed as opportunities of compensation for start-up and growth emerge. Up to the year 2000, this expectation was actualized. Thus we can suggest the following proposition.

Proposition 1: The start-up of entrepreneurial ventures is positively influenced by the government support policy if and only if Venture Capitals employ a high level of investment in the early stage and entrepreneurs expect an opportunity for positive compensation in the exit market.

This proposition, derived from the data of Korean venture industry for the last decade, might give practical lessons to government authorities of developing countries in charge of supporting venture industry. In order to activate the start-up of venture firms, inducing venture capital's investment in the early stage is a key strategy. Many studies have already emphasized the importance of government support in the establishment of venture firms. Moreover various policies promoting venture industry may be created, such as the certification system for venture firms employed in Korea. However the venture capitalists' high level of investment in the early stage was a greater inducement force for the start-ups than the effect of government policy. Therefore government authorities in charge of venture industry should consider various inducement tools that could encourage investment in the early stage.

Venture Capital

Norton and Tenenbaum (1993) and Gupta and Sapienza (1992) find that venture capital firms focusing on the early stage investment are specialized in industry sectors. In other words, they focus their investment on few industries rather than distributing funds across various industries for risk diversification. The reason for this high-risk investment pattern is that they could control risks through networking among investors, sharing information and learning (Ruhnka and Young, 1991).

Gompers (1996) compared the venture capital firms with more business experience to the less experienced ones in exit strategy. The results show that the firms with less experience try for an IPO in a shorter span of time than the ones with more experience due to their insufficient reputation in the market. Firms with insufficient reputation need to show good results in fund operation before the dismissal of investment partnerships. Consequently, they exhibit less interest in early stage firms at relatively high risks, and their investment portfolios tend to be comprised of later stage firms.

Moreover the investment patterns of venture capital firms change according to the ups and downs of stock markets. Gompers (1998) finds that the size of fund per investment and the investment in later stage firms increased in the late 1980s with the rapid inflow of capital. In other words, at the time of booming stock markets and venture capital markets funds flow easily in the market and the size of the investment fund increases. This increase begets the augmentation of the size of fund per investment due to the limitation in investment opportunities. This leads to more investments in late stage firms. On the other hand, during the slump the size of investment fund and the size of fund per investment are cut down as the flow of funds into venture capital declines. This means that the investment in early stage firms will be relatively increased.

Black and Gilson (1998) suggest that countries with active venture industry either have a well-developed secondary stock market besides the primary stock market, as in the USA and UK, or make practical use of the secondary market of other countries, as in the case of Israel and Ireland. According to Gompers and Lerner (1999), the venture industries in Europe and the USA have developed their unique features because of their differences in whether the stock market recovered rapidly enough to provide the secondary market for venture capital, since the crash in 1987. In the countries with less-developed venture industry, such as Japan and Germany, the return on venture capital is not fully guaranteed. Studies conducted by Black and Gilson (1998) and Milhaupt (1997) show that the development lag in venture capital industry in Japan and Germany is due to their peculiarities in the financial structure centered on banks. They point out that this fact cannot be amended in a short period of time, nor can it simply be modified exclusively for the venture industry alone.

Bygrave and Timmons (1992) find a positive relationship between the amount of fund supplied through the IPO and the size of fund flowing into venture capital from 1969 to 1986. They show that the NASDAQ stock index and the size of fund flowing into venture capital markets move in the same direction. Gompers (1995) also finds a strong positive relationship between the movement of the market and the size of fund flowing into venture capital markets. The correlation between the NASDAQ composite index and the inflow of funds into venture capital was as high as 0.79, based on the data from the 1970s to the early 1990s. Namely, investments in venture capital increase when the stock market booms. The booming market and the increasing inflow of funds into venture capital surely activate business venturing by entrepreneurs.

According to our observation, during the last decade in Korea investments in early stage firms have declined since the KOSDAQ stock market reached its peak in 2000. On the other hand, up until the present, which is the time of market slump, investments in mid and late stage firms have continuously increased. As shown in Figure 7.8, the Korean venture capitalists invest more than 60 per cent of their total funds mainly into the information technology sector. However Figure 7.7 shows a decline in investment in the early stage right after the market boom in 2000, which contradicts the pattern that the existing studies suggest. Many venture capital firms in Korea were newly established in the late 1990s, and engaged in extensive investments in early stage firms at the peak of the market. Those venture capital firms that have enough business experience by now are trying to avoid investment in early stage firms.

These investment patterns are quite contrary to that of the USA in the 1980s. The discrepancy found here can be explained as follows. First, it is due to the attributes of the KOSDAQ stock market from 1999 to 2000. At that time venture capital firms could profit from the investment in early stage firms due to weak regulations in the IPO process. Therefore the firms with less business experience did not need to make investments in later stage firms in order to build up a reputation.

Second, investment in early stage firms by the less experienced firms was possible due to the rapid inflow of funds into venture capital firms, as shown in Figure 7.4. After all, venture capital firms with almost no experience could raise plenty of funds. So they did not need to build up a reputation in haste through the IPOs for additional fund raising.

Third, the Korean venture capital firms do not need to show the results to external investors in a hurry because the investment does not necessarily come from external funds. During the boom of the KOSDAQ market they had very few reasons to show the results as they could internally supply 70 per cent of the funds they needed.

Fourth, it is due to the results of rash investments led by the novice firms at the time. Investment at this time was made under circumstances in which the standard operating procedures of traditional venture capital firms were almost ignored. Young venture capital firms invested aggressively and heavily in early stage firms at high risks as they could hardly separate the grain from the chaff in choosing an investment vehicle due to investment fever initiated by the Internet bubble.

Fifth, at the current time of the stock market slump, investments in mid and later stage firms have increased while investments in early stage firms have decreased due to the overflowing funds employed by an insufficient number of venture capitalists. Investments in later stage firms have increased as more funds flow into venture capital markets. At present, there are about 100 venture capital firms in Korea, but only 30 of them can be described as a substantially functioning investor. This is a very small number, compared to the amount of actual funds they employ. This phenomenon is due to the fact that the focused supply of government funds started to play a major role in venture capital markets since the inflow of private funds to venture capital markets was virtually suspended after 2001. These unique observations lead to the following propositions.

Proposition 2: A higher level of venture capital investment in early stage companies is possible when government regulations on the exit market are weaker.

Proposition 3: Regardless of the market slump, abundant investment funds employed by a relatively small number of venture capitalists lead to increased investments in later stage companies.

These propositions, derived from the data of the Korean venture industry for the last decade, gives practical advice to government authorities of developing countries in charge of venture industry. First, in order to induce more venture capital investment in the early stage, regulations for the IPO market should be more flexible. As seen in the existing studies, the opening of a stock market for venture firms is the first to come. Interestingly, as the regulation for IPO in the stock market for ventures gets weakened, venture capital investment in the early stage increases resulting in more establishment of venture firms. However one other important fact to consider is that the level of regulation should be determined concerning both the development of venture industry and the protection of stockholders participating in the stock market.

Second, government's more active involvement in the efficient supply of the venture capital market is another way to induce venture capital investment in the early stage. In the case of developing countries like Korea, the government is usually a key investor in the venture capitalists market. Therefore the government should always be aware of all the number of venture capitalists including the ones privately owned and prevent any unnecessary future investment.

Industrial Development

Throughout the history of mankind, technology has always been the core factor that produces the wealth of a nation and elevates the quality of human lives. Technological advancement is now considered as the most important determinant of the continuity of a nation's economic development. The critical effect of technology and knowledge on economic development has been acknowledged for quite a long time. The studies on the relationship between technology and economical efficiency include Schumpeter's (1934) who insisted that in terms of economic development of capitalist societies, firms' continuous pursuit of profit is a driving force of a nation's economic development.

From the 1950s to the 1970s exogenous development theories in which technology is considered as an exogenous factor were dominant in the field of study. However, starting from the 1980s, indigenous growth theories emerged, which state that when economic subjects intentionally increase the development of technology, economic development is stimulated. New growth theories emphasize the increasing returns of accumulated knowledge based on new technology and human capital. And the efficiency of production is generated mostly by the external effect, the collection of learning and knowledge. Therefore the new growth theory views accumulation of knowledge as the base of economic growth and emphasizes economic mechanisms or systems that such knowledge could be applied. In particular, the new growth theory derives the political importance of knowledge, technology and government involvement (Dosi, 1982; Nelson and Winter, 1977; Romer, 1986; Albach, 1989).

Stoneman (1987) suggests that the government involvement in the technology sector is initiated by market failure. He also states that due to its longevity and mobility, the technology market is quite fragile so its failure mostly results from the difference in the concept of time between individual and whole economic subjects, and difficulties in production and shipment, and semi-optimal information activity (Gielow et al., 1985). Roobeek (1990) emphasizes the effect of government involvement in R&D and other innovative activities, and also suggests that governments in developing countries should successfully function as entrepreneurs.

Freeman (1974) insists that rapid development of industries, such as chemical, pharmaceutical, electrical, computer, aircraft and nuclear energy, in the twentieth century was based on a systematized series of scientific studies. Therefore he suggests that in order to gain national competitive advantage, the government should not keep their hands off and let entrepreneurs independently play their role. In the late 1980s policies promoting technological advancement had been considered extremely crucial in the USA as a way to strengthen the competitive power of manufacturing industry in the global market (Branscomb, 1992). Khalil (2000) points out that the main cause of the Southeast-Asian economic crisis in 1997 was the lack of policies promoting technological advancement.

Many European countries post-World War II acknowledged the indispensability of policies promoting technological advancement for the growth of the economy, the elevation of national competitiveness, and also considered them as resolutions for other socioeconomic problems. The areas in need of government involvement include high technology industry,

R&D of small and medium firms and also R&D for traditional industry (Hauff and Scharpf, 1975; Gielow et al., 1985; Bruder and Does, 1986; Fritsch et al., 1993; Chung, 1996). According to existing studies, conglomerates spend twice as much on R&D per patent, but the return on investment is greater for small and medium firms. In fact the successful innovators were small and medium firms in most cases. The small and medium firms in Germany show excellence in innovative ability in technology and they are called the hidden champions of the global market (Simon, 1992, 1996).

Freeman and Perez (1988) explain the economic development since the industrial revolution in terms of five stages of long-term bullish waves. The main player of the fifth wave is the telecommunications industry and the leading countries are Japan, USA, Germany, Switzerland, Taiwan and Korea. The most important fact is that such a long-term wave contributes to the success of not only a particular industry but also its nation as a whole. Therefore government vision and active involvement in these industries is strongly demanded.

Foreseeing the potential growth of the information technology industry in the mid 1990s, the Korean government established the KOSDAQ market and employed a unique policy titled 'Certification System for Venture Firms'. Along with the Internet boom during the late 1990s and the early 2000s, venture industry had been greatly advanced, and despite the decrease in the number of venture start-ups after the KOSDAQ stock market crisis (see Figure 7.2) and perceived conservatism in venture capital (see Figure 7.7), the information technology industry as Korea's representative industry with national competitiveness stayed firm (see Figure 7.13). The proportion of information technology venture firms among all venture firms, the proportion of information technology companies among venture capital invested firms, and the percentage of information technology companies among all the KOSDAQ listings remained continuously high. The growth of information technology in Korea continued through the dramatic rise and fall of KOSDAQ in the mid 1990s, even up until now when venture industry is in its stage of contraction (see Figure 7.13). It is clear that the government's supportive policies greatly contributed to the nonstop development of the information technology industry itself. These observations lead to the following proposition.

Proposition 4: Government support policy is effective in fostering and sustaining entrepreneurial firms in an industry with national competitiveness.

To make a strategic fostering of a particular industry, inducing venture firms in that industry and reallocating resources are two of the most effective ways. The existing studies emphasize the governments' role in developing countries that could generate advancement in technology. In particular, the wave of economic growth along with industrial change is a critical factor of a nation's success. Also small and medium firms are considered relatively more effective compared with conglomerates in terms of technological development. The Korean government predicted the endless possibilities of the information technology industry during the period of the fifth wave, and continued to employ support policies. Consequently, venture industry went though a vast amount of development and the concentration of information technology among all venture industries remained even at the stage of contraction. Therefore government authorities in charge of venture industry are encouraged to induce the establishment of venture firms in the industry of their focus. In order to do so, they need to make policies concerning mechanisms of concentrated resource allocation for venture firms in their nation's core industry.

CONCLUSIONS

In this chapter we reviewed the processes of emergence, take-off and restructuring in the Korean venture industry over the last ten years. The growth of the information technology sector and the financial crisis in combination provided strong stimuli to the government in restructuring strategic industrial sectors at the national level. This change effort favored high technology ventures for a while. The Korean venture capital system started to function properly from 1999. From 1999 to 2000 entrepreneurial start-ups dramatically increased along with venture capital funding for early stage companies.

Since the crash of the KOSDAQ market in 2000, however, the rate of start-ups and the growth of venture firms have remained depressed. Venture capital inclined more toward conservatism in investment with a significant contraction in the inflow of private capital money. Nevertheless, the KOSDAQ market started to function as an efficient capital market, encouraging the growth of competitive industries such as the information and communications industry.

Considering the existing literature, the case of Korea provides some interesting propositions. First, more investments in early stage venture firms can be made by venture capital when government regulations on the IPO market are weaker. Second, regardless of the IPO market slump, the abundant investment funds employed by a relatively small number of venture capitalists lead to increased investment to later stage venture firms.

Third, government support policy can be effective in fostering entrepreneurial firms in an industry with national competitiveness.

Propositions derived from the data of the Korean Venture Industry for the last decade present some practical lessons to the government authorities of developing countries in charge of venture industry. First, for the active establishment of venture firms, inducing investment in the early stage is extremely essential. Second, in order to induce more venture capital investment in the early stage, regulations for the IPO market should be more flexible. Third, government's more active involvement in the efficient supply of the venture capital market is another way to induce venture capital investment in the early stage. Lastly, to make a strategic fostering of a particular industry, inducing venture firms in that industry and reallocating resources are two of the most effective ways.

NOTE

1. The research assistantships of Ji Sun Lim, Yeo Im Hwang and Na Sung Pyo are acknowledged.

REFERENCES

Acs, Z.J. and D.B. Audretsch (1990), *Innovation and Small Firms*, Cambridge, MA: MIT Press.

Albach, H. (1989), 'Innovationsstrategien zur verbesserung der wttbewebsfahigkeit', *Zeitschirft fur Betriebswirtschaftslehre*, **59** (12), 1338–52.

Bae, Y. (2002), 'Redefining venture administration', research report no. 373, Samsung Research Institute, Seoul.

Baum, J.R., E.A. Locke and K.G. Smith (2001), 'A multidimensional model of venture growth', *Academy of Management Journal*, **44** (2), 292–303.

Birch, D.L. (1990), 'Sources of job growth and some implications', in J.D. Kasarda (eds), *Jobs, Earnings, and Employment Growth Policies in the United States*, Norwell, MA: Kluwer Academic Publishers, pp. 71–6.

Black, B. and R. Gilson (1998), 'Venture capital and the structure of capital markets: banks versus stock markets', *Journal of Finance Economics*, **47** (3), 243–77.

Bollinger, L., K. Hope and J.M. Utterback (1983), 'A review of literature and hypotheses on new technology-based firms', *Research Policy*, **12** (1), 1–14.

Branscomb, L.M. (1992), 'Does America need a technology policy?', *Harvard Business Review*, **70** (2), 24–31.

Brown, C.V. and P.M Jackson (1990), *Public Sector Economics*, Oxford, UK: Basil Blackwell.

Bruder, W. and N. Dose (1986), *Forscchungs und Technologiepolitik in der Bundersrepublik Deutschland*, Opladen: Westdeutscher Verlag.

Bygrave, W.D. and J.A. Timmons (1992), *Venture Capital at the Crossroads*, Boston, MA: Harvard Business School Press.

Carree, M.A. (1997), 'Market dynamics, evolution and smallness', Doctoral Dissertation, Erasmus University Rotterdam.
Carree, M.A., A. Stel, R. Thurik and S. Wennekers (2002), 'Economic development and business ownership: an analysis using data of 23 OECD countries in the period 1976–1996', *Small Business Economics*, **19** (3), 271–90.
Casson, M. (1995), *Entrepreneurship and Business Culture: Studies in the Economics of Trust*, vol. 1, Aldershot, UK and Brookfield, USA: Edward Elgar.
Cheon, M. (2000), 'A study on international competitiveness and industrial policy: Korean and Japanese automobile industries', Master's thesis, Baejae University, Seoul, Korea.
Cho, D. (1994), *Korea-Japan Industry Policy*, Seoul: Institute of International Corporate Strategy Research.
Choi, B., K. Choi and K. Park (2002), 'Economic spillover effects and political implications of the IT industry', research report, Korea Institution for Industrial Economics and Trade, Korea.
Chung, K. (2003), 'Support policy towards ventures by the Korean government', Master's thesis, Yonsei University, Seoul, Korea.
Chung, S. (1996), *Technologiepolitik fur Neue Produktionstechnologien in Korea und Deutschland*, Heidelberg: Physica-Verlag.
Cole, A.H. (1959), *Business Enterprise in its Social Setting*, Boston, MA: Harvard University Press.
Cooper, A.C. (1993), 'Challenges in predicting new firm performance', *Journal of Business Venturing*, **8** (3), 241–53.
Covin, J.G. and D.P. Slevin (1991), 'A conceptual model of entrepreneurship as firm behavior', *Entrepreneurship Theory and Practice*, **16** (1), 7–25.
Dosi, G. (1982), 'Technological paradigms and technological trajectories: a suggested interpretation of the determinants and directions of technical change', *Research Policy*, **11** (3), 147–62.
Freeman, C. (1974), *The Economics of Industrial Innovation*, Harmondsworth: Penguin.
Freeman, C. and C. Perez (1988), 'Structural crises of adjustment: business cycles and investment behavior', in G. Dosi, C. Freeman, R. Nelson, G. Silverberg and L. Soete (eds), *Technical Change and Economic Theory*, London: Pinter Publishers, pp. 38–66.
Fritsch, M., T. Wein and H.J. Ewers (1993), *Markversagen und Wirtschaftspolitik: Mikrookonomische Grundlagen Staatlichen Handels*, Munchen: Verlag Franz Vahlen.
Gartner, W.B. (1985), 'A conceptual framework for describing the phenomenon of new venture creation', *Academy of Management Review*, **10** (4), 696–706.
Gielow, G., H. Krist and F. Meyer-Krahmer (1985), *Industirelle Forschungs und Technologieforderung-Diskussion Theoretischer Ansatze und Ihrer Empirischen Evidenz*, Karsruhe: FhG-ISI.
Gompers, P.A. (1995), 'Optimal investment, monitoring, and the staging of venture capital', *Journal of Finance*, **50** (5), 1461–89.
Gompers, P.A. (1996), 'Grandstanding in the venture capital industry', *Journal of Financial Economics*, **42** (1), 133–56.
Gompers, P.A. (1998), 'Venture capital growing pains: should the market diet?', *Journal of Banking and Finance*, **22** (6), 1089–104.
Gompers, P.A. and J. Lerner (1999), *The Venture Capital Cycle*, Cambridge, MA: MIT Press.

Gupta, A. and H. Sapienza (1992), 'Determinants of venture capital firm's preferences regarding the industry diversity and geographic scope of their investments', *Journal of Business Venturing*, **7** (5), 347–62.
Hamermesh, D.S. (1993), *Labor Demand*, Princeton, NJ: Princeton University Press.
Hauff, V. and F.W. Scharpf (1975), *Modernisierung der Wirtschaft: Technologiepolitik als Strukturpolitik*, Frankfurt am Main: Europaische Vergsanstalt.
Johnson, C.A. (1982), *MITI and the Japanese Miracle: The Growth of Industrial Policy 1925–1975*, Stanford, CA: Stanford University Press.
Kang, W., T. Chung and K. Kim (2004), 'Conditions for the revitalization of the venture ecosystem', CEO Information Report no. 471, Samsung Economic Research Institute, Seoul.
Khalil, E.L. (2000), 'Making sense of Adam Smith's invisible hand: beyond pareto optimality and unintended consequences', *Journal of the History of Economic Thought*, **22** (1), 49–63.
Kim, S., C. Kim and S. Kim (2000), *The Government and the Firm: The Political Economy of Industry Policy*, Seoul: Dae-Young.
Klepper, S. and K.L. Simons (1999), 'Dominance by birthright: entry of prior radio producer and competitive ramifications in the U.S. television receiver industry', *Strategic Management Journal*, **21** (10), 997–1016.
Koh, B., S. Yong and S. Lee (2003), 'Empirical study on determinatives of success in venture firms', research report, Korea: The Korean Venture Management Review.
Korea Institute for Industrial Economics and Trade. (2006), 'The key indicators of major industries', Seoul, Korea.
Kyläheiko, K. and A. Miettinen (1995), 'Technology management and entrepreneurship: a critical view', in S. Birley and I. MacMillan (eds), *International Entrepreneurship*, London: Routledge, pp. 39–58.
Lee, B. (2002), 'Growth of venture firms and policy challenges', Economic Research Institute, Seoul, Korea.
Lee, I. (2002), 'Revitalizing the venture firm: the role of market and government', Korean Venture Capital Association, Seoul, Korea.
Lee, I. (2003), *Venture Capital Industry of Korea*, Seoul: In-Sung.
Lee, I., K. Lee, S. Park and W. Kim (1998), 'Analysis on the factors for success for each growth stage and principal policy', Information Society Development Institute, Seoul, Korea.
Lynn, R. (1991), *The Secret of Miracle Economy: Different National Attitudes to Competitiveness and Money*, London: Social Affairs Unit.
Milhaupt, C.J. (1997), 'The market for innovation in the United States and Japan: venture capital and the comparative corporate governance debate', *Northwestern University Law Review*, **91** (3), 865–98.
Nelson, R. and S. Winter (1977), 'In search of a useful theory of innovation', *Research Policy*, **6** (1), 36–77.
Norton, E. and B.H. Tenenbaum (1993), 'The effects of venture capitalists characteristics on the structure of the venture capital deal', *Journal of Small Business Management*, **31** (4), 32–41.
Reynolds, P.D. (1997), 'Who starts new firms? Preliminary explorations of firms-in-gestation', *Small Business Economics*, **9** (5), 449–62.
Robbins-Roth, C. (2001), *From Alchemy to IPO: The Business of Biotechnology*, Cambridge, UK: Perseus Books.

Romer, P. (1986), 'Increasing returns and long run growth', *Journal of Political Economy*, **94** (5), 1002–37.

Roobeek, A.J.M. (1990), *Beyond the Technology Race: An Analysis of Technology Policy in Seven Industrial Countries*, Amsterdam: Elsevier Science Publishers.

Rosen, R.J. (1991), 'Research and development with asymmetric firm sizes', *Rand Journal of Economics*, **22** (3), 411–29.

Ruhnka, J.C. and J.E. Young (1991), 'Some hypotheses about risk in venture capital investing', *Journal of Business Venturing*, **6** (2), 115–33.

Scherer, F.M. (1991), 'Changing perspectives on the firm size problem', in Z. Acs and D.B. Audretsch (eds), *Innovation and Technological Change: An International Comparison*, Ann Arbor, MI: University of Michigan Press, pp. 24–8.

Schumpeter, J.A. (1934), *The Theory of Economic Development: An Inquiry into Profit, Capital, Credit, Interest and Business Cycle*, London: Oxford University Press.

Simon, H. (1992), 'Lessons from Germany's Midsize Giants', *Harvard Business Review*, **70** (2), 115–23.

Simon, H. (1996), *Hidden Champion: Lessons from 500 of the World's Best Unknown Companies*, Boston, MA: Harvard Business School Press.

Small and Medium Business Administration (2001, 2005), 'Annual report of the research on the actual conditions of a venture business', Korea: Small and Medium Business Administration.

Stoneman, P. (1987), *The Economic Analysis of Technology Policy*, Oxford, UK: Clarendon Press.

Verheul, I., S. Wennekers, A. David and R. Thurik (2001), 'An eclectic theory of entrepreneurship: politics, institutions and culture', discussion paper 01-030/3 Tinbergen Institute Amsterdam, The Netherlands.

Zysman, J. and L. Tyson (1983), 'American industry in international competition: government policies and corporate strategies', in J. Zysman and L. Tyson (eds), Ithaca, NY: Cornell University Press, pp. 10–12.

PART 4

Firm level responses to entrepreneurial opportunities in emerging regions

8. The founding conditions of entrepreneurial firms as a function of emerging institutional arrangements in China

Atipol Bhanich Supapol, Eileen Fischer and Yigang Pan

INTRODUCTION

In developed markets institutional structures typically change relatively slowly and incrementally. As developing economies emerge, however, institutional change is pervasive and frequently profound: rapid reform in political philosophies, regulatory environments and laws and policies governing how business may operate are common (for example, Park et al., 2006; Tan, 2005). Clearly, these changes are likely to present both challenges and opportunities for the managers of entrepreneurial ventures in such economies as feasible and attractive means of doing business may shift from one era or period to the next. Equally, these kinds of evolving economies also represent a challenge for entrepreneurship scholars accustomed to studying how new and growing firms compete in relatively slowly evolving macro-institutional environments. In order to study entrepreneurship in emerging economies, we need theories that are appropriate to such dynamic environments.

In this chapter we address this challenge by drawing on the evolutionary economic perspective. We conceptualize the evolving macro-institutional conditions in such economies as founding conditions that may interact with firm choices at the time of founding to affect firm performance over time. There is a growing body of evidence that suggests there is a significant impact on firm performance of both initial choices and industry specific founding conditions (for example, Bamford et al., 1999). We build on the various theoretical perspectives that explain such findings to consider how changing macro-institutional contexts in emerging economies may constitute differing founding conditions that impact emerging organizations' abilities to benefit from the strategies and tactics they adopt.

As indicated above, there are both managerial and theoretical reasons why it is particularly important to investigate this matter. From a managerial point of view such a study is important for those interested in fostering successful organizational emergence and development in evolving economies. It is essential to know the extent to which there may be choices that are likely (or unlikely) to be successfully implemented, irrespective of the environmental founding conditions at startup, versus choices that are likely to result in superior performance only for firms founded in contextual conditions with particular imprinting characteristics (Davies and Walters, 2004).

From a scholarly perspective, such a study is important for those interested in broadening their understanding of how to study entrepreneurial phenomena in emerging economies. Our chapter offers a theoretical perspective on such economies as a dynamic series of founding conditions. It adapts a perspective that has previously been focused on considering whether industry specific founding conditions and initial choices impact survival and performance (for example, Bamford et al., 1999; Boeker, 1989; Eisenhardt and Schoonhoven, 1990; Tolbert and Zucker, 1983). It highlights that in emerging economies the broader set of macro-institutional conditions can constitute founding conditions, and considers whether these conditions, in conjunction with firm level choices, limit the extent to which firms can benefit from the choices that they make.

To apply and explore this perspective we develop specific arguments related to the performance of organizations founded in three different eras of China's emergent economy, looking at differences in the extent to which those founded at different times appear to benefit or suffer from making the same choices. The chapter develops two hypotheses and examines both using data on enterprises founded and operating in the Chinese economy. China's is a particularly appropriate emerging economy in which to examine changing founding conditions, as its macro-institutional environment has evolved profoundly over time (Tan, 2005), creating quite contrasting conditions for firms founded in distinct eras of economic reform.

To delimit the scope of this study we highlight that this chapter does not address the question of whether economic conditions influence the nature of specific strategic choices that firms are likely to make. This research question has recently been addressed specifically in the context of China (Tan, 2005). Instead, this study is concerned with whether the macro-institutional contextual conditions at the time of founding imprint on the enterprise and affect its ability to benefit from choices made either initially or later. This question has received no direct attention (Davies and Walters, 2004; Park et al., 2006; Tan and Tan, 2003), and it is an important key goal of this study, to advance our understanding of how to study entrepreneurship in dynamic emerging economies.

In the following sections of this chapter we first review the theoretical rationale that addresses why founding conditions affect organizational survival and performance. We next develop specific hypotheses regarding the likely impact on the performance of Chinese firms founded in one of three eras having made or of making specific choices (that is, ownership structure, operating in an industry receiving state support). We then describe our methods and results, and discuss our conclusions and implications.

FOUNDING CONDITIONS AND THE FUTURE PERFORMANCE OF THE FIRM

Many scholars have mounted arguments that initial choices and external founding conditions matter to the nature and future performance of firms. One of the earliest to do so was Stinchcombe (1965), who argued that external conditions and initial choices at the time of founding constitute 'imprinting forces' that define initial characteristics and create internal consensus around the appropriate way of organizing and conducting business. He also argues that events shortly after founding are 'traditionalizing forces' that tend to preserve initial organizational characteristics. Subsequent theorists have elaborated in various ways on the idea that founding environments and choices matter.

One particularly influential school of thought comes from population ecologists who have stressed that the environment at founding determines the resources available to the firm, and thereby shapes their future prospects (for example, Aldrich, 1999). For example, the rate of growth of a market at the time it is founded appears to influence the size firms are likely to achieve (Eisenhardt and Schoonhoven, 1990) and the scale of entry of new firms (McDougall et al., 1994). Likewise, the concentration of firms in an industry at the time a new firm is founded influences its likelihood of survival (Carroll and Hannan, 1989). And the munificence in an environment influences the likelihood that firms founded in it will survive (Swaminathan, 1996).

Institutional theorists have examined the more social and cognitive aspects of founding conditions, further contributing to our insights into the ways that initial contexts of operations and initial choices matter. One important notion developed in this tradition is that of isomorphism, which suggests that organizations conform to the accepted norms of their populations (DiMaggio and Powell, 1983). In effect, an environment legitimates certain ways of organizing, which in turn have certain consequences. For example, Tolbert and Zucker (1983) found that city governments adopted certain organizational structures and processes more than others because the environment legitimated those structures and processes. And while institutional theorists acknowledge that some firms choose to deviate from institutional norms, they stress that once legitimated within an organization, certain cognitions and beliefs become taken for granted and perpetuated within the firm (cf. Oliver, 1997).

Scholars concerned with strategic inertia have shed further light on the impact of initial strategies and assumptions. Miller and Friesen (1984), for example, discussed strategic momentum, a tendency in organizations to persist in certain courses of action or classes of strategies and to adapt slowly to changes in their environments. Others have noted that strategy influences structures of organizations (for example, Chandler, 1962) and may limit the consideration or implementation of future new strategies (Freeman and Boeker, 1984; Boeker, 1989). More broadly, organizational researchers have argued that an organization's history is crucial to its future development, and that organizations can only be understood in light of their early phases. For example, Selznick (1957) described how early political and social processes shaped organizations' subsequent patterns of activities.

These views are consistent with the perspective that has been developed in the evolutionary economics literature (for example, Nelson and Winter, 1982, 2002). This body of work typically characterizes firms as having organizational routines – recurring patterns of organizational processes – that are based on the repeated interaction of several firm members and that enable coordinated activity because they make the behavior of firm members mutually predictable. It is argued that firms 'remember by doing', with the organizational routines embodying the firm's 'memory'. Knowledge reflected in organizational routines is typically tacit in nature; the firm members holding it may be unable to express it verbally.

A key point of relevance to the arguments developed below is that this tacit knowledge can be highly adaptive at the point in time in which it develops, but less so later. Firm behavior is frequently maladapted to its current environment because it becomes outdated: changes in socioeconomic systems often impose exogenous changes that lead to new and unfamiliar problems for firms. Yet learning is inhibited because routines are deeply routed in behavior practices and constitute a largely tacit knowledge. Evolutionary economists argue that when there is a contrast between current challenges or opportunities and those that featured in earlier contexts in which organizational processes became routine, organizations that once were competent can be much less so (for example, Nelson and Winter, 2002, p. 30).

Our perspective builds on the common logic articulated across these approaches. Whereas the institutional and evolutionary economic perspectives typically emphasize industry level influences on early organizational characteristics we argue, however, that in the case of emerging economies it is important to consider the broader macro-institutional context as contributing to the imprinting conditions at the time of founding. That is, we argue that organizations founded in differing stages of the evolution of an economy are likely to be affected by normative assumptions, reflected, for example, in political philosophies, regulatory structures, laws and policies that prevail across the economy within the period. This argument is consistent with the core premise of institutional theory, which is that rules, norms and beliefs surrounding economic behavior define that which is socially acceptable or socially preferable firm level behavior (Oliver, 1997, p. 698). We further assert, consistent with the evolutionary economic perspective (for example, Nelson and Winter, 2002), that the impact of these assumptions, combined with initial firm level choices, can shape organizations' routines and tacit knowledge such that firms are limited in the extent to which they adapt as the economy changes.

To make our arguments more concrete and to provide a specific example of what is meant by dynamic macro-institutional founding conditions, we describe in the next section changes in the founding conditions that faced firms as the Chinese economy evolved. We then develop hypotheses regarding particular choices made either at founding or later, the impact of which is susceptible to the influence of the founding context.

FOUNDING CONDITIONS IN THE EVOLVING CHINESE ECONOMY

Since China began the reform of the Chinese planned economy in 1978, the nation has gone through three distinctive stages (Lin et al., 1996; Tan, 2005). The first stage of reform focused on introducing the incentives system to farmers and workers and decentralizing the decision making responsibility to households in farming villages and factories in the cities (Tan, 2005). This stage existed between 1979 and 1986 (Lin et al., 1996). State-owned enterprises began to experience some management autonomy, such as to produce outside the mandatory state plan, and to enjoy performance-related incentives. Towards the end of this stage, most state-owned enterprises (SOEs) were allowed to sell outputs in excess of quota that was negotiated with the state in the market. Township, village and private enterprises began to compete in the market. SOEs began to experience market uncertainties and turbulence. As a result, the profitability of SOEs in general declined and government subsidies increased despite improvements in productivity. We call this stage the 'Incremental Reform Stage'.

The second stage began in 1987, when the state began to enter into contracts with managers of SOEs, in an attempt to clarify the authority and responsibilities in a formal manner. This stage ended in 1996, when the state began to introduce the modern corporate system to large SOEs. We call the second stage the 'Structural Reform Stage'. During this period township, village and private enterprises grew at a faster pace (Luo, 1999). The share of industrial output from non-state enterprises grew from 22 per cent in 1978 to 56.9 per cent in 1993 (Lin et al., 1996). These non-state enterprises existed outside the state's plan. They acquired supplies in the competitive market and sold their products in the competitive markets (Tan, 2005). At the same time, SOEs were still receiving varying amounts of state subsidized resources. On the whole, during this era, SOEs appeared to react to changes in the market environment, instead of being proactive, innovative and future oriented (Tan, 2005; Tan and Tan, 2003).

The third stage, which we call the 'Ownership Reform Stage', began in 1997 and ended in 2002 when China entered the World Trade Organization (Tan, 2005). For the first time, the private sector was no longer considered simply a supplement to the state-owned economy. It was now recognized as an important component of the socialist market economy (Kanamori and Zhao, 2004). During these three stages of market development many different reform policies were introduced, and the ideology regarding the role of state-owned enterprise dramatically evolved. Clearly, in the last stage state-owned firms were not given the eminent and protected status that they once received. They now appear to compete on a much more level playing field with private firms.

CHOICES MADE BY FIRMS

As the review in the previous section suggested, contextual conditions at founding within an emerging economy should exert a significant imprinting effect in conjunction with initial choices made by firms. We now identify two specific choices of particular relevance in the case of China's evolving economy, and develop hypotheses regarding the impact these choices will have on firm performance depending on the founding conditions in place at the time of startup. A model is developed which systematically relates firms' performance to: (1) ownership structure (private versus state-owned enterprise; and (2) whether the firm operates in an industry that is supported by the government, taking into account the eras in which firms were founded.

Ownership

There are numerous discussions on the effects of different ownership forms on performance, especially regarding the impact of state versus private ownership and the economic benefits of privatization. Yet to date the empirical evidence has been inconclusive (Dewenter and Malatesta, 2001). In part, the reason for mixed results is related to the difficulty in teasing out the separate effects of ownership status and market environments. Many of the studies have also been restricted by the number of SOEs available for comparison, or by the number of industries for which data may be available.

A number of recent studies have tried to separate out the pure effect of ownership and the associated agency problem from the effects of environment factors, such as the firm's financing decision and the degree of competition or rivalry in the marketplace (Bartel and Harrison, 2005). Ramaswamy (2001), for example, explored the interactive role of ownership status and competitive rivalry in influencing performance differentials between SOEs and private firms. Based on 110 Indian manufacturing firms (55 private firms and 55 state-owned firms), the study found that SOEs in general do not perform as well as private firms, and that the performance differential increases with increasing competitive intensity – as measured by the Herfindahl index. When return on investment was used as a measure of performance the authors found that competitive intensity on its own is not statistically significant. However there appeared to be a moderating effect on performance when an interaction between state ownership and rivalry was considered, having controlled for size, foreign ownership and industry. The poorer performance of state-owned firms was significantly more pronounced when the environment within which they operate was characterized as having a lower level of competition, all else being equal.

Based on the foregoing, and given our premise that founding conditions will significantly affect the institutionalized assumptions regarding how to operate a private versus a state-owned enterprise, we expect privately-owned firms to exhibit superior performance compared with state-owned firms to the extent that those enterprises were founded during a time when state-owned firms were protected from the full competitive forces of the market. The argument put forward is that private enterprises, created in the period where there are significant barriers to entry and a great deal of protection for state-owned firms, must have in their possession competitive resources which would allow them to successfully participate in markets where they otherwise would have been precluded. The competitiveness of these resources combined with the mindsets of the managers (cf. Oliver, 1997) would translate into a firm likely to perform well over time. In the absence of such resources, overcoming regulatory barriers would have been prohibitively difficult, and private firms would then choose not to enter, or would fail to survive.

Analogously, state-owned firms created during such a period would not, because of the available regulatory support and financial subsidies provided by the government, be as efficient, or as profit oriented as their private counterparts. We would expect to see a larger performance gap in the case where firms were founded in a period of relatively non-market oriented institutional environment.

Over time, however, both China's state-owned firms and private firms have had to contend with more market discipline as the country continues to liberalize many of its key sectors and allow for greater competition. Newly created government firms need to be more responsive to new sets of competitive forces, and these agencies are now more accountable and are often held to the standards of private sector performance. With the new economic surrounding where there are fewer economic supports and less preferential treatment for state-owned firms, entry conditions have become relatively more leveled (Tan, 2005). For this reason, we would expect to see a reduction in the profitability gap between private and state-owned firms founded during a more liberalized and competitive environment. That is, overall we expect to see a positive and significant effect associated with private ownership status; however the performance superiority associated with private ownership is expected to be smaller in the case where the initial founding conditions are more level and more competitive. We expect an interaction between the time period in which the firm was founded and the choice of ownership structure such that the magnitude of the difference in performance between state and privately owned firms is less for firms founded later in the evolution of China's economy. Thus:

Hypothesis 1: The magnitude of performance difference between state-owned firms and privately owned firms is less for firms founded later in the evolution of China's economy than for firms founded earlier.

That is, as more recently established state-owned enterprises are forced to meet the efficiency of private sector firms, we would expect to see smaller performance differentials. Thus, like the initial choices studied by Bamford et al. (1999) in their study of banks' initial choices and founding conditions, in the present study the choice of state versus private ownership is likely, along with environmental contextual variables, to significantly influence the firm's subsequent development.

Operating in a Government Supported Industry

The second firm level choice we examine is whether or not the firms have operations in a state-supported sector. Firms may choose the industries in which they operate either at startup or at some later point in their life cycle, although there is a tendency for the initial choices regarding industry of operation to influence subsequent choices regarding industry (for example, Porter, 1980).

In China a number of industries have, in recent times, been particularly promoted via state support. These so-called 'pillar' industries include machinery, electronics, petrochemicals, automobiles and construction (Pan et al., 1999, p. 91). Industry specific support began in the mid 1990s. All else being equal, we would expect firms operating in these pillar industries to face a more munificent environment and therefore to be able to outperform firms not operating in pillar sectors.

However we posit that founding conditions will differentially condition firms to take advantage of the munificence provided by state support. That is, firms operating in these pillar industries, but created prior to the introduction of state support, are likely to have developed routines and assumptions that make them less able to take advantage of state sectoral support than firms founded during or after the point in the mid 1990s when state support was introduced. As such, it is reasonable to expect a different relationship between state support and firm performance for firms founded prior to the introduction of support compared with firms founded later. In short, we expect that there is an interaction between founding conditions and the choice of operating in a government supported industry. Therefore:

Hypothesis 2: The magnitude of performance difference between firms in government supported industries and those in non-government supported industries is greater for firms founded later in the evolution of China's economy than for those founded earlier.

DATA, MEASURES AND METHOD OF ESTIMATION

Data

China, unlike many other economies, has a significant population of SOEs which were created at various stages of economic liberalization. To date, SOEs remain pervasive and their operations cut across all industries and span over many different product sectors. By any reasonable measure, state-owned firms have and continue to represent a colossal economic force within the Chinese economy.

Over the 25 years since China began to switch to a market oriented economy, however, private companies have rapidly emerged and have now overtaken SOEs, both in terms of numbers and in economic output. It should be noted, thought that the degree of private ownership and the contribution of private firms varies across product sectors and remains a function of government regulatory constraints. Similarly varied is the level of state support for certain sectors.

The data used in our study come from the Industrial Census of China conducted annually by the State Statistical Bureau of China (SSB). The census includes all firms operating in China, both domestic and foreign, that have annual sales revenues above US$ 25 000 (RMB 200 000 Yuan). In other words, it includes all manufacturing and service firms except very small businesses. For the reporting year 2002, after deleting extreme and missing values, we have for our analysis over 100 000 firm level observations spanning nearly 500 different product sectors. The database contains the following information on each firm: identification number, ownership type, product sector, age, geographic location, number of employees, revenues, profit, different types of assets, short-term and long-term debt, and costs. The database is a comprehensive and reliable data source; China's industrial census has been used in research published in leading academic journals, including the *Quarterly Journal of Economics* (Chow, 1993), *American Journal of Sociology* (Walder, 1995) and the *Journal of International Business Studies* (Buckley et al., 2002; Pan et al., 1999).

Measures

Performance

While there is some support for using a multidimensional construct to measure performance, there is no consensus as to what are the appropriate variables for performance, especially in the case of government or collective ownership where there is an implied or explicit multiplicity of objectives and goals. Given the nature of our data set and in keeping with previous studies, we choose to use the standard profitability measure or the enterprise returns on assets (ROA) as an indicator of performance. The ROA measure is calculated as profits divided by total assets, and the latest available and most complete data is for the reporting year 2002.

Ownership

The transition from a command to a market based economy has resulted in new forms of private enterprise ownership. In our study firms were classified into nine separate ownership structures, namely SOEs, privately owned enterprises, collective enterprises, foreign wholly owned enterprises, overseas Chinese wholly owned enterprises, foreign equity joint ventures, overseas Chinese equity joint ventures, foreign contractual joint ventures and, finally, overseas Chinese contractual joint ventures. We utilized a series of binary variables to capture this spectrum of ownership form that prevailed in China during the transitional period. In our definition of private firms we include wholly owned private firms and private cooperatives, and exclude private shareholding and listed firms. Using this strict definition of private ownership precludes us from examining the potential differences between shareholding and non-shareholding private firms. Nonetheless, it enables us to get around the problematic issue of dealing with 'privatized' SOEs.

Operating in government supported industry

Chinese economic development policy included governmental support for selected industrial sectors deemed to be the backbone of the economy and growth. A sector that received a great deal of policy support in the 1990s was the automobile sector. Other sectors that received preferential treatment in China include machinery, electronics, petrochemicals and construction (Pan et al., 1999). A government support variable was constructed as a binary variable, equal to one if the firm was operating in a supported sector, and zero otherwise.

Control Variables

To take into account the effects of the market environment on performance, in particular the degree of market competition, we use the traditional and well accepted measure – the Hirshman-Herfindahl (HHI) index. The primary focus is the impact of product sector concentration on performance, and whether state-owned firms and private firms' performance differential depends on the degree of market concentration. Again the empirical evidence on the influence of market concentration on performance is still inconclusive.

Two other measures of market rivalry included in our study are shares of total shipments accounted for by SOEs and by foreign firms. Presumably, the higher the shares accounted for by the government, the lower the market orientation of the sector and, thereby, the lower the overall level of performance. On the other hand, the higher the share of total shipments generated by foreign owned and/or affiliated firms, the higher the degree of foreign/market competition and, hence, the higher the performance of firms in general.

One product sector specific variable is also included as an additional determinant of performance. This is the rate of market growth, measured as the total product sector shipment growth over the period 1998 to 2002. Industry demand growth is added to the model to account for inter-industry differences in growth opportunities. Firms should find it easier to realize superior performance on average in markets where the demand is growing more rapidly. The existing literature has provided strong empirical support for this positive association.

We also wished to take into account other predictors of firm performance. We included in our model two financial measures – the debt to asset ratio and the working capital balance. There are many dimensions to the capital structure of SOEs and privately owned firms. The financial sector in China has historically been dominated by a few large state-owned banks. Traditionally, these banks provided debt financing to SOEs and were often pressured politically to lend to losing state enterprises (Perkins, 2001; Qian, 2001). Decades of government directed bank lending has resulted in state-owned and private enterprises having different terms and conditions governing their debt capital. Only recently has the banking sector been liberalized and alternative private banking institutions have been licensed to operate in China. While ideally in an empirical analysis, a distinction ought to be made regarding debt financed by the government or privately, we only have information on the firm's total debt level, and not on debt source. As such, the leverage variable is broadly defined as the ratio of a firm's total debt to total assets in the observed year.

In an earlier and highly cited paper, Altman (1968) identified working capital to total assets ratio as one of the major financial factors determining the likelihood of business success. Following Altman's work, it is assumed that working capital balance and performance should be positively correlated, all else being the same.

A number of possible measures can be used as proxies for a firm's ability to realize economies of scale and scope. The variables typically used in the literature are the firm's sales revenue or the firm's market share in a specified product sector. These variables are, however, highly correlated and we elected to use the firm's market share variable as it also provides a measure of relative market power. Market share in our study is measured as the sales volume of the firm relative to the sector total value of shipments. We expect a positive relationship between firm performance and market share.

The asset turnover ratio, defined as the ratio of sales to total assets, is included to account for a firm's resource deployment efficiency. The measure shows how efficiently firm's assets are being put to use to generate income. A high ratio compared with other firms in the same industry could indicate that the firm is working close to capacity and at a higher level of efficiency. We also employed two additional efficiency measures to capture firm's productivity, namely, the ratio of total employment to total assets and the ratio of total fixed assets to total assets. We would expect negative associations between performance and these factors (Pan et al., 1999).

We also entered a number of market specific factors designed to account for the effects of product sector maturity and minimum efficiency scale. We use, respectively, the mean age of firms and the average sales of firms for the product sectors in question.

Descriptive Statistics and Method of Estimation

Our data included 102 578 firms taken from 454 different manufacturing product sectors of the Chinese economy, each observed in the year 2002. The firms in our sample were founded in different years under different ideological and economic conditions (1979–86, 1987–96 and 1997–2002). Over the three sub-periods of interest there was a marked shift in the distribution of ownership. Between 1976 and 1986 there were 9225 enterprises created, of which nearly 44 per cent were private (not including stock limited or listed firms) and about 15 per cent were SOEs (Table 8.1). For the 45 079 firms founded between 1997 and 2002, the proportion of privately owned firms was over 68 per cent, while the proportion of state-owned firms created during this same period was slightly over 3 per cent. Among all the firms examined, about 39 per cent received government support, and this proportion was relatively stable over the three sub-periods.

The performance of firms, as measured by returns on assets, is also summarized in Table 8.1. The returns on assets on average were slightly higher for firms founded in the latter two sub-periods. It appears, however, that the average returns were much higher for private firms than the returns of state-owned firms founded across all three sub-periods. Generally, the returns on assets appear to be greater for firms operating in government supported sectors and this was consistent throughout the three sub-periods.

Other attributes of the firm and product sectors examined over the three periods are also presented in Table 8.1. On average the ratio of debt to assets tended to be higher for firms created during the first sub-period (66.42 per cent), and smaller amongst firms founded over the next two periods (58 per cent). Also, the average share of state-owned firms' contribution to product sector total shipments in 2002 was greater for those firms created in the first sub-period than for firms created during the subsequent periods. As for the environment within which the firms were operating, it would appear that firms created in the first sub-period operated in product sectors that experienced lower growth rates on average (57 per cent) than firms created between 1987–96 and 1997–2002 (63 per cent).

In order to test whether ownership and government support have statistically significant relationships with the performance of firms, and whether founding conditions as reflected in the period of founding significantly moderate these relationships, we specify and estimate a regression model:

$$Y_i = f(X_{1i}, X_{2i}, \ldots, X_{ni})$$

where Y_i denotes the performance of ith firm measured as the ith firm's return on assets in the year 2002 and $X_{1i}, \ldots X_{ni}$ denotes the first through

Table 8.1 Descriptive statistics

Variable	1976–86		1987–96		1997–2002	
Total number of enterprises founded	N = 9 225		N = 48 274		N = 45 079	
			Number (Proportion in %)			
Ownership						
Privately owned enterprise (PE)	4 023 (43.61%)		18 551 (38.43%)		30 774 (68.27%)	
State-owned enterprise (SOE)	1 408 (15.26%)		2 861 (5.93%)		1 500 (3.33%)	
Collectives (CE)	3 179 (34.46%)		8 285 (17.16%)		2 806 (6.22)	
Government Support						
Firms receiving government support (GOV)	3 655 (39.62%)		19 395 (40.18%)		17 504 (38.83%)	
			Mean (Standard Deviation)			
Returns on Assets (ROA)						
Returns on assets (ROA) for all firms	3.47% (7.49)		4.64% (8.59)		4.66% (9.15)	
Returns on assets (ROA) – private and state-owned firms	**Private** 4.32% (7.14)	**State-Owned** 0.12% (6.91)	**Private** 5.33% (7.94)	**State-Owned** 1.13% (7.54)	**Private** 4.61% (8.36)	**State-Owned** 0.55% (7.30)

Returns on assets (ROA) – by government support	With Support	Without Support	With Support	Without Support	With Support	Without Support
	4.15% (7.43)	3.03% (7.52)	5.32% (8.86)	4.16% (8.34)	5.46% (9.68)	4.16% (8.77)
Other Variables						
Debt to total assets (DTA)	66.42% (34.59)		57.72% (8.60)		58.08% (28.15)	
Private sector share of total product sector sales	7.67% (4.55)		7.53% (4.48)		7.67% (4.54)	
State sector share of total product sector sales (STMS)	10.54% (10.06)		7.85% (8.38)		8.35% (8.40)	
Foreign share of total product sector sales (FFMS)	26.29% (16.26)		33.19% (17.80)		31.35% (17.43)	
Product sector sales growth between 1998 and 2002 (GSales)	57.00% (40.58)		62.62% (60.08)		62.88% (41.25)	

Notes: All firms included are from the manufacturing sector (excluding tobacco products). Observations with missing variables and outliers have been excluded.

Source: Authors' findings; primary data source is the Statistical Bureau of China, Industrial Census (1998 and 2002).

*n*th factors, some of which are firm specific and others are product market specific. X_{1i} represents a vector of mutually exclusive binary variables indicating the type of ownership of the firm in question. There are nine different classes of ownership represented by eight dummy variables, the default status being state-owned firms. X_{2i} is the government industrial support binary variable, which is equal to one if the firm was operating in a supported sector, and zero otherwise. X_{9i} to X_{ni} are product sector specific and other firm specific covariates. The list of control variables included in the model is defined in Table 8.2, and correlation measures are presented in Tables 8.3a, 8.3b and 8.3c for founding periods 1979–86, 1987–96 and 1997–2002, respectively.

Included in the pooled regression model are interaction terms between founding period dummies and the private enterprise dummy, and between founding period dummies and the government support variable. If founding period conditions are to change the underlying relationships, then the coefficients of the interaction terms would be statistically significant, and this implication can be tested using the standard significance test.

Because our firm level observations were collected from 454 different product classes, the data can be said to be nested within product sectors, and the structure is therefore hierarchical. For hierarchically nested data, the conditionally independent observations assumption required for the Ordinary Least-Squares (OLS) regression does not apply, and the least-squares estimates of the standard errors are biased. Statistical models must take this into account, and tests of statistical significance of coefficients must rely on the asymptotically consistent variance-covariance estimates. The heteroscedasticity problem that arises as a result of observations coming from different product classes has been widely discussed, and a number of techniques appropriate for estimating heteroscedastic regression models have been proposed (White, 1980; Green, 2003).

One approach for dealing with nested hierarchical data and group-wise heteroscedastic residuals is to allow the regression coefficients to vary over product sectors. Each specific sector will have its own intercept and regression coefficients. A problem with specifying all coefficients in a model as random is that the model can become unstable and convergence criteria may not be realized. As such, we allow only the coefficients of the intercepts for each product sector and the coefficients of the private ownership variable, a key variable of interest, to vary across product sectors. Other covariates, both at the firm and sector levels, are assumed to have a fixed effect on performance. We also assume the model residual to be normally and independently distributed with mean zero and homoscedastic variance.

Table 8.2 A list of variables included in the multivariate regression model

Variable	Code	Definition
1. Return on Assets	ROA	= profit (amounts in RMB) over total assets (amounts in RMB) reported in the year 2002
2. Ownership	PE	= 1 if privately owned (defined as private firms – including wholly owned, private cooperative and excluding private shareholding and/or listed firms)
		= 0 otherwise
	SOE	= 1 if state-owned (defined as state wholly owned enterprise)
		= 0 otherwise
	CE	= 1 if the firm is classified as a Chinese collective enterprise
		= 0 otherwise
	FWHO	= 1 if the firm is a foreign wholly owned subsidiary
		= 0 otherwise
	HKWHO	= 1 if the firm is an overseas Chinese wholly owned subsidiary
		= 0 otherwise
	FEJV	= 1 if the firm is a foreign equity joint venture
		= 0 otherwise
	HKEJV	= 1 if the firm is an overseas Chinese equity joint venture
		= 0 otherwise
	FCJV	= 1 if the firm is a foreign contractual joint venture
		= 0 otherwise
	HKCJV	= 1 if the firm is an overseas Chinese contractual joint venture
		= 0 otherwise
3. Government Support	GOV	= 1 if the firm operates in a government supported 'pillar' sector (machinery, electronics, petrochemicals, automobiles and construction) as outlined in the article by Pan et al. (1999, p. 9)
		= 0 otherwise
4. State enterprise market share	STMS	= the share of total product sector shipment value accounted for by state-owned firms in the year 2002

Table 8.2 (continued)

5. Foreign firm market share	FFMS	= the share of total product sector shipment value accounted for by all foreign firms in the year 2002
6. Hirshman-Hirfindahl index	HHI	= sum of the square of individual firm's market share calculated in 2002
7. Product market growth	GSales	= growth rate of the product sector total sales revenue over the period 1998–2002
8. Debt to asset ratio	DTA	= the firm's total debt in 2002 divided by the reported total asset value in 2002
9. Working capital	WC	= the firm's total reported working capital divided by total assets of the firm in 2002
10. Turnover		= the firm's total sales over total assets in 2002
11. Inventory	Inv	= the firm's total fixed assets over total assets in 2002
12. Employment	Employ	= the firm's total employment over total assets in 2002
13. Firm market share	MS	= total sales revenue of the firm over the total product market sales revenue in 2002
14. Industry maturity	Mage	= average age of firms in each product class in the year 2002
15. Size of the industry	Msales	= mean value of total product sector sales revenue in 2002

Notes: FWHO (wholly owned foreign firm); HKWHO (wholly owned Honk Kong firm); FEJV (foreign equity joint venture firm); HKEJV (Hong Kong equity joint venture firm); FCJV (foreign contractual joint venture firm); HKCJV (Hong Kong contractual joint venture firm).

Source: State Statistics Bureau of China, Industrial Census (2002).

Table 8.3a Correlation matrix for selected variables (1979–86)

	ROA	PE	SOE	GOV	STMS	FFMS	HHI	GSales	DTA	WC	Turnover	Inv	Employ	MS	Mage	Msale
ROA	1.000	**0.102**	**−0.191**	**0.072**	**−0.056**	**0.050**	**0.029**	**0.059**	**−0.331**	**0.074**	**0.393**	**−0.094**	0.004	**0.065**	**−0.033**	**−0.076**
PE		1.000	**−0.383**	**.078**	**−0.081**	**−0.033**	−0.003	**0.048**	**−0.063**	0.004	**0.119**	**−0.045**	**−0.053**	−0.006	**−0.064**	0.000
SOE			1.000	**−0.078**	**0.205**	**−0.045**	**0.056**	**−0.059**	**0.119**	**−0.112**	**−0.209**	**0.080**	−0.019	0.004	**0.126**	**0.046**
GOV				1.000	0.013	**0.043**	**0.102**	**0.209**	**−0.057**	**0.227**	0.004	**−0.216**	**−0.070**	**0.053**	**0.067**	**0.051**
STMS					1.000	**−0.485**	**0.141**	**−0.202**	0.017	**−0.120**	**−0.088**	**0.135**	−0.008	**0.047**	**0.660**	**0.086**
FFMS						1.000	−0.010	**0.254**	**−0.077**	**0.118**	**0.061**	**−0.126**	−0.010	−0.010	**−0.512**	**0.134**
HHI							1.000	**0.108**	−0.016	**0.046**	−0.019	**−0.070**	−0.018	**0.431**	**0.066**	**0.092**
GSales								1.000	**−0.041**	**0.122**	**0.034**	**−0.107**	**−0.061**	**0.024**	**−0.264**	**0.131**
DTA									1.000	**0.022**	**−0.116**	0.010	**0.042**	**−0.030**	**0.028**	**0.059**
WC										1.000	**0.178**	**−0.581**	**0.030**	−0.004	**−0.061**	**−0.059**
Turnover											1.000	**−0.029**	**0.237**	**0.026**	**−0.112**	**−0.054**
Inv												1.000	**0.098**	**−0.039**	**0.093**	**0.030**
Employ													1.000	**−0.050**	**0.024**	**−0.127**
MS														1.000	0.011	**0.028**
Mage															1.000	**−0.148**
Msales																1.000

Notes: Bold indicates the correlation measure is significant at 95% or greater. See Table 8.2 for list of abbreviations.

Table 8.3b Correlation matrix for selected variables (1987–95)

	ROA	*PE*	*SOE*	*GOV*	*STMS*	*FFMS*	*HHI*	*GSales*	*DTA*	*WC*	*Turnover*	*Inv*	*Employ*	*MS*	*Mage*	*Msale*
ROA	1.000	**0.064**	**−0.103**	**0.067**	−0.007	−0.000	**0.015**	**0.034**	**−0.278**	**0.070**	**0.348**	**−0.103**	**0.011**	**0.061**	**0.030**	**−0.025**
PE		1.000	**−0.202**	**0.057**	**0.018**	**−0.131**	−0.005	0.000	**0.044**	**−0.019**	**0.133**	**−0.131**	**0.025**	**−0.048**	**0.034**	**−0.019**
SOE			1.000	−0.008	**0.209**	**−0.133**	**0.025**	**−0.037**	**0.127**	**−0.057**	**−0.127**	**0.030**	−0.001	−0.005	**0.156**	**0.011**
GOV				1.000	**0.109**	**−0.029**	**0.105**	**0.178**	−0.002	**0.162**	**−0.023**	**−0.156**	**−0.124**	**0.041**	**0.198**	**0.147**
STMS					1.000	**−0.538**	**0.110**	**−0.127**	**0.059**	**−0.097**	**−0.083**	**0.066**	**−0.073**	**0.038**	**0.651**	**0.117**
FFMS						1.000	**0.016**	**0.150**	**−0.100**	**0.118**	**0.044**	**−0.047**	**0.093**	**0.014**	**−0.503**	**0.171**
HHI							1.000	**0.114**	**−0.010**	**0.032**	**−0.023**	**−0.043**	**−0.046**	**0.468**	**0.067**	**0.172**
GSales								1.000	**−0.029**	**0.061**	**−0.010**	**−0.058**	**−0.072**	**0.027**	**−0.190**	**0.116**
DTA									1.000	**0.132**	**−0.048**	**−0.122**	0.006	**−0.024**	**0.049**	**0.031**
WC										1.000	**0.183**	**−0.632**	**0.018**	−0.002	**−0.052**	**−0.016**
Turnover											1.000	**−0.096**	**0.314**	**0.027**	**−0.090**	**−0.015**
Inv												1.000	**0.062**	**−0.013**	**0.024**	**0.011**
Employ													1.000	**−0.074**	**−0.048**	**−0.112**
MS														1.000	**0.034**	**0.046**
Mage															1.000	**−0.081**
Msales																1.000

Notes: Bold indicates the correlation measure is significant at 95% or greater. See Table 8.2 for list of abbreviations.

Table 8.3c Correlation matrix for selected variables (1997–2002)

	ROA	***PE***	***SOE***	***GOV***	***STMS***	***FFMS***	***HHI***	***GSales***	***DTA***	***WC***	***Turnover***	***Inv***	***Employ***	***MS***	***Mage***	***Msale***
ROA	1.000	−0.008	**−0.084**	**0.070**	**−0.058**	**0.065**	0.006	**0.040**	**−0.267**	**0.110**	**0.308**	**−0.087**	−0.001	**0.047**	**−0.027**	**−0.037**
PE		1.000	**−0.276**	−0.005	**0.063**	**−0.171**	**−0.019**	**−0.040**	**0.070**	−0.006	**0.094**	−0.004	**0.026**	**−0.058**	**0.064**	**−0.019**
SOE			1.000	**−0.019**	**0.152**	**−0.101**	0.015	**−0.041**	**0.060**	**−0.059**	**−0.114**	**0.079**	−0.002	0.007	**0.114**	**0.017**
GOV				1.000	**0.098**	0.014	**0.098**	**0.158**	0.006	**0.165**	**−0.036**	**−0.139**	**−0.104**	**0.044**	**0.208**	**0.137**
STMS					1.000	**−0.562**	**0.067**	**−0.156**	**0.032**	**−0.094**	**−0.112**	**0.101**	**−0.065**	**0.036**	**0.658**	**0.166**
FFMS						1.000	**0.020**	**0.195**	**−0.047**	**0.114**	**0.065**	**−0.124**	**0.049**	0.004	**−0.506**	**0.137**
HHI							1.000	**0.126**	**−0.019**	**0.021**	**−0.033**	**−0.028**	**−0.036**	**0.456**	**0.050**	**0.166**
GSales								1.000	**−0.030**	**0.040**	−0.001	**−0.052**	**−0.055**	**0.040**	**−0.237**	**0.104**
DTA									1.000	**0.233**	0.014	**−0.141**	**0.028**	−0.004	**0.040**	**0.021**
WC										1.000	**0.249**	**−0.681**	**0.049**	**0.010**	**−0.029**	**−0.016**
Turnover											1.000	**−0.136**	**0.236**	**0.013**	**−0.126**	**−0.034**
Inv												1.000	0.003	**−0.016**	**0.054**	0.005
Employ													1.000	**−0.053**	**−0.046**	**−0.088**
MS														1.000	**0.032**	**0.026**
Mage															1.000	**−0.047**
Msales																1.000

Notes: Bold indicates the correlation measure is significant at 95% or greater. See Table 8.2 for list of abbreviations.

RESULTS

Table 8.4 presents the results of the mixed model regression for the three distinct founding periods. The reported chi-square statistics indicate that the variability in the returns on assets of firms can be successfully explained by the independent variables included taken as a group as it is statistically significant at the standard acceptable confidence level of 95 per cent or higher. The model is also successful to some extent in disentangling the joint effect which these factors and the era of founding exert on the profitability of firms as demonstrated by the generally significant *t*-statistics.

Consistent with hypothesis 1, our results indicate that ownership type is important and that founding conditions have a moderating influence on the relationship between ownership type and performance. The interaction terms indicates that the magnitude of this performance differential in favor of private enterprises (PEs) is significantly smaller for firms founded later in the evolution of the Chinese economy, all else being equal. Findings from the pooled regression model indicate that there is a statistically significant decline in the performance gap between private and state-owned enterprises when we compare firms found in the first and the second sub-periods, as well as the second and third sub-periods. Specifically, private firms created earlier in the Chinese economic evolution would realize on average a return on assets which is about 1.58 per cent higher than that of state-owned enterprises created at the same time. Private firms created in the period 1987–96 would realize a smaller premium of about 1.18 per cent over their state-owned firms of the same vintage. Finally, private firms created in the period 1997–2002 would observe a much smaller margin of about 0.30 per cent over the state-owned firms found at the same time.

This finding provides some support for the hypothesis that condition at the time of firm founding leaves long lasting effects on firms' performance. That is, the performances of SOEs appears to be closer to that of PEs as the operating conditions at the time of firm founding became more competitive and more market oriented. By the same token, it would appear that SOEs created in a more liberalized environment must contend with market competitive forces much more than their counterparts which were created earlier under a more governmentally supportive and regulated situation, as argued by Ralston et al. (2006).

Next consider the effect of the Chinese industrial support policy. We found that the marginal effect of the Chinese industrial support variable was statistically and significantly larger for firms founded within the latest economic period than those founded earlier. Consistent with hypothesis 2, it appears that firms found in the latter periods of the Chinese economic

Table 8.4 Fixed effects regression model: determinants of firm profitability (1979–2002)

Independent Variable	**Pooled**
Hypothesis 1: Ownership Variables	
Private Enterprise (PE)	1.5798[b]
	(8.22)
PE * D1	−0.3977[a]
	(−2.18)
PE * D2	−1.2794[b]
	(−6.84)
Hypothesis2: Government Support	
Government Support (GOV)	0.0921
	(0.44)
GOV * D1	0.5309[b]
	(2.92)
GOV * D2	0.8577[b]
	(4.67)
Covariates	
Collective Enterprise (CE)	1.7979[b]
	(15.35)
Foreign Wholly Owned	0.7637[b]
	(4.88)
OS-Chinese Wholly Owned	−0.8931[b]
	(−6.22)
Foreign Equity JV	1.4686[b]
	(10.48)
OS-Chinese Equity JV	0.2922[a]
	(2.13)
Foreign Contractual JV	1.4353[b]
	(4.78)
OS-Chinese Contractual JV	−0.7038[b]
	(−3.24)
State Enterprise Mkt Share (STMS)	0.4310
	(0.56)
Foreign Firm Mkt Share (FFMS)	1.8575[b]
	(4.52)
Hirshman-Herfindahl Index (HHI)	−5.9678[b]
	(−4.05)
Product Market Growth (GSales)	0.2544[a]
	(2.31)
Debt to Asset Ratio (DTA)	−0.0791[b]
	(−93.88)
Working Capital (WC)	0.0044[b]
	(2.97)

Table 8.4 (continued)

Independent Variable	**Pooled**
Turnover	0.0417[b]
	(111.00)
Inventory (Inv)	−0.0203[b]
	(−18.29)
Employment (Employ)	−0.3335[b]
	(−22.04)
Firm Mkt. Share (MS)	27.0084[b]
	(13.39)
Industry Maturity (Mage)	0.0879[b]
	(4.08)
Industry Size (Mean Industry Sales Revenue: Msales)	−2.0E–6[b]
	(−6.51)
Constant	2.6937[b]
	(7.03)
D1	0.3203[a]
	(2.28)
D2	0.3890[a]
	(2.57)
N, AIC	102 578, 710 960
Chi-Square LR Test	1 254[b]

Notes: [a]$p<0.05$; [b]$p<0.01$; *t* statistics in parentheses.

Source: Authors' findings; primary data source is the Statistical Bureau of China, Industrial Census (1998 and 2000).

evolution were able to make better use of the available governmental sectoral support, other things being equal.

The coefficients of the other variables are generally as expected and are consistent with findings in prior literature (Pan et al., 1999). As shown in Table 8.4, the efficiency variables as measured by the turnover ratio and the market share of the firm have a positive and significant effect on performance. Fixed asset to total asset and total employment to total asset ratios both show significant and negative impact on performance as expected.

Regression results indicate that firms' profitability was generally higher where the market was relatively more mature and lower where the market average scale of operation was higher. As expected, the market share variable has a highly significant and positive effect on profitability. As for the degree of rivalry, as indicated by the product sector specific Hirshman-Herfindahl index, higher performance appears to be associated with a higher degree of competition within the product space. The share of market sales accounted for by foreign firms has a positive and significant effect on profitability, while the share of market sales accounted for by SOEs had no discerning effects on performance, controlling for other factors.

Firms performed significantly worse when they were burdened with a higher debt load. Unfortunately, our database does not distinguish between different sources of debt, and we cannot test directly whether government sponsored debt and private commercial debt has the same influence on firm performance. It could also be argued that performance and debt to asset ratio may be jointly determined and that a single equation model will result in biased estimates. However it is plausible that given the pervasiveness of relationship based lending in China, lending may not be as dependent on performance as it would elsewhere, and that the direction of causality would more likely be from the leverage ratio to performance. The effect of working capital balances on performance is positive and statistically significant as expected.

Several different specifications of the main profitability model were tested. We included in the above model the age of the firm as an additional explanatory variable to take into account possible firm specific maturity or experience effects, and the results on key variables of interest are consistent, and parameter estimates are not significantly different from those of the main model. Allowing the time frame to vary by a year on either side did not significantly change the result. Consistently, the most important set of determinants are the ones presented here. Overall, it would appear that recently created state-owned enterprises seem to more closely mimic the operations of private firms. They typically have a greater managerial prerogative and, in most cases, investment decisions of state-owned firms founded recently are no longer driven predominantly by government non-commercial objectives.

DISCUSSION AND FUTURE RESEARCH

Although our findings must be regarded as speculative in nature owing to limitations of our data set, the empirical results indicate that prevailing macroeconomic, political, regulatory and institutional factors at the time of founding seem to have a prolonged effect on firm performance. Institutional drivers at different time periods may have a positive or a negative effect on a firm's performance, depending on how they interact with the firm's strategic choices. We show that there is a significant difference in the performance differential between private and state-owned firms across founding periods, and that the decline in the performance advantage of private firm is statistically significant. The implication of this finding, although preliminary, is that the superior effect associated with private ownership is moderated significantly by the environmental conditions at the time of founding. Similarly, government support positively affects returns on assets, and the magnitude of the effect appears to be a function of the initial conditions at the time support recipients were created.

Our study serves to highlight the need for entrepreneurship research to take into account changing contextual factors at the time of founding when examining drivers of performance, survival and viability of entrepreneurial firms and startup organizations. Such considerations are particularly likely to be critical in contexts where environmental circumstances and macro-institutional conditions are changing rapidly and profoundly. In China, as in many other dynamic transitional economies, political, social and regulatory structures have changed dramatically over the last few decades. There have been several major shifts in the political and economic philosophy within these economies, and subsequently, changes in the legal infrastructure, bureaucracy and market liberalization. As argued in the earlier sections of this chapter, strategic choices that an organization selects under prevailing institutional regimes, although appropriate at the time, may indeed be sub-optimal and ineffective under an alternate regime. We surmise further that ostensibly optimal choices adopted at the time of founding can in fact be liabilities for the firm under different environmental and institutional circumstances.

This study makes contributions to both the literature on emerging economies and the literature on founding conditions as they affect startups. With regards to the former, it suggests – though more evidence in support of this view is definitely required – that the social, political and economic conditions that comprise an emerging economy are likely to shape the future course of firms founded at specific points in time. Firms doubtless vary in their dexterity and ability to adapt to changes in economic conditions, but there will be a general tendency for firms born into particular founding conditions to be a product of those founding conditions. Those who manage firms in such economies and those who are engaged in transforming policies in such economies need to be sensitive to the imprinting effect of founding conditions on firms that are started in different politico-economic eras.

Complementing and extending the existing literature concerning the effect of founding conditions, this study suggests that economic policies in place at the time of founding, in interaction with firm level choices, may have a lasting impact on firms. Prior empirical work by evolutionary economists, while theoretically consistent with the viewpoint elaborated in this chapter, has tended to focus within industries on the effects that firm experience (and presumably initially institutionalized routines) prior to market entry on post-entry size, strategy and survival (for example, Klepper and Simons,

2000; Klepper, 2002; Mitchell, 1991; Tripsas and Gavetti, 2000). This study argues and offers preliminary evidence that across industries, the macro-institutional context, coupled with firm level choices, has an imprinting effect that shapes future performance. Both the choice of type of ownership and the choice of whether to operate in a state supported industry would appear to interact with the environmental conditions in place at the time of founding to influence later outcomes. While more research is needed on a wider range of economic contexts and initial choices, this research suggests that researchers and managers take into account the pervasive economic and political assumptions that are in place when firms are founded if they are to fully understand the future performance of firms.

Naturally, this study has limitations which suggest these conclusions should be regarded as tentative. The cross-sectional nature of the data set used for this research, and the limited number of variables available, considerably constrain the extent to which hypotheses regarding founding conditions and initial choices can be fully tested.

Because of these limitations, several potential biases arise. One possible bias with using the Chinese Statistical Bureau 2002 database is that it includes only those enterprises that were incorporated or created during the three identified periods which are 'still in operation' in 2002. That is, there is an inherent sample selection bias due to the fact that only surviving entrants, and to some extent relatively more successful firms, are included in the study. Nonetheless, the sample allows us to compare the impact of founding conditions on surviving firms, which is consistent with the theoretical focus of this investigation.

Ideally, we would estimate the above model comparing the performance of firms in the same time window (measured relative to the year of founding) for each of the three sub-periods of interest to account for possible bias related either to the liabilities of newness and/or other experience and age related factors. However, in order to do this we would need annual data for the years 1979 to 2002. This is problematic as the industrial census started in China much later than the economic reform of China. To partly address the bias problem from using a single year performance measure for all firms, we incorporated the variable of industrial maturity – the mean age of firms in the industry.

In addition, it may also be argued that lagged measures of certain explanatory variables rather than current values should be used in the firm performance model. Using current as opposed to lagged measures poses an econometric problem to the extent that the explanatory variables are deemed to be stochastic and contemporaneously related to the disturbance term. In this case, least squares estimators are inconsistent. Unfortunately, we are limited to the 2002 industrial census data and must utilize current values, and if these independent variables are correlated with the disturbance term, our estimates would be unreliable.

A number of extensions suggest themselves for future research. Clearly, institutional conditions and the strategic choices that firms make at the time of founding are important, and variables capturing these factors need to be specifically incorporated into future models. A natural extension would be to examine the underlying relationship of these factors in a way that allows for the assessment of performance using the same time window since founding, and by using several performance indicators.

An econometric complication which typically arises in this type of analysis is that one or more of the explanatory variables may be endogenous. As such, future research may usefully model the underlying structural relationship as a system of simultaneous equations, and take into account the possible endogeneity bias. Another extension would be to examine how commercial performance of privatized state-owned firms may be affected by founding conditions, and to what extent the efficacy of government support measures is determined by firms' initial operational choices and their surrounding institutional governance. This, however, requires more in-depth information to be gathered on certain firms' strategic dimensions and institutional forces at the time of founding as well as current factors affecting performance. Insights from this line of research may be highly relevant to policy makers, managers and business practitioners in emerging or transitional economies.

We recognize that in some instances the assumptions made in the analysis are dictated by the availability of information and the nature of the data used, and this in turn may have a significant impact on our results. Notwithstanding these limitations, this study provides a valuable theoretical perspective regarding how to study entrepreneurship in emerging economies, and makes a complementary contribution to the literature on the impact of founding conditions.

REFERENCES

Aldrich, H. (1999), *Organizations Evolving*, Thousand Oaks, CA: Sage Publications.

Altman, E.I. (1968), 'Financial ratios, discriminant analysis and the prediction of corporate bankruptcy', *Journal of Finance*, **23** (4), 589–609.

Bamford, C.E., T.J. Deans and P.P McDougall (1999), 'An examination of the impact of initial founding conditions and decisions on the performance of new bank start-ups', *Journal of Business Venturing*, **15**, 253–77.

Bartel, A.P. and A.E. Harrison (2005), 'Ownership versus environment: disentangling the sources of public-sector inefficiency', *Review of Economics and Statistics*, **87** (1), 135–47.

Berger, A.N. and E. Bonaccorsi di Patti (2006), 'Capital structure and firm performance: a new approach to testing agency theory and an Application to the Banking Industry', *Journal of Banking and Finance*, **30** (4), 1065–102.
Boeker, W. (1989), 'Strategic change: the effects of foundings and history', *Academy of Management Journal*, **32** (2), 489–515.
Buckley, P., J. Clegg and C. Wang (2002), 'The impact of inward FDI on the performance of Chinese manufacturing firms', *Journal of International Business Studies*, **33** (4), 637–56.
Carroll, G.R. and M.T. Hannan (1989), 'Density delay in the evolution of organizational populations: a model and five empirical tests', *Administrative Science Quarterly*, **34**, 411–30.
Chandler, A. (1962), *Strategy and Structure*, New York: Doubleday.
Chow, G.C. (1993), 'Capital formation and economic growth in China', *Quarterly Journal of Economics*, **103** (3), 809–42.
Davies, H. and P. Walters (2004), 'Emergent patterns of strategy, environment and performance in a transition economy', *Strategic Management Journal*, **25** (4), 347–64.
Dewenter, K.L. and P.H. Malatesta (2001), 'State-owned and privately owned firms: an empirical analysis of profitability, leverage, and labor intensity', *American Economic Review*, **91** (1), 320–34.
DiMaggio, P. and W. Powell (1983), 'The iron cage revisited: institutional isomorphism and collective rationality in organizational fields', *American Sociological Review*, **48**, 147–60.
Eisenhardt, C. and C.B. Schoonhoven (1990), 'Organizational growth: linking founding team, strategy, environment, and growth among U.S. semiconductor ventures, 1978–1988', *Administrative Science Quarterly*, **35**, 504–29.
Freeman, J. and W. Boeker (1984), 'The ecological analysis of business strategy', *California Management Review*, **26** (3), 73–110.
Green, W.H. (2003), *Econometric Analysis*, 5th edn, Englewood Cliffs, NJ: Prentice Hall.
Kanamori, T. and Z. Zhao (2004), *Private Sector Development in the People's Republic of China*, Tokyo: Asian Development Bank Institute.
Klepper, S. (2002), 'The capabilities of new firms and the evolution of the US automobile industry', *Industrial and Corporate Change*, **11**, 645–66.
Klepper, S. and K. Simons (2000), 'Dominance by birthright: entry of prior radio producers and competitive ramifications in the U.S. television receiver industry', *Strategic Management Journal*, **21**, 997–1016.
Lin, Justin Yifu, Fang Cai and Zhou Li (1996), 'The lessons of China's transition to a market economy', *Cato Journal*, **16** (2), 201–31.
Luo, Yadong (1999), 'Environment-strategy-performance relations in small businesses in China: a case of township and village enterprises in Southern China', *Journal of Small Business Management*, **37** (1), 37–52.
McDougall, P., J.G. Covin, R.B. Robinson Jr. and L. Herron (1994), 'The effects of industry growth and strategic breadth on new venture performance and strategy content', *Strategic Management Journal*, **14**, 137–53.
Miller, D. and P. Friesen (1984), *Organizations: A Quantum View*, Englewood Cliffs, NJ: Prentice-Hall.
Mitchell, W. (1991), 'Dual clocks: entry order influences on incumbent and newcomer market share and survival when specialized assets retain their value', *Strategic Management Journal*, **12**, 85–100.

Nelson, R. and S. Winter (1982), *An Evolutionary Theory of Economic Change*, Cambridge, MA: Harvard University Press.
Oliver, C. (1997), 'Sustainable competitive advantage: combining institutional and resource based views', *Strategic Management Journal*, **18** (9), 697–713.
Pan, Y., S. Li and D.K. Tse (1999), 'The impact of order and mode of market entry on profitability and market share', *Journal of International Business Studies*, **30** (1), 81–104.
Park, S.H., S. Li and D.K. Tse (2006), 'Market liberalization and firm performance during China's economic transition', *Journal of International Business Studies*, **37**, 127–47.
Perkins, D.H. (2001), 'Industrial and financial policy in China and Vietnam: a new model or replay of the East Asian experience', in J.E. Stiglitz and S. Yusuf (eds), *Rethinking the East Asian Miracle*, New York: Oxford University Press, pp. 247–94.
Porter, M. (1980), *Competitive Strategy*, New York: Free Press.
Qian, Y. (2001), 'Government control in corporate governance as a transitional institution: lessons from China', in J.E. Stiglitz and S. Yusuf (eds), *Rethinking the East Asian Miracle*, New York: Oxford University Press, pp. 295–322.
Ralson, D., J. Terpstra-Tong, R. Terpstra, X. Wang and C. Egri (2006), 'Today's state-owned enterprises of China: are they dying dinosaurs or dynamic dynamos', *Strategic Management Journal*, **27**, 825–43.
Ramaswamy, K. (2001), 'Organizational ownership, competitive intensity, and firm performance: an empirical study of the Indian manufacturing sector', *Strategic Management Journal*, **22**, 989–98.
Selznick, P. (1957), *Leadership in Administration*, New York: Harper & Row.
Stinchcombe, Arthur L. (1965), 'Social structure and oganizations', in James G. March (ed.), *Handbook of Organizations*, Chicago, IL: Rand McNally, pp. 142–93.
Swaminathan, A. (1996), 'Environmental conditions at founding and organizational mortality: a trial-by-fire model', *Academy of Management Journal*, **39** (5), 1350–78.
Tan, J. (2005), 'Venturing in turbulent water: a historical perspective of economic reform and entrepreneurial transformation', *Journal of Business Venturing*, **20**, 689–704.
Tan, J. and D. Tan (2003), 'Maximizing short-term gain or long-term potential: entry, growth, and exit strategies of Chinese technology start-up', *Journal of Management Inquiry*, **13** (1), 1–7.
Tolbert, P. and L. Zucker (1983), 'Institutional sources of change in the formal structure of organizations: the diffusion of civil service reform, 1880–1935', *Administrative Science Quarterly*, **28**, 22–39.
Tripsas, M. and G. Gavetti (2000), 'Capabilities, cognition, and inertia. Evidence from digital imaging', *Strategic Management Journal*, **21** (10–11), 1147–61.
Walder, A. (1995), 'Local governments as industrial firms: an organizational analysis of China's Transitional Economy', *The American Journal of Sociology*, **101** (2), 263–301.
White, A. (1980), 'A heteroscedasticity-consistent covariance matrix estimator and a direct test for heteroscedasticity', *Econometrica*, **41**, 817–38.

9. The entrepreneurial role of border traders in Laos and Thailand

Edward Rubesch

INTRODUCTION

Prahalad (2005) suggests there is a 'fortune at the bottom of the pyramid' for firms who can reach the market comprised of the 4 billion people living in developing countries on less than US$ 2 per day. Indeed, many multinational firms face stagnant sales in mature markets, and supplying products to emerging markets in developing countries represents the largest potential for future sales growth (Czinkota and Ronkainen, 1997). However, while firms recognize the potential of emerging markets in developing countries, these markets also present many challenges. Distribution infrastructure is often limited (Arnold and Quelch, 1998). At the same time, using intermediaries can be risky as firms are uncertain about how well their distributors will honor agreements, or it may be difficult to monitor channel partners to ensure they are not also participating in gray marketing or counterfeiting of the firm's branded products (Li, 2003). Emerging markets often have underdeveloped legal systems, with inconsistent law enforcement. Government officials may participate in illegal activities to such an extent that those activities become accepted as 'normal', and engaging in such activities may, in fact, be a requirement in order to do business (Reid et al., 2001). According to Prahalad (2005), 'entrepreneurship on a massive scale is the key' (p. 2) to overcoming barriers to doing business and unlocking the potential of emerging markets in developing countries.

This research investigates the activities of one type of entrepreneur who operates in an emerging market: traders on the Thailand-Laos border who travel to Thailand on a daily basis, and buy branded household consumer products to re-sell in and around Vientiane, the capital city of Laos. The fact that these border traders exist is easily observable to anyone who spends time at one of the border crossings near Vientiane. Numerous traders travel with large quantities of instant noodles, snack items, cooking oil and other ingredients for food preparation, soap and shampoo, and other convenience products that are used on a daily basis. A more perplexing question, and one

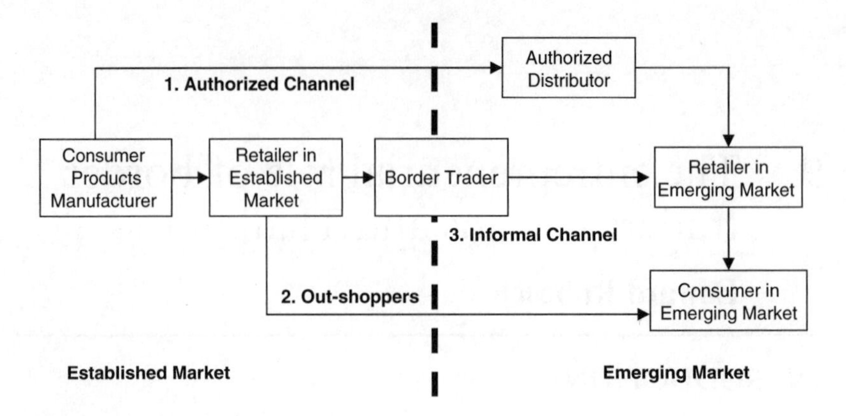

Source: The author.

Figure 9.1 Three product flows into emerging markets

that motivated this research, is, with the authorized distributors serving the market, why do the border traders exist at all?

Three flows of consumer products into the Vientiane market can be identified (Figure 9.1). First, the border is relatively easy to cross for those who have the wherewithal and inclination, and numerous Lao residents travel in their own vehicles to the adjoining provinces in Thailand to buy household goods. Second, multinational corporations have set up authorized distributors to supply the Vientiane market. These distributors import and stock products, run promotional programs and build a market share for the brands they represent. Finally, border traders cross the border to buy consumer products from retailers in Thailand to supply shops and mini-marts in and around Vientiane.

A prevailing view in the literature is that a border trader's opportunity comes from either avoiding duties (Gillespie and McBride, 1996) or free-riding on the efforts of authorized distributors (Cespedes et al., 1988). These reasons, however, are not sufficient to explain the situation on the Thailand-Laos border. Border traders must endure numerous hassles to ply their trade. Crossing the border itself involves a level of bureaucratic annoyance. The border traders studied in this research do not own their own cars, and must change between various modes of transport to get to Thailand and back, carrying their goods with them along the way. Duties average only 10 per cent for most convenience goods traded between Laos and Thailand (ASEAN, 2006). Given the fact that the border traders' buy from retailers in Thailand, it is hard to understand how they can compete with authorized distributors who receive trade discounts to import and sell products in Laos.

One possible explanation could be the fact that a developing country like Laos, with a large percentage of the population living in economically disadvantaged conditions, has an ample supply of laborers willing to undertake the job in order to improve their situations. Yet, if this were true, it would seemingly preclude the existence of the authorized channel, which would be unable to compete with large numbers of people who are willing to work for meager profits.

What cannot be explained by these hypotheses is the coexistence of both the authorized channel and the informal channel of border entrepreneurs. This research, therefore, attempts to answer the question: In what ways does an informal channel of border traders provide value to customers in an emerging market, and how does it relate to authorized distributors?

THEORETICAL BACKGROUND

Entrepreneurial Opportunities from Supplying Emerging Markets

Entrepreneurship is the combination of both a lucrative opportunity and an enterprising individual with the right capabilities to exploit it (Venkataraman, 1997). The academic field of entrepreneurship 'involves the study of sources of opportunities; the processes of discovery, evaluation, and exploitation of opportunities; and the set of individuals who discover, evaluate, and exploit them' (Shane and Vekataraman, 2000, p. 218). The border trader recognizes opportunities that are created by asymmetrical flows of people, goods and information across international borders, and calls upon a range of skills to exploit them.

Border traders introduce new products to a market, or find new sources of supply, and in this way they are 'carrying out new combinations' (Schumpeter, 1934, p. 66). They facilitate exchanges that would not otherwise occur, by finding buyers who are willing to pay a premium for products that are not otherwise available in the marketplace. In doing so, these border traders gain an entrepreneurial profit (Kirzner, 1973).

However there must be some barrier, which keeps buyers and sellers apart, that gives the entrepreneur their opportunity. Burt (1992, p. 18) devised the concept of a 'structural hole', a barrier or insulator between parties that prevents them from interacting directly. The entrepreneur fills the structural hole and becomes a broker of information, which allows the interaction to take place:

> The information benefits of access, timing, and referrals enhance the application of the [structural hole] strategy . . . . Having access to information means being

able to identify where there will be an advantage in bringing contacts together and is the key to understanding the resources and preferences being played against one another. (Burt, 1992, pp. 33–4)

Barriers come in many forms. For border traders, the barrier, or structural hole, that keeps buyers and sellers apart is formed by the border itself, and the bureaucratic burdens of passing through border posts, doing paperwork and having it checked, paying duties and other fees and the other associated tasks of crossing an international boundary. In addition, in emerging markets, transportation infrastructure is often underdeveloped, increasing barriers to the flow of people, goods and information, making the structural hole bigger.

Nijkamp et al. (1990) developed a typology of barriers to reflect the way different flows might be obstructed, including:

- physical barriers
- congestion barriers
- fiscal barriers
- institutional barriers
- technical barriers
- market regulation barriers
- time-zone differences
- cultural, language and information barriers.

Physical barriers may be natural, such as mountains and rivers, or manmade, like the Berlin Wall. Congestion barriers result from limitations in infrastructure, such as insufficient roads, telephone systems or Internet nodes. Fiscal barriers add costs to trade flows through the use of visas, import duties or tariffs. Institutional barriers relate to the costs of traveling from one jurisdiction to another, and may be in the form of border crossing paperwork (and the delays in preparing it) or in the fluctuation of currencies, which might require hedging. Institutional barriers also include regulations that vary from country to country, such as accounting practices. Technical barriers are similar to institutional barriers, in that they involve compliance with industrial standards, which may vary in different areas, and may require testing to prove conformance. Similarly, they include technological differences in cell phone protocols, or different railway gauges, causing delays in information flows and transportation. Cultures and languages can also be barriers, and do not necessarily coincide with national borders, since people on either side of a border may have more in common with each other than with their compatriots in other parts of the same country. In addition, each of these barriers may be asymmetric in nature, allowing flows in one direction more than the other. For example, fiscal barriers, such as duties or tariffs, are regularly used to protect local industries or to promote domestic policies, such as the attempt to curb cigarette smoking in the UK with higher excise taxes.

Borders differ in their permeability, or in the ease with which people and goods flow across them. While policies set by countries sharing a common border affect its permeability, as a practical matter no border can be completely closed. As Anderson (1996) pointed out, even the most tightly controlled borders allow a small number of exchanges to occur: 'The completely closed frontier has always been an aspiration and, except for brief periods, during wars or other exceptional circumstances, it has scarcely existed. Transfrontier transactions and individual flight across the frontier occur despite the policies of authoritarian regimes' (p. 6). Readings in the popular press provide interesting examples. For instance, North Korea maintains one of the world's most tightly controlled borders along the Yalu River, which separates the country from China, and yet

> here and there, shadowy figures can be seen on both sides of the misty river quietly carrying out an illegal – but thriving – trade in women, endangered species, food and consumer appliances that makes a mockery of North Korea's reputation as a tightly controlled and internationally isolated state. (Watts, 2004, p. 17)

Instead, a less permeable border merely results in increased cost and time required for goods and people to cross it. Losch (1954) viewed international boundaries as discontinuities, which, in effect, increase the artificial distance between two locations on either side. Borders that are relatively more closed, or which have more impediments to their crossing, increase the apparent overall distance. Banomyong (2000) operationalized this by analytically comparing time versus distance, and cost versus distance, for multimodal freight shipments from Southeast Asia to Europe. Overall travel time and overall cost were used as variables to measure the artificially expanded distance caused by a variety of barriers, including a shortage of skilled labor, underdeveloped infrastructure, onerous regulations or the need to pay bribes to local officials (Banomyong, 2000).

For the entrepreneur who is able to do so, overcoming barriers at international borders represents a value-added activity. Crossing such barriers incurs cost, which may be prohibitive. However, if there are customers willing to pay the price, and there are no other available channels, an exchange occurs and value is created. Viewing market entry as value creation is the key to understanding the role of entrepreneurial border traders. In an example from the criminal world, Naim (2005) argued that the ability to add value by crossing the border was exactly what led to the shift in power from Columbians to Mexicans in the supply of illicit drugs to the USA:

> [Mexican organizations] possess the most enviable situational advantage of all: territorial control of the approaches to the U.S. border, the single most lucrative bottleneck in the drug supply chain, the point where the most value is added. (p. 75)

However, while the activities of entrepreneurial border traders add value, they are rarely captured in official statistics. The fields of sociology and anthropology have developed a rich literature of the activities of entrepreneurs in developing regions, especially those that go unrecorded as part of the informal economy. While a range of occupations has been attributed to the informal sector, an important distinction is drawn between illegal activities, which produce illicit goods, and informal activities, which largely involve legal goods supplied outside the frameworks provided by governments or authorities. This distinction is especially necessary given that social norms are locally defined, and what is illegal differs greatly across cultures and nations (Portes and Haller, 2005, p. 405). Castells and Portes (1989) define three economic types: criminal, formal and informal, which are determined by a combination of two characteristics: (1) the nature of the final product, and (2) the way it is produced and distributed. The formal economy is made up of legal products, which are produced and distributed according to established laws and regulations. Criminal activities deal with illegal products, which are produced and distributed illicitly. Informal activities, finally, cover the supply of legal goods and services, delivered through channels that are not legally established. Staudt's (1998) research on the US-Mexican border contrasted crime, such as the drug trade, which harms people, with informal activity where 'while the means of informality do not comply with all regulations, its ends (goods and services) are legitimate' (p. 19).

CHARACTERISTICS OF THE BORDER ENTREPRENEUR

Not everyone who recognizes an opportunity necessarily becomes an entrepreneur. As Burt (1992) pointed out,

> behavior of a specific kind converts opportunity into higher rates of return. The information benefits of structural holes might come to a passive player, but control benefits require an active hand in the distribution of information. Motivation is now an issue. Knowing about an opportunity and being in a position to develop it are distinct from doing something about it. . . . You enter the structural hole between two players to broker the relationship between them. Such behavior is not to everyone's taste. A player can respond in ways ranging from fully developing the opportunity to ignoring it. (p. 34)

In developing countries realizing profit potential depends on the entrepreneur's ability to manage risk and hassles (Fadahunsi and Rosa, 2002). The management of risk is an intrinsic characteristic of entrepreneurship, albeit a somewhat controversial one. Schumpeter was emphatic that the entrepreneur never took risks, the burden falling, instead, on whoever provided credit or capital for the undertaking (Schumpeter, 1934). A more recent view (Shapero, 1985) is that while an entrepreneur takes risks, he or she 'sees the odds being affected by his or her knowledge, personal intelligence, creativity, dedication and persistence' (p. 3). The successful entrepreneur, therefore, not only willingly accepts risks, but also finds ways to minimize them. For border traders, risk includes an additional dimension: there is no overriding authority to which one can turn to settle disputes and seek redress of grievances. 'Because the rules of behavior in cross-border trade are ambiguous and formal state institutions are frequently not in a position to enforce what rules there are, the possibilities for real or perceived malfeasance on the part of either or both parties to a transaction are great' (Spener, 1995, p. 21). Authority ends on either side of the border. Those whose activities straddle the border cannot rely solely on legal safeguards to protect their interests. Part of the risk is the amount of effort required to overcome hassles. While the term 'hassles' perhaps lacks technical elegance, it nonetheless captures accurately the nature of the multitude of small, persistent irritations that the entrepreneur in an emerging market faces on a regular basis, including bureaucratic red tape or poor transportation infrastructure. As Staudt (1998) observed,

> whichever side of the border they work on, self-employed informals are the quintessential free traders. They distance themselves from regulatory embrace. They do not burden state-subsidized credit, for start-up capital comes from their own savings or that of relatives. What are informals' special skills? They are able to maneuver on both sides of the border, without calling official attention to themselves, using uneven enforcement of policies and regulations to lower their business costs. (p. 73)

Border traders who have the special skills to effectively manage both risk and hassles will be more successful. One way entrepreneurs acquire such special skills is by observing and learning from other successful entrepreneurs (Minniti and Bygrave, 1999; Bygrave and Minniti, 2000). Border traders who have access to role models, such as friends or family, who have already succeeded in the occupation, will have advantages in finding ways to manage risk and hassles. Moreover entrepreneurs often do not own or control all of the necessary resources needed to successfully capitalize on a particular opportunity. Instead 'most of these resources have to come from other people and institutions. Thus, the entrepreneur has to assemble, organize and execute the market development and value-chain infrastructure before potential profits can be realized' (Venkataraman, 1997, p. 125). Like all entrepreneurs, border traders who can build networks of supporting individuals will increase their chances of success.

THEORETICAL FRAMEWORK

The preceding discussion leads to the development of a theoretical framework to explain the role of border traders who supply emerging markets (Figure 9.2). The border trader's opportunity comes from overcoming barriers on and around the international borders of developing countries that contain emerging markets. The barriers can come in a variety of forms, and include stifling bureaucratic processes, or underdeveloped transportation or communications infrastructure. These barriers lead to asymmetries in the flows of people, products or information that prevent all participants from benefiting equally.

The border trader calls upon a number of entrepreneurial capabilities to take advantage of these opportunities once they are recognized. They can develop ways of overcoming the risks and hassles associated with border trade. These methods include developing creative solutions, but border traders also rely on learning from others, such as relatives or friends who already participate in the occupation, or building networks which provide resources that the entrepreneur does not directly possess.

RESEARCH DESIGN

The Setting of the Research

This framework applies to a case, studied by the author, involving border traders who purchased products from border provinces in Thailand, and supplied Vientiane, the capital city of Laos. This geographical area was selected because the metropolitan area in and around Vientiane represents an emerging market of nearly 1 million people lying in close proximity to Thailand, an established consumer market.

While Laos has little manufacturing capability, a number of authorized distributors supply consumer products manufactured in the neighboring countries of Thailand, Vietnam and China. Residents of Vientiane, seeking products not available locally, also travel to shop in the nearby provinces of Udorn Thani and Nong Khai in Thailand.

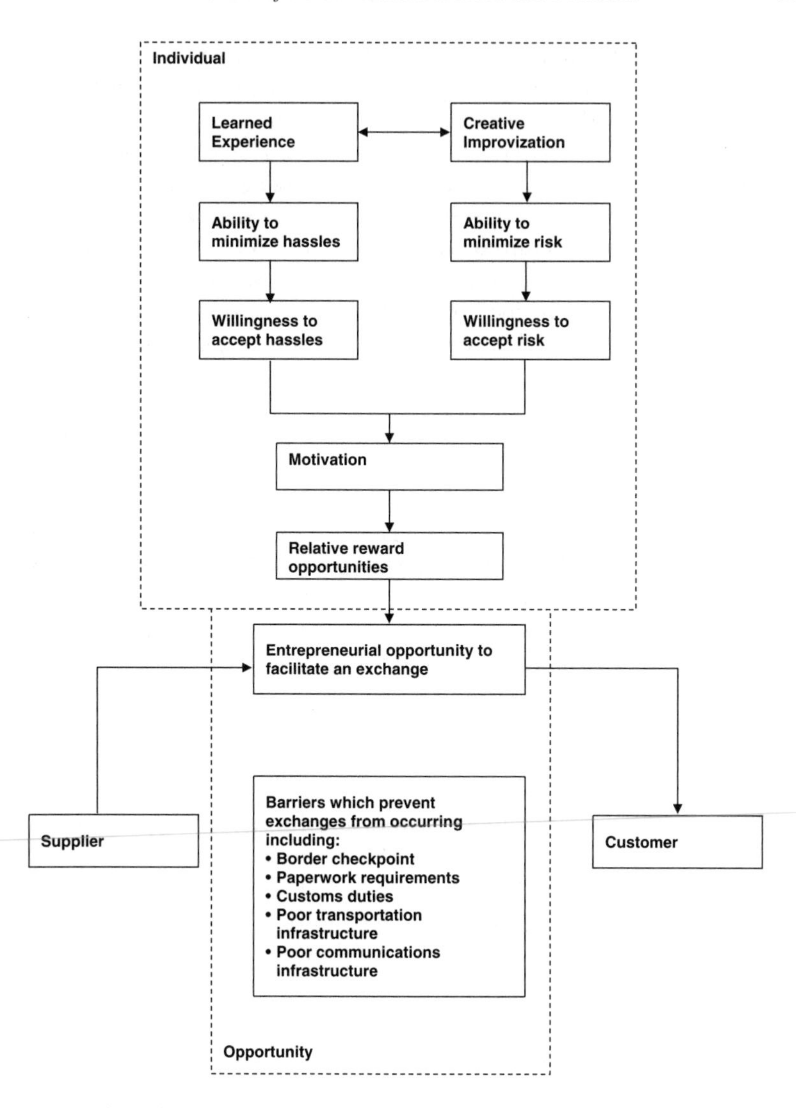

Source: The author.

Figure 9.2 Theoretical framework for entrepreneurial border traders: combination of individual and opportunity characteristics

In addition, there is an informal channel made up of entrepreneurial border traders, locally known as 'the ant army', who have identified opportunities in the Vientiane market which are not met by individual outshoppers traveling to Thailand, and which the authorized distributors have not found or are unwilling to address. The way these entrepreneurs create value and facilitate exchanges, by connecting customers in Vientiane to suppliers in Thailand, is the focus of this research.

The border traders sell to retailers in Vientiane. The retailers can be characterized as one of two types: local open-air markets or mini-marts. Open-air markets fulfill the needs of most residents of greater Vientiane, and are similar to markets found throughout the rest of the country. These markets are usually housed under one large roof, with or without walls. The area underneath the roof is divided into corridors, and stalls are set up to sell goods. Some markets swell beyond the size of the original structure, with tents and tarpaulins used beyond the limits of the roof to keep out the sun and rain.

Mini-marts are more numerous within central Vientiane, and target a smaller but more affluent group of customers. Around the very center of town, these small shops are located on every block, supplying the needs of locals, foreign residents and tourists. They are more likely to stock a wide range of imported products, especially foreign food items, such as imported cookies and biscuits, salami and cheese, and ingredients for preparing foreign dishes. The more up-market shops carry a range of imported liquor, in addition to locally produced beer and whisky.

Border traders buy products in Thailand from retailers in Nong Khai or Udorn Thani. These retailers may be located in open-air markets or in shophouses. In most provincial towns in Thailand, as in Laos, the open-air market continues to be a large supplier of basic household needs for many residents. Ready-prepared food is available from shops specializing in noodle soup, fried food, or barbecued pork or chicken. Customers eat at the stall, or buy the food to take home in plastic bags secured by rubber bands. Shops sell fresh vegetables, fresh fruit, fresh meat, fresh fish, dried fish, rice, dried chilies, and dried beans and grains, each concentrating on one product category. Other shops sell packaged and canned food. Still others are dedicated to branded personal care products, such as soap, shampoo or haircare products. Several shops offer the same product selection, including shops sitting side by side. In addition to selling to consumers, shops in Nong Khai and Udorn Thani sell to wholesale customers such as border traders.

The shophouse is the mainstay of traditional commerce throughout Southeast Asia, especially in cities and larger towns. A 'typical' shophouse includes a large open first floor for displaying products. The front of the shop has a wide opening to make it easy to carry products in and out, which is secured at night by a roll-up door. The shophouse may have several stories, with upper floors reserved for accommodation for the shop owner and their family. Cement structures are replacing older shophouses made of wood. A description of a 'typical' shophouse would also include an abundance of goods stacked floor to ceiling, hanging from hooks and spilling out onto the sidewalks in front.

Methodology

For this research a qualitative approach was necessary given the fact that informal trade across borders, in the context of how entrepreneurs facilitate exchanges that would not otherwise occur, has not been studied in any depth. Miles and Huberman (1994) noted that, while qualitative research may come in many varieties, it contains certain 'recurring features' (p. 5):

- It is conducted through prolonged contact in the field, studying individuals, groups, societies or organizations.
- The researcher's role is to get an integrated or 'holistic' understanding of the situation under study.
- Data are gathered from the perceptions of participants inside the situation.
- A major objective is to understand how the participants act or manage their 'day to day situations'.

Lincoln and Gupa (1985) suggested qualitative inquiry requires a different scheme for justifying results than internal and external validity, the commonly used measures for quantitative methods. Instead of internal validity, the credibility of results becomes the requirement for the qualitative researcher. Similarly, results from qualitative studies should be evaluated for their usefulness in applying them to other settings, rather than by using a strict measure of external validity (Lincoln and Gupa, 1985).

'Triangulation' (Denzin, 1989) or 'multiple methods' (Miles and Huberman, 1994) are often suggested as ways to demonstrate credibility for qualitative research. Denzin (1989) suggested there are four types of triangulation:

1. Data triangulation utilizes data from different, ideally diverse, sources.
2. Investigator triangulation draws on the observations from multiple researchers in the field.
3. Theory triangulation attempts to explain existing data with different hypotheses.
4. Methodological triangulation applies two or more research methods, including the possible mix of quantitative and qualitative methods, to reach credible conclusions.

The author adopted an experimental design that combined different sources of data and different methods of obtaining those data. This occurred in roughly two phases an exploratory phase and an in-depth phase.

1. An exploratory phase comprised open-ended, unstructured interviews with a range of participants connected to either the border trade between Nong Khai and Vientiane, or the sale of consumer products in Nong Khai or Vientiane, including:
   - A manager from the regional office of a large multinational consumer products company, who worked in the group that oversaw the IndoChina markets of Laos and Cambodia.
   - The store manager of a large hypermart retailer located in Nong Khai.
   - The owner of three retailers in Vientiane.
   - A manager from a distributor of consumer products in Laos.
   - Three border traders, including one who also owned a shop in a market in Vientiane.
   - The owners of two retailers in Nong Khai who regularly sold to border traders.
2. The in-depth phase consisted of structured, in-depth interviews with 47 border traders.

The goal of the exploratory phase was to try to understand the scope and constraints of the border trade situation, and to develop a structured questionnaire to be used in the in-depth phase.

The research took place at a single location on the Thailand–Laos border. It is, therefore, a case study, a research strategy that focuses on understanding a single setting (Eisenhardt, 1989), and can involve single cases or multiple cases (Yin, 1994). Yin (1994) suggested choosing single individuals as the unit of analysis instead of other case study designs:

> Of course, the 'case' also can be some event or entity that is less well defined than a single individual. Case studies have been done about decisions, about programs, about the implementation process, and about organizational changes . . . Beware of these types of topics – none is easily defined in terms of the beginning or end points of the 'case'. For example, a case study of a specific program may reveal (a) variations in program definition, depending upon the perspective of different actors, and (b) program components that existed prior to the formal designation of the program. Any case study of such a program would therefore have to confront these conditions in delineating the unit of analysis. (p. 22)

While any number of legal and illicit goods cross the border between Thailand and Laos, the design of the research specifically targeted consumer products that people use on a regular basis for themselves or in their homes, including snack foods, cooking oil and other food ingredients, shampoo, soap, cosmetics, fresh seafood and dried fruit. The sampling frame for interviewing was determined with this goal in mind; while there are also traders who deal in auto parts, furniture, bedding and industrial products, these traders were excluded from this research.

There are two official border-crossing points from Nong Khai province to Vientiane. The Friendship Bridge allows road vehicle traffic between Thailand and Laos, and a shuttle bus travels back and forth between terminals at the border checkpoints on either side. Also, a river ferry crosses between Tha Deau, Laos and Tha Sadet in downtown Nong Khai. Border traders were selected for interview from the population of people crossing the border at these two locations, while waiting at the shuttle bus station at the bridge or at the ferry terminal. The author selected convenient and quiet points at both locations for the in-depth interviews.

FINDINGS

General Findings

Interviewed border traders had experience ranging from over 20 years for those who had started their business before the opening of the 12 year old bridge between Vientiane and Nong Khai, to one month for a trader who was just getting started. Most traders, especially those crossing at the bridge, supplied shops in and around Vientiane. Traders crossing at the Tha Sadet river ferry generally supplied local villages opposite Nong Khai, 25 km from Vientiane. Two border traders in the survey traveled to and from Vang Vieng, approximately 150 km from Vientiane, to supply shops in that town.

Border traders went to Thailand to purchase goods nearly every day, and spent a large part of their daily lives traveling. To travel the 25 km from central Vientiane to the bridge crossing the Mekong River took approximately one hour by tuk-tuk (a motorized three-wheeled vehicle found in Thailand and Laos). Crossing the river, which is also the border, entailed utilizing a shuttle bus, which ran between the immigration points on either side. Paperwork was checked and border passes were issued at checkpoints on both sides. Traveling from the bridge into Nong Khai town required another 45 minutes. The same journey was repeated on the return trip, with whatever goods the trader purchased being transferred at each stage from one mode of transport to the next. This daily routine, of long travel times and numerous vehicle transfers, was one of the aggravations that the border trader endured in her occupation:

> I leave home most days around 8 am and go to the market to receive money for the products that I delivered previously. Then, I go to the border by tuk-tuk, or sometimes bus, and get a border pass. In Thailand, I hire a tuk-tuk (motorized rickshaw) to buy goods and then return the same way. (Trader, 15 years)

> I ride a motorcycle and leave it at the border checkpoint on the Laos side. Next, I get my border pass stamped and come across the bridge, hire a tuk-tuk, and go buy goods. (Trader, three years)

Traders who saw opportunities to buy from the larger city of Udorn Thani (50 km from Nong Khai) or to supply far-flung destinations outside Vientiane, such as Vang Vieng (150 km away), faced even longer distances and additional time spent in transit.

Transporting goods entailed many costs. Tuk-tuks were hired both to take the border trader around to suppliers and customers, and also to carry products. Porters were often utilized to carry goods between vehicles at the shuttle bus stations on either side of the bridge, and were a necessity for moving goods up and down the steep stairs that led from the river ferry terminals to the boat landings:

> There is no way to avoid using porters at Tha Sadet [ferry terminal]. I have to get my goods down to the boat, and up again on the other side. I cannot do it myself. Porters cost me a lot of money each day, but I have no choice. (Trader, two years)

Every border trader interviewed in the research bought to order, and only purchased additional items to test the market for a new product. Many border traders gave responses similar to that of one trader of 16 years: 'If the shops don't order, I don't buy. I want to be sure that I can sell what I buy'. When a trader visited a customer to receive an order, she also tried to collect outstanding debts. For many traders the ordering and collection processes were linked, and determined how frequently those traders could buy more goods. Several traders explained that they could not go to Thailand every day, and instead went every other day, or two to three times per week, because there would be at least a day's lag between product delivery and receipt of funds.

Border traders, therefore, overcame numerous barriers to ply their trade. While immigration procedures took time, crossing the border was only one of the barriers that had to be surmounted. All border traders encountered long transportation times to and from Thailand, and if a particular individual was seeking greater product selection by traveling to Udorn Thani, or was supplying unmet demand in an area outside the city, the level of hassle was even greater. This leads to the next question: What was the opportunity that made such trips worthwhile?

Opportunities

Border traders supplied a wide range of products. Food and cooking ingredients were the most frequently purchased items by traders interviewed in the survey. Personal care products, such as soap and shampoo, were also common. A small number of traders in the sample specialized in goods, such as shoes and clothing, cosmetics and beauty products, or agriculture products, while others bought a broad range of products depending on orders from customers.

And yet authorized distributors also supplied the same product categories and brands provided by border traders. More significantly, the authorized channel was able to supply products at lower cost. This was confirmed in interviews:

> The distributor will supply those goods. I can't buy them because the distributor will sell them cheaper than me. (Trader, four years)

> If the distributor has it, I don't supply it. I have different products. I don't sell products that are supplied by the distributor. (Trader, 15 years)

> If the distributor has the product, the distributor is cheaper. If the distributor does not import the product, then I can bring it in and make a profit. (Trader, 15 years)

Similarly, one shop owner confirmed the price advantage offered by distributors, and also explained one of the benefits she received from buying from border traders:

> Border traders are more expensive than distributors. Usually, I buy from border traders only when I sell out of an item. They can get it to me fast. If I order today, I will receive it tomorrow morning. Even if it is more expensive it is better than not having the product to sell. (Shop owner, Vientiane)

Border traders recognized this competitive advantage:

> They [shops] buy from us because we are faster than the distributor. The distributor will deliver once or twice a week, but we come everyday. If goods sell out, they will buy from us because it's faster than waiting for the distributor even though the distributor's price is a little lower than ours. We are faster and deliver on-time. (Trader, 20 years)

Border traders, therefore, competed not on price, but by providing more frequent deliveries than authorized distributors, who typically delivered every two to three days. This gave shops some valuable benefits. For example, it allowed them to place regular orders with authorized distributors, and buy additional items from border traders if they sold out during the time between deliveries. This translated into inventory cost savings for the shops. Without the border traders, shops would be faced with the choice of increasing stock to a level which was always sufficient to meet expected demand, or keeping stock at an average level and losing sales during peak periods.

Daily delivery also provided advantages for customers buying fresh products:

> For goods like soap and toothpaste, they [shops] don't buy so often. But, if you eat Chinese sausage today, and want to eat it again tomorrow, you have to buy it everyday. (Trader, eight years)

> If a restaurant buys seafood from a trader, they know that the trader will supply it immediately. The restaurant does not stock fresh seafood, and they don't have to buy frozen food. (Trader, 15 years)

Daily delivery of fresh products is essential in developing countries, where few people have refrigerators in their homes, and where refrigeration is limited even in shops and restaurants. For authorized distributors, who were only willing to provide two to three deliveries per week of dry goods, the additional step of daily, refrigerated deliveries would require a large increase in expense and risk.

Another benefit that border traders gave to shop owners was that border traders (who were buying from retailers in Thailand) could buy exactly the quantity desired by the customer, whereas the distributor required customers to buy in case sizes:

> Some of the goods I buy from the distributor don't sell very well and I have to throw them away when they expire. Like chicken soup. It's hard to sell. If I buy from the distributor I have to buy a full case, so I don't want it. If I buy from a border trader I can order just 2 or 3 servings, which I can sell in a couple of days, and then order more. (Shop owner, Vientiane)

In other cases, product varieties were not offered in Laos by authorized distributors, but could be found just across the border in Nong Khai, within reach of the border traders' distribution capabilities. For example, one shop owner sold fish sauce in 1-gallon containers because many of her customers ran restaurants and used larger quantities of the product. Since the authorized distributor only sold fish sauce in the smaller, standard-sized bottles to consumers, the shop owner relied on the services of border traders to bring the larger size from Thailand.

Finally, one of the border traders' biggest advantages was in being able to supply new products to consumers in Vientiane, as they became available in Thailand. Thai television spills over the border and is popular in Vientiane. Consumers in the Lao capital have current knowledge of available products and promotions from watching Thai television commercials, and have learned that they can rely on retailers in Vientiane to get products that are being sold in Thailand:

> All goods are advertised on television. In Laos we can see Channels 3, 5, and 7 [from Thailand]. In Vientiane, customers watch television and they know what is being sold, and they will go to the store and ask, 'Do you have that product? Do you have it yet? If you do not have it, you should buy it to sell'. (Trader, 20 years)

One trader pointed out that some shopkeepers watched Thai television advertisements, and if they saw new products they would order them immediately from traders. Another trader explained that watching television was an important way of learning new opportunities: 'I cannot miss television. I have to catch up on new products. In selling goods, we have to see what is available' (Trader, 12 years).

Some traders continually looked for new products when they were in Thailand buying goods: 'If I go and see something different, I will buy it and see if it will sell' (Trader, four years). Another explained, 'Shops [in Nong Khai] tell us when there are new items and ask us if we want to buy' (Trader, three years).

The opportunity for border entrepreneurs, therefore, came from offering a number of advantages to retailers in Vientiane not available from authorized distributors. Products were supplied on a daily basis, allowing shops to quickly buy additional items when they ran out of stock. Products could be purchased in individual quantities instead of by the case, as required by authorized distributors. Furthermore border traders could buy products in Thailand not offered by authorized distributors, such as those in flavors, fragrances or pack sizes that were demanded in smaller volumes. Finally, border traders could respond to new products launched in Thailand, and supply them to consumers in Vientiane on a timely basis. The border trader's advantage, therefore, came not from doing the job more cheaply, but from performing at a higher service level.

While individual border traders provided valued services to customers in the Vientiane market, a more significant result was that combined they composed an entire channel that supplied products in parallel to authorized distributors. This informal channel did not compete against authorized distributors; it complemented them.

Authorized channels supplied products at lower cost and, consequently, the intermediaries making up those channels were free to set prices and determine exactly the level of service that they were willing to provide to meet customers' needs while still generating a profit. If a distributor wanted to capture sales lost to the informal channel, it needed only to increase its service level. To do so undoubtedly meant greater investment in delivery trucks, personnel, warehousing space and stock, and, consequently, greater overall business risk. However the authorized distributor need not, as suggested in the literature, view the informal channel as a competitive threat.

Instead, allowing the informal channel to meet some of the market's demand removed the most costly and risky aspects of the distribution challenge: supplying low volume and new products, on a frequent basis, to a large number of small retail points. Authorized distributors were then free to focus on more profitable, higher volume product lines. An owner of a mini-mart in Vientiane offered an example of this. Border traders buying from Nong Khai originally supplied a popular brand of disposable diaper to the Vientiane market. Once the market grew to a sufficient level (over a period of three years in the estimation of the shop owner) an authorized distributor added the brand to its line of products that it was supplying to Laos.

In some cases, therefore, informal channels do not threaten authorized channels and instead provide a number of benefits from their complementary activities. This suggests that firms would benefit from incorporating informal channels into strategies for entering emerging markets, by harnessing the entrepreneurial capabilities of border traders.

Motivation

Like many entrepreneurs, the initial motivation for some border traders was the desire or need to make money. Some traders reported having no other skills and, like one trader of three years, 'just stayed at home' or 'during the growing season worked on a farm', before starting this work. This exemplifies the dim employment prospects offered to many border traders. Most had no formal schooling, and had worked primarily as farm laborers. Their employment choices included the option of continuing as farm laborers, pursuing some other form of manual labor or entering government service at low pay. For many of those who had an entrepreneurial inclination it was easy to find better opportunities by trading:

> I had a job washing dishes in Thailand. I received a monthly salary of Baht 1,200 [about US$ 30; 1 US$ is equal to approximately Baht 40] and my hands were falling apart [laughs]. Then, I went with a friend to buy some goods to sell in Laos. I found I could make Baht 1,200 in three days, so I didn't wash dishes any more. On a day that I get a big order, I can make Baht 1,000 at one time. Even days with small orders, I make Baht 200 or 300. On days that I am tired, I can rest. (Trader, 20 years)

> The people who are traders usually come from families who already did this business. They have parents who first brought them to work with them. For me, my mother was also a trader, before passing it on to me. When I was a teenager, and my mom first took me to sell goods at the market, I felt very shy and embarrassed. But when I finished school and went to work as an employee and received a monthly salary, the salary was low, and did not cover my expenses. I saw my mother sell, and saw she had money to use everyday – she touched money everyday. In the beginning, before buying goods in Thailand, I sold fruit to get a regular cash flow. Now, if I spend Baht 1,000, I get profits of Baht 200, 300, and keep anything that is leftover. (Trader, 15 years)

In addition to the opportunity to make additional money, trading provided another benefit to those with an entrepreneurial bent: flexibility. Flexibility meant being able to choose one's own hours, or to decide whether to work on a particular day:

> This is a more comfortable job, meaning when I feel tired, and I think that it's enough, I can rest. When I don't feel tired, I work. If I worked for the government, I would need to work all the time. I wouldn't have time to stop or rest. Doing this work, when I want to stop, I can stop. (Trader, 20 years)

It also meant being able to combine occupations to increase income: 'When I have free time, I also fish or do farm work' (Trader, 15 years). Ultimately, trading allowed each person to be her own boss, a benefit valued by entrepreneurs the world over. A husband of a trader illustrated his entrepreneurial outlook by explaining his contingency plan: 'If there is a problem with this job, I can work in my garage repairing cars. My wife can be a hairdresser at home'.

Despite seemingly limited choices, traders portrayed the entrepreneur's desire to be in control of their situation, and many expressed that desire in terms of wanting 'to touch money':

> Doing other work is not enough. I see that this occupation has good cash flow. I can cover expenses. If I work with the government I get paid a monthly salary; it's not enough to feed me [laughs]. I want to touch money. When I get money, I can use it to pay for household expenses. Being a trader is easier and more convenient. (Trader, 20 years)

The decision to become a border trader was made easier by observing parents, other family members, friends or others successfully involved in trading. This introduced them to the occupation, gave them a support network and, in some cases, even helped provide the start-up capital. A number of traders came from families that had already been doing the occupation for several years:

> My mother has done this job for ten years, since the border checkpoint at the bridge was first opened. Nowadays, my mother and sister work together, and I work separately. (Trader, four years)

> My mother showed me how to do this job, and she still does this work the same as before. As a child, on Saturdays and Sundays when I didn't go to school, I went to sell goods with my mother. When my mother gave me money from working with her, I saved it. When I went to sell with her, I got a lot of money and my mother said to keep it for my own investment. When I grew up, I started selling by myself. (Trader, 20 years)

Those who did not have access to a family already doing the job had a greater challenge in learning the business on their own. Some learned from friends who were already border traders:

> I knew a friend of the family who used to do this. I saw that she made good money, so I tried this work too. (Trader, seven years)

> The first day, I came with a friend. She bought here and there, and I followed her. (Trader, three years)

Others learned by watching people already trading, or by teaching themselves:

> On the first day, I came to the border checkpoint and saw some border traders. I went up to one and asked where they sold their products. I asked if they were buying to re-sell somewhere else. (Trader, three years)

Traders, therefore, relied heavily on help from others to get started. Those whose parents were traders, or those who had friends doing the work, had an advantage in learning the job and its benefits. Without those initial connections, a new border trader had to learn the job on her own:

> I always saw traders from Vientiane buying things, and I thought I would like to buy things to sell near my house, which is 1.5 hours from Vientiane. . . . I went to the main market and talked with the Chinese and Vietnamese people who have shops there. I cannot remember their names. I asked them if they wanted to buy anything from Thailand. (Trader, three years)

> The first time, I went to Nong Khai and walked around to find each shop, and learned about the shops and prices myself . . . Then I went to shops [in Vientiane] and asked if they wanted anything from Nong Khai, and I already knew the prices so I let them know. (Trader, 15 years)

Traders usually started small, and then gradually built up their business. Many traders reported initial investments of approximately Baht 1000–2000.

The newest trader, operating for only one month, declared an initial investment of 'Baht 5000–6000'. Two traders said they started with Baht 10 000. A number of traders explained they had been able to expand their businesses over a period of time:

> I started with about Baht 10,000 of my savings. Today, I now buy Baht 30,000 to 40,000 each trip. (Trader, six years)

> I started with Baht 1,000 to buy 10 bolts of cloth. I sold them and made a profit, and built up from there. Today, I make a profit of Baht 200 per day. (Trader, 20 years)

In order to have an approximate understanding of the size of the return that traders received for their efforts, they were asked about their expenses and the profit they made on their trips to Thailand. However no effort was made by the author to tally a comprehensive income statement for the traders. Whether the profit was acceptable was a subjective judgment that depended on each trader's family situation, prospects for employment, money available for investment, ability to manage the various hassles of the occupation and the person's general outlook. The fact that they were motivated to do this work was evidence that the occupation supplied a reasonable return in the trader's calculation. The interviews, therefore, did not attempt to produce valid answers with respect to reported expenses and profits, but reliable ones, consistent with a trader's responses about overall investment.

According to responses in the interviews, border traders received a gross profit of approximately 15 per cent of their sales. From this they needed to cover their transportation costs, duties and other expenses. Traders who reported spending Baht 2000–3000 per trip claimed to receive net profits of approximately Baht 300–400. Traders spending more, in the range of Baht 10 000 per day, reported netting approximately Baht 1000 after expenses. These responses indicate that border traders earned a profit of approximately 10 per cent on each trip.

Managing Risk

Customs duties had a significant effect on the profitability of any particular trip, and border traders attempted to manage inconsistent enforcement to their advantage. However assessing the impact of import duties was a particularly challenging area of this research. Border traders were often reticent to answer directly, since the topic is sensitive and border traders could not be sure of the interviewing team's intentions. In the interview protocol this issue was raised last, once a rapport had been built up with the person being interviewed, and border traders were promised strict anonymity. Most traders stated they paid duties, and many claimed to pay 'full' duties:

> I pay full duties according to the law. Some products that Laos produces have high duties, such as cosmetics. When I buy them, I unpack them and keep only the cream tube. I do not bring the box back to Laos. Powdered detergents and agriculture tools pay lower duties. If I am only carrying a few products, sometimes duties are reduced. (Trader, 20 years)

This trader's statement reflects the contradictions in understanding the effects of duties on border traders. The trader believed she paid full duties 'according to the law', yet had strategies for decreasing duties by discarding external packaging and packing products more densely. Another trader referred to the opportunity to negotiate with customs officials:

> The bigger the size of the goods, the higher the duties we pay. But sometimes we can reduce duties on items such as food, consumer goods, and small products by negotiating with the customs official. (Trader, 11 years)

In the developing world the term 'negotiation' can often take on an illicit quality. However the author witnessed the negotiation of duties on several occasions while observing the general activity at the customs posts during the research. At the border checkpoint at the bridge people carried products through immigration where their travel documents were checked, then continued past the customs station. Customs officials pulled certain people out of the flow to check their goods. The author observed that those carrying the most were obvious targets for the officials.

If a border trader was singled out by customs officials, a general discussion, or negotiation, took place. Traders and officials spoke with familiarity with each other, indicating that, while they may not have been directly introduced, both sides knew each other professionally. The negotiation occurred for all to hear, with no overtures to any secret or 'under the table' agreement. In the cases the author observed traders adopted a strategy of combining general friendliness and humor, with claims that they were carrying products of little value. When more charming methods failed, traders appealed to pity. The author captured one trader's final pleas to a customs official:

> I won't pay. I can't. If I have to pay, I won't accept it. I will throw everything away. I have to pay everyday. I'm just trying to feed myself. I can't cover it. How can you make me pay so much every time? I've paid many days already, so don't make me pay today. (Trader, unknown experience)

The above was delivered in what appeared to be a fairly good natured way, without detracting from the seriousness of the overall meaning. On that particular occasion the official waved the trader through with a smile and, it appeared to the author, resignation at having to negotiate with a persistent individual. This trader mentioned to the author afterward, 'no matter what happens we have to fight, and then move on'.

After the negotiation between a trader and customs official was concluded the amount of duties was written up on an official receipt and paid at the customs counter, in plain view of everyone present. From the author's observations on several occasions at the bridge there was no clear or obvious way that a particular official benefited from the levied amount. Instead of the 'climate of illegality' observed by Fadahunsi and Rosa (2002) in Nigeria, traders seemed simply to be making use of every advantage they had over customs officials, who had the challenge of monitoring many individuals with limited resources, and probably limited motivation, to catch any but the more serious offenders.

Traders had other methods of minimizing customs payments. Smaller traders enlisted the help of those carrying large amounts by combining their shipments. The combined batch was not likely to be much bigger, and the customs official was apt to arrive at the same overall levy, saving the smaller trader the expense:

> If I have a small amount of goods, they can be combined with other traders with bigger shipments, to pay less. Small items can be placed in boxes among big items to save money. (Trader, two months)

This form of help was observed on a more general social level, as fellow passengers on the shuttle bus, who were not traders, were enlisted to carry part of the load past the customs inspection table. This fact was confirmed in interviews:

> We give the goods to people who are not traders. These people are just travelers, and they take one or two bags of something for us; that is the technique. They will carry for us because they know who we are, and what we are doing. (Trader, 20 years)

Traders faced risks in trusting others, but in doing so gained competitive advantages. Traders who were unwilling or unable to trust others to help them controlled every aspect of their occupation, but also found the scale of their activities limited. These border traders received orders, traveled to each supplier to buy goods and carried those items personally to customers.

Border traders who formed trusting relationships with other parties were able to minimize the occupational hassles that had to be endured, but which also gave the trader her entrepreneurial opportunity. For example, hiring a tuk-tuk to deliver the goods separately from the trader allowed the trader greater flexibility, and provided an opportunity to increase her business, because products could be in transit from several suppliers at once. Similarly, some traders hired porters to carry, pack or store goods on a temporary basis. This process was a basic requirement for the traders who worked at Tha Sadet, and needed to load and unload goods on the steep stairs leading down to boats crossing the river. However porters could also be used to collect and watch over goods being sent from suppliers to the ferry terminal, without the need for traders to supervise them directly. As one porter of three years at Tha Sadet explained:

> Traders who use porters and tuk-tuks don't have to waste time to come back before they have finished buying their goods. Traders hire the tuk-tuks to bring goods here. They call porters who they work with regularly to pick up their goods and wait for them. Tuk-tuk drivers know all of the porters. The hired porters wait for the tuk-tuk, and then they pack, organize, and count the goods, and write the name of the trader on the side of the box.

The case of one 20 year veteran trader showed how a relatively complex and lucrative chain could be set up by forming relationships with suppliers, trusted tuk-tuk drivers and porters. She was able to conduct all of the purchasing of products from the shuttle bus station, placing orders by mobile phone. This freed her from the necessity of traveling from store to store. A trusted tuk-tuk driver, one she had worked with for over ten years, was given cash and relied upon to pay for and deliver goods from shops in Nong Khai to the border shuttle bus station. To illustrate the relative magnitude of this reliance on trust, it must be noted that both tuk-tuk drivers and porters received anywhere from Baht 50 (US$ 1.25) to Baht 200 (US$ 5) for their services, yet they often carried goods several times that value. Most border traders purchased Baht 3000–5000 (US$ 75–125) worth of goods. Traders who were willing entrusted the goods, and sometimes the cash to pay for them, to the porter or tuk-tuk driver. This reliance on trust was even more significant when considering that the activity took place across an international border, where legal jurisdictions are often unclear. Most border traders were from Laos; the tuk-tuk drivers were from Thailand; porters could come from either country. A border trader who found her goods absconded with by an unscrupulous tuk-tuk driver could probably count on little legal protection. Instead, trust between drivers, porters and traders who had worked together over a long period of time, and trust among members of the greater social community involved in border trade who worked together every day, provided the only security against malfeasance.

CONCLUSION

The Entrepreneurial Opportunity

The most significant finding of this research is that border entrepreneurs on the Thailand–Laos border do not compete by avoiding duties or undercutting the prices of authorized distributors. Instead, they succeed by finding new customers or by identifying new products which are not currently available in the Vientiane market. The activities of these entrepreneurs complement those of authorized distributors. The border traders find and evaluate new opportunities, and provide service levels that authorized distributors are unable or unwilling to provide. This research indicates that border traders are more expensive, so the authorized distributors can choose to take over any business once it develops to a lucrative level.

Value Provided by the Network of Border Entrepreneurs

Individually, each border trader, motivated by her search for profit opportunities, overcomes barriers and facilitates exchanges between a small number of suppliers and customers. Combined, the individual border traders form an informal distribution network that gives two advantages over the authorized channel: (1) it makes more frequent deliveries, and (2) it provides products that the authorized channel does not. This provides benefits to the market. Customers are served more quickly and retailers are able to control their inventory more precisely with less need to build additional safety stock to meet peak demand. The network provides convenience goods in flavors, fragrances and pack sizes that are demanded by certain customers in quantities too small to entice authorized distributors. Moreover it supplies new products as they become available in Thailand. Entrepreneurs in the informal channel undertake the functions of recognizing a new product opportunity, testing the new product in the market and finding initial customers.

The Entrepreneurial Characteristics of Border Traders on the Laos–Thailand Border

Poor infrastructure and limited transportation options mean most people crossing the border need to endure long travel times to and from the border checkpoints and must change vehicles several times en route. This creates a barrier so that customers in Laos have incomplete information about where goods can be purchased and what current prices are, while suppliers in Thailand have limited knowledge about the market situation in Laos. Border traders call on a number of social skills, attributed to entrepreneurs elsewhere, to overcome these barriers, linking suppliers on one side of the border with customers on the other. Many of the respondents interviewed in this research came from families which had other family members who were already border traders, and who provided knowledge, experience, a support network and often start-up capital. Others relied on friends, who were already border traders, to build up expertise. Border traders also worked in a social community that included other border traders, porters, tuk-tuk drivers and other participants. Social forces and trust within this community, which straddles both sides of the border, help ensure that its members do not take advantage of one another. The combination of these social linkages gives border traders advantages in minimizing hassles, building trust and increasing their entrepreneurial capabilities.

Limitations of the Study and Areas of Future Research

In case studies like the research presented here the question arises about how well the results can be used to draw more general conclusions. Vientiane, Laos, is an emerging market with relatively easy access to a large consumer market, Thailand. The results of the research, therefore, would be most applicable to situations where a large consumer market exists just over the border from a source of supply of desired goods. The cases studied by Staudt (1998) and Spener (1995), on the US–Mexico border, and Piron (2002), on the Singapore–Malaysia border, share characteristics with the research described here, and have produced similar results. However, with increasing efficiency of transportation, and liberalization of trade policies, this situation will continue to become more common throughout the world. Further research should explore entrepreneurial traders on other international borders to understand their role in opening emerging markets, and to find out if their activities are consistent with the findings here.

The results also showed that border traders are all women. Further research should consider the economic and social factors which lead to this outcome. Are women better suited for this work? Or are they left with no other options?

This case study also highlights how entrepreneurs overcome the barriers that are found in many emerging markets, where low purchasing power means few people own vehicles to enable them to travel long distances to shop, and few have refrigeration in their homes. Customers in emerging markets, therefore, must buy for their immediate needs, and cannot afford to buy a week's supply of food or consumer products. For an authorized distributor this results in a daunting challenge: how to meet the needs of consumers with lower purchasing power, who nonetheless have higher service requirements.

In their search for opportunities individual entrepreneurial border traders are able to:

- Find a large number of customers, who have low purchasing power.
- Supply products to them more frequently.
- Deliver in smaller, less profitable quantities.
- Make products available to more numerous retail locations, close to consumers' homes.

This suggests the potential for firms to incorporate entrepreneurial border traders into their strategies for entering emerging markets. Prahalad (2005) offers as an example the Shakti Project initiated by a subsidiary of Unilever in India. Shakti encourages small groups of women to help each other build small enterprises by providing entrepreneurial development training. These self-employed women form a network that sells directly to rural villagers, educating consumers on the benefits of the company's products and delivering the goods to them. However, to develop such strategies on a broader scale, future research must investigate the benefits, and challenges, of such interactions.

REFERENCES

Anderson, M. (1996), *Frontiers*, Cambridge, UK: Polity Press.

Arnold, D. and J. Quelch (1998), 'New strategies in emerging markets', *MIT Sloan Management Review*, **40**, 7–20.

ASEAN (2006), Association of Southeast Asian Nations, consolidated 2005 CEPT package by country, www.aseansec.org/17693.htm, 8 January 2006.

Banomyong, R. (2000), 'Multimodal transport corridors in Southeast Asia: a case study approach', unpublished doctoral dissertation, Cardiff University, Cardiff, UK.

Burt, R. (1992), *Structural Holes: The Social Structure of Competition*, Cambridge, MA: Harvard University Press.

Bygrave, W. and M. Minniti (2000), 'The social dynamics of entrepreneurship', *Entrepreneurship, Theory and Practice*, **24**, 25–36.

Castells, M. and A. Portes (1989), 'World underneath: the origins, dynamics, and effects of the informal economy', in A. Portes, M. Castells and L. Benton (eds), *The Informal Economy*, Baltimore, MD: Johns Hopkins University Press, pp. 11–37.

Cespedes, F., E. Corey and V. Rangan (1988), 'Gray markets: causes and cures', *Harvard Business Review*, **66**, 75–82.

Czinkota, M.R. and I.A. Ronkainen (1997), 'International business and trade in the next decade: report from a Delphi study', *Journal of International Business Studies*, **28**, 827–44.

Denzin, N. (1989), *The Research Act: A Theoretical Introduction to Sociological Methods*, Englewood Cliffs, NJ: Prentice-Hall.

Eisenhardt, K. (1989), 'Building theories from case study research', *Academy of Management Review*, **14**, 532–50.
Fadahunsi, A. and P. Rosa (2002), 'Entrepreneurship and illegality: insights from the Nigerian cross-border trade', *Journal of Business Venturing*, **17**, 397–430.
Gillespie, K. and J. McBride (1996), 'Smuggling in emerging markets: global implications', *Columbia Journal of World Business*, **31**, 40–54.
Kirzner, I. (1973), *Competition and Entrepreneurship*, Chicago, IL: University of Chicago Press.
Li, L. (2003), 'Determinants of export channel intensity in emerging markets: the British experience in China', *Asia Pacific Journal of Management*, **20**, 501–16.
Lincoln, Y. and E. Gupa (1985), *Naturalistic Inquiry*, Newbury Park, CA: Sage.
Losch, A. (1954), *The Economics of Location*, New Haven, NJ: Yale University Press.
Miles, M. and M. Huberman (1994), *Qualitative Data Analysis: An Expanded Sourcebook*, Thousand Oaks, CA: Sage.
Minniti, M. and W. Bygrave (1999), 'The microfoundations of entrepreneurship', *Entrepreneurship, Theory and Practice*, **23**, 41–52.
Naim, M. (2005), *Illicit*, New York: Doubleday.
Nijkamp, P., P. Rietveld and I. Salomon (1990), 'Barriers in spatial interactions and communications', *Annals of Regional Science*, **24**, 237–53.
Piron, F. (2002), 'International outshopping and ethnocentrism', *European Journal of Marketing*, **36**, 189–210.
Portes, A. and W. Haller (2005), 'The informal economy', in N. Smelser and R. Swedberg (eds), *The Handbook of Economic Sociology*, Princeton, NJ: Princeton University Press, pp. 403–25.
Prahalad, C. (2005), *The Fortune at the Bottom of the Pyramid: Eradicating Poverty Through Profits*, Upper Saddle River, NJ: Wharton.
Reid, D., J. Walsh and M. Yamona (2001), 'Quasi-legal commerce in Southeast Asia: evidence from Myanmar', *Thunderbird International Business Review*, **43**, 201–23.
Schumpeter, J. (1934), *The Theory of Economic Development*, Cambridge, MA: Harvard University Press.
Shane, S. and S. Venkataraman (2000), 'The promise of entrepreneurship as a field of research', *Academy of Management Review*, **25**, 217–26.
Shapero, A. (1985), 'Why entrepreneurship? A world-wide perspective', *Journal of Small Business Management*, **23**, 1–5.
Spener, D. (1995), 'Entrepreneurship and small-scale enterprise in the Texas border region: a sociocultural perspective', DAI 57/01, July 1996 (AAT 9617350), p. 473.
Staudt, K. (1998), *Free Trade: Informal Economies at the U.S.-Mexican Border*, Philadelphia, PA: Temple University Press.
Venkataraman, S. (1997), 'The distinctive domain of entrepreneurship research: an editor's perspective', in J. Katz and R. Brockhaus (eds), *Advances in Entrepreneurship, Firm, Emergence, and Growth*, Greenwich, CT: JAI Press, pp. 119–38.
Watts, J. (2004), 'Frozen frontier where illicit trade with China offers lifeline for isolated North Koreans', *Guardian*, 9 January, p. 17.
Yin, R. (1994), *Case Study Research: Design and Methods*, Thousand Oaks, CA: Sage.

10. The value of social capital to family enterprises in Indonesia

Michael Carney, Marleen Dieleman and Wladimir Sachs

INTRODUCTION

We hypothesize that in poorly developed institutional environments family firms enjoy a competitive advantage over professionally managed ones, as family links and tacit business arrangements provide the means for coping with contextual hostility, lack of trust and imperfect information. Because family firms with simple organization structures may more readily respond to the exigencies of hostile environments they can outperform non-family firms that are endowed with greater resources and more sophisticated structures (Mintzberg, 1979). Family firm owner-managers have greater discretion than professional managers to make 'risky deals' (Miller and Breton-Miller, 2005), commit the firms assets 'on a handshake' (Blyler and Coff, 2003), exercise a 'capacity to trust' (Redding, 1990) and cultivate 'guanxi' (Xin and Pearce, 1996). These arguments suggest that social capital is a key resource and the basis of competitive advantage where formal contracts are otherwise difficult to enforce (Carney, 2005).

Social capital may be especially advantageous in transitional and emerging economies due to uncertainties inherent in dynamic and sometimes hostile conditions. However many analysts predict that once that transitional/emergent phase has passed social capital will decrease in value and firm success will increasingly rely upon the creation of proprietary techno-organizational competencies (Kock and Guillen, 2001; Peng, 2003). Established firms who are immersed in previous stage conditions will become increasingly out of tune with more codified institutional contexts (Tan, 2005). Meanwhile newcomer firms are better attuned to emerging conditions and more willing to invest in competence destroying innovations that will form the basis for advantage in the next stage of development (Tan and Tan, 2005). Hence in these stage perspectives of development social capital is the basis of temporary, not long-term, advantage.

In this chapter we argue that 'stage perspectives' assume, at least implicitly, that economic and social development is an ordered process through which societies converge upon certain institutional models. Due to its optimistic assumption of social and economic progress this view of institutional development has a deep appeal. While some may argue the correctness of this view from a very long historical perspective (for example, Fukuyama, 1992), in management studies we are interested in a time frame not exceeding a few decades and in this respect the 'stages' view oversimplifies questions of institutional development. For instance, economic development may occur without any corresponding social progress or institutional development. In other cases risk and uncertainty remain latent and unstable institutional conditions pose continuing threats. If so, social capital may be of lasting advantage.

Below we examine the endemic nature of hostility and the continuing benefits and costs of social capital in developing economies. In this light we explore the question why family firms enjoy an advantage over non-family firms. To illustrate our exploration we narrate a paradigmatic case of a successful multi-generation family firm, the Salim Group, located in the chronically hostile environment of Indonesia. The case opens with the foundation of the firm during the World War II Japanese occupation and traces the adaptations of the firm through a colonial war of independence, and successive eras of socialist rule, state-led industrialization, crony capitalism, financial crisis and a precarious democracy all punctuated by sporadic outbreaks of violence against the ethnic Chinese minority.

We discuss the managerial implications for firms that operate for many years 'living dangerously'. The stages model of institutional development urges firms to learn economic-techno capabilities that will be required to succeed in the next developmental stage (Kock and Guillen, 2001). But if initial hostile conditions persist then such advocacy may encourage firms to prepare for a future that may not materialize. If so, then firms should rather enrich and even deepen their social capital capabilities that are adapted to institutional under development.

ENVIRONMENTAL HOSTILITY

We distinguish between the scarcity-munificence (Staw and Szwajkowski, 1975) and the benign-hostility dimension of firm environments. Acquiring resources in scarce environments is difficult due to intensively competitive structural conditions, such as low barriers to entry, powerful buyers or sellers, or market decline (Harrigan, 1980). These low opportunity conditions produce high firm mortality rates, low profitability and poor employee remuneration (Covin and Slevin, 1989). In contrast, benign-hostility stems not from intensive competition but from non-market threats emanating in the macro-institutional environment. Specific threats to firms or certain entrepreneurial groups arise from institutional underdevelopment (North, 1990) and a heightened risk of asset confiscation, rent expropriation, corruption, discrimination, regulatory restriction, physical destruction of property, personal abduction and even physical harm. Hostile environments may also be scarce but often they are highly munificent, characterized by ample natural resources, good physical infrastructure, an abundance of opportunities and limited competition.

Some countries in the Middle East, Africa, the former Soviet Union and Southeast Asia exhibit hostility that requires actors to develop special skills and resources for coping with non-market threats. If countries show signs of convergence on high-quality institutions, such as greater transparency and the rule of law, then they may be considered developmental or transitional (Peng, 2003). However some remain crisis prone with the potential for political discontinuity, instability and dramatic reversals in their development. Countries as diverse as Iran, Russia and Venezuela appeared to be set upon a path of benign development only to experience disruptions and a snap back to prior conditions (Rajan and Zingales, 2001).

Environments can also be hostile for firms if they evince low levels of trust, collective social capital or civic traditions (Putnam et al., 1994). Low trust societies often have a long history of social conflict or state predation that engenders defensive behaviour in its citizens who seek to encase themselves behind a thick social wall that separates reliable insiders from untrustworthy outsiders (Fukuyama, 1995). Exclusionary behaviour is an understandable response to chronic conflict or state predation but it can impede the development of more open and universalistic values needed to support the diffusion of liberal market institutions. If the state has been historically poor in delivering collective goods, such as security and justice, then secret societies, cliques and criminal organizations may emerge to provide basic levels of security, but such organizations later retard institutional development (Nozick, 1974).

Otherwise benign environments may be hostile for some sub-sectors of the population, such as firms owned by ethnic or religious minorities. Market dominant minorities are pervasive in many developing economies, such as the Chinese in Southeast Asia, Indians in East Africa, Lebanese in Francophone West Africa and the Tutsi in Rwanda, and frequently generate antipathy in the general population (Davis et al., 2001; Kotkin, 1993). Minorities may initially occupy marginal social positions yet prosper despite official discrimination and restrictions on their activities (Bonacich, 1973).

Because contextual hostility is often latent, it is a long-term phenomenon and the 'transitional/emergent phase' may be regarded as permanent from a managerial perspective. Accumulated learning and skill in perceiving, interpreting and managing active and latent threats can be a valuable capability. Competitive advantage in these contexts may rest on various forms of social capital, such as social solidarity (Granovetter, 1994), network resources (Gulati, 1998), contact capabilities (Guillen, 2000), political resources (Frynas et al., 2006) or guanxi (Xin and Pearce, 1996). However social capital is heterogeneous and the differences in its various forms should be clarified since they may have different uses or benefits and entail different costs for individuals and social aggregates.

FORMS OF SOCIAL CAPITAL

The basic notion is that 'the goodwill others have toward us is a valuable resource' (Adler and Kwon, 2002, p. 18) and there is substantial agreement that social capital facilitates individual, organizational and collective action. Social capital has been defined in different ways and for different purposes in various social sciences and has become an 'umbrella' concept that encompasses both micro and macro constructs, such as trust, networks, relational contracting, goodwill, firm reputation and the capacity for society-wide collective action. While this may lead to the critique that the concept is diluted, it is precisely this heterogeneity that renders it a useful lens for examining the bases of competitive advantage (Nahapiet and Ghoshal, 1998).

To differentiate between these perspectives Adler and Kwon (2002) distinguish between bonding and bridging forms of social capital. Research on bonding capital emphasizes the collective good qualities of social capital, suggesting that it resides at the level of organizations, social groups or whole societies. We focus on social capital as the basis of solidarity or identity that bind groups into cohesive entities. Bonding capital accrues from membership in groups or closed networks that have defined boundaries, a common identity or a shared solidarity, based on factors, such as kinship, religion, political affiliation, ethnicity or school attendance. Membership offers an inheritance of goodwill to an individual if the group is disposed towards shared norms and values that generate group trust and mutuality. Not all social aggregates share such norms but those that do typically enjoy certain advantages over those that do not.

In contrast, bridging capital accrues from creating new linkages across social boundaries in relatively open networks. From this perspective social capital resides in relationships that may be 'appropriated' by individuals or by firms and utilized for their own private purposes. In this ego centric view of social capital scholars emphasize the uses, effects and benefits conferred on individuals from linkages that are autonomously and deliberately constructed (Woolcock, 1998). The focus is upon an individual's capacity for brokerage among 'friends, colleagues, and more general contacts through whom you receive opportunities to use your financial and human capital' (Burt, 1992, p. 9). Cohesive bonds, or trust, may develop in such linkages due to frequent positive interactions but there is not necessarily an inheritance of goodwill, so bridging social capital must be actively cultivated and maintained.

Researchers have tended to emphasize the value of social capital, but there is growing recognition that there are disadvantages and costs associated with it (Locke, 1999), for example, due to lack of network diversity, costs associated with redundancy or dependency (Steier and Greenwood, 2000) or through costs and risks associated with decoupling of existing ties. In certain situations the costs of the generation and maintenance of social capital may outweigh the benefits. Benefits and costs may also shift over time due to changing circumstances that alter the value of relationships. Therefore an appraisal of social capital as a source of competitive advantage must be made based on the balance of benefits and costs, as well as an assessment of specific contingencies.

THE BENEFITS AND COSTS OF BONDING AND BRIDGING CAPITAL

The primary benefit of bonding capital is reduced transactions costs among group members who trade with one another. The dense social networks that arise within cohesive groups provide channels for the transmission of fine grained information about member reputation and conduct. The costs of searching and screening potential trading partners and enforcing contracts are less among group members due to their capacity to identify and apply binding social sanctions on opportunistic behaviour. Moreover the inheritance of shared norms and values represent a bank of social credit and shared understandings facilitate problem solving and dispute resolution. Credible information freely transmitted within the group can serve as a functional substitute to more formal screening, search and enforcement mechanisms, such as credit rating bureaus, executive placement agencies or arbitration services (Xin and Pearce, 1996; Poppo and Zenger, 2002).

Hostile environments often lack efficient formal institutions that enforce contracts and provide reliable data about potential partners (Khanna and

Palepu, 1997). In these contexts transactions costs will be high for the general population but cohesive groups with substantial bonding capital may enjoy a transaction costs advantage, especially with regard to uncertain medium- and long-term contracts (Williamson, 1985). In low trust societies lower transaction costs provide advantages to subgroups that can draw upon their communities to mobilize resources and identify opportunities that are unavailable to the general population (Granovetter, 1995; Portes and Sensenbrenner, 1993).

However the transaction cost benefits of bonding capital may be diminished if the group exerts strong pressure for compliance with its norms. Such pressures may result in over-embeddedness, which can dampen experimentation, innovation and blind members to external information. They may isolate actors from external influences and resources that reside beyond the group (Uzzi, 1997). If this occurs, the group incurs an economic penalty by foregoing opportunities that are available in the general population. Over-embeddedness can also produce excessive claims by group members on successful individuals for gifts, jobs, favours and nepotism. If such claims are not limited in some fashion, they may overwhelm the firm's resources.

The primary benefit of bridging capital is an increased capacity for brokering and facilitating transactions between people and organizations that are otherwise unconnected. Actors in brokerage positions may be better at identifying opportunities, getting things done, and thus may appropriate rents for themselves (Blyer and Coff, 2003; Coff, 1999). Bridging capital need not rest upon prior goodwill; indeed individuals from distinct social groups may be suspicious and mistrustful of one another. Rather, bridging capital is based upon an instrumental rationality and enforced by a continuing mutuality and the expectation of reciprocity. Such ties between actors are self-enforcing so long as both parties perceive their shared interests (Telser, 1980).

In environments where access to resources is restricted the capacity to cultivate linkages with gatekeepers is advantageous. The value of bridging capital accrues from quid pro quo exchange, for example, an entrepreneur securing a licence from a government official in exchange for a seat on the entrepreneur's board. The skills required to cultivate such exchanges may be relatively rare. Unconcealed reciprocity between officials and entrepreneurs may be either illegal or socially proscribed and significant social skill is required to effect these transactions. Social skill inheres in the capacity to generate trust and reputation, in diplomacy, persuasion, and the ability to aggregate interests, convince others and frame situations Such skills are necessary corollaries of wheeling and dealing (Fligstein, 1997).

Where contract enforcement mechanisms are weak a reputation for transactional integrity and fair dealing may be more valuable than in more benign institutional contexts. Reputation is generated through numerous interactions with a variety of stakeholders and opportunism in one domain may negatively impact reputation in others (Rindova and Fombrun, 1999). The self-binding quality of a favourable reputation can aid in attracting resources from third parties and from external sources. In markets where actors, including government, have an unknown propensity for opportunism actors with known reputations can intermediate or go between two non-trusting parties by acting as a guarantor to both. For example, Guillen (2000) suggests that diversified business groups in emerging markets derive some of their dynamic growth from their ability to access, mobilize and combine foreign and domestic resources under conditions of restricted foreign trade and investment. Such firms intermediate between foreign firms and the state, which have little knowledge of one another. Reputational assets generate economies of scope because scarce resources, such as high level contacts, may be shared among group firms across industries. Reputation has a 'virtuous circle' quality with regard to resource acquisition and foreign direct investment will tend to flow to the largest and most visible domestic firms.

Individuals establish linkages beyond their social boundaries in expectation of private benefits, but there are associated costs and obligations. There is potential for unbalanced expectations of reciprocity: a favour granted at one point in time may be expected to be repaid in multiple at a later point. The quantification and accounting for the value of obligations incurred through bridging is an inexact and vexing process. A focal actor must maintain a mental calculus about the terms of trade (Redding, 1990). The potential for unbalanced reciprocity presents the actor with a delicate decision regarding maintaining or defecting on a particular link. Pure brokerage may be viewed as opportunism and held in disdain in certain cultures and brokers may be resented and vulnerable. Given the benefits and costs of bridging and bonding capital, we now investigate the relationship between social capital and family firms.

THE FAMILY FIRM AND SOCIAL CAPITAL IN HOSTILE ENVIRONMENTS

La Porta et al. (1999) have shown that family control is ubiquitous in the world; and that the larger business families in weak institutional contexts are often organized in corporate groups. We adopt a loose definition that sees the family as a coalition of economic interests that exercises control over its economic wealth rather than identifying family firms in terms of stock concentration (Anderson and Reeb, 2003) or some combination of ownership, control rights, owner participation in management and generational succession (Amit and Villalonga, 2006). Family firms may adopt different forms in different institutional contexts due to factors, such as death duties, inheritance laws, property rights and the status of women, each of which will stimulate different family succession patterns and the probability of a family firm incorporating as a public entity (Colli et al., 2003). In hostile environments families may seek to conceal their identity and shift assets beyond the boundaries of a legally defined firm. Indeed, family firms in these contexts are likely to adopt structures and practises that protect their wealth from the arbitrary exercise of power.

Nevertheless, family firms may possess a competitive advantage in hostile markets not because they are small (Covin and Slevin, 1989) but due to their governance structure that facilitates owner-managers' discretion in generating and appropriating rents from their social capital. Empirical studies reveal that large family business groups are often organized in the form of an ownership pyramid (Morck and Yeung, 2003). The pyramid structure achieves managerial control by a key shareholder with limited capital and facilitates the diversion or insertion of funds by the key shareholder, often called, respectively, tunnelling and propping (Friedman et al., 2003). Family firm governance frees owner-managers to adopt an entrepreneurial posture (Covin and Slevin, 1989), act decisively (Mintzberg, 1979) or engage in questionable acts (Dyer and Mortensen, 2005) and exercise greater discretion in selecting business partners (Uzzi, 1997). In this regard, family governance structures may be viewed as an instrument for exercising centralized family control and strategic flexibility.

Uniting ownership and control obviates the classic agency problem because owners have both managerial control and a direct claim on the firm's profits, thereby establishing high-powered incentives. However in hostile environments accounting standards are typically undeveloped and capital markets are often non-existent or illiquid. Hence the effects of strong incentives on firm performance may not be reflected by either accounting measures or capital market valuations. More importantly, family control permits owners to appropriate rents before they show up in the bottom line (Coff, 1999). For families in hostile contexts the capacity to redirect rents is a vital wealth preservation mechanism.

Additionally, uniting ownership and control consolidates authority in the person of the owner which frees them from the oversight placed on executives of non-family firms. These strong command characteristics (Miller and Breton-Miller, 2005) provide owner-managers with broad discretion as they generally do not have to justify their actions to outsiders. Lack of capital market oversight also facilitates the conduct of secretive and non-transparent transactions. The capacity to engage into transactions with unspecified obligations and uncertain time horizons is a useful mechanism for engaging in reciprocity based contracts that are essential to the generation of social capital. Family firm governance structures allow owner-managers to apply both particularistic and highly calculative criteria to transactions. In his research Uzzi (1997) found that owner-managers typically blend both highly calculative arms-length transactions with trusting and forgiving relations with a subset of their trading partners. The application of particularistic criteria by owner-managers may manifest in selecting partners on the basis of social criteria, which is instrumental in building bonding and bridging capital.

In contrast, due to the checks and balances inherent in managerial firms it is more difficult for professional managers to engage in extra-contractual commitments on behalf of their principals. Moreover professional managers cannot easily favour their friends or kin as it is likely to be viewed as illegitimate 'nepotism', which is not normally considered illegitimate in the context of family firms. Whether, when and to what extent families are able to maximize the benefits and minimize the costs of different forms of social capital remains an unexplored question. To advance this exploration we narrate a case history of Liem Sioe Liong who founded one of Southeast Asia's largest and most successful ethnic Chinese family business groups, the Salim Group of Indonesia. At its peak the Salim Group companies had revenues in excess of $20 billion and employed over 200 000 people. Over the 60 year period covered by this case study Indonesia was an extremely hostile and dangerous environment.

CASE STUDY METHODOLOGY

The threats posed to firms by hostile environments are often complex and nuanced and responses to them are equally complex and often covert. The case study method is useful since it permits thick descriptions of threat response processes (Redding, 2005). However there are several inherent problems with case studies, including limited generalizability, reliability and researcher bias. Propositions based upon a single case can be overly complex and idiosyncratic if researchers become immersed in too much detail (Eisenhardt, 1989). However if the researcher is sensitive to these problems then cases offer significant contributions to theory development due to researcher proximity to the field. In particular, generalizability can be increased by the careful selection of 'extreme, critical or deviant cases', which may reveal much more about the phenomenon under investigation than a typical case or a random sample. Flyvbjerg (2001) suggests that extreme cases may be 'paradigmatic' because they epitomize their type and serve as a metaphor or referent for more general exemplars of the category. Extreme cases, such as Perrow's (1984) use of the Three Mile Island incident to develop ideas about disasters, can offer insight into previously shielded social systems, a quality especially valuable when studying 'elite and powerful groups in societies where they practice their skills' (Pettigrew, 1990, p. 275). Pettigrew advises researchers conducting longitudinal field research to select cases that represent 'extreme situations, critical incidents and social dramas' (p. 275), because such situations may make phenomena 'transparently observable'.

Data for the case study were collected between 2003–5 from diverse sources including 56 personal interviews of which two with Anthony Salim, the Salim Group's Chief Executive, fourteen with other Salim Group executives, seven with academic experts, nine with executives of the Indonesian Bank Restructuring Agency, four with former cabinet members, two with financial analysts, four with journalists, five with executives of other conglomerates and three with executives of foreign partners. Other sources included annual reports, news analysis and archival data. Interviews focused upon the Salim Group's corporate strategy, actor's mutual perceptions and interactions between company and the institutional environment, domestic competitors and international partners. The case study does not seek to either prove or falsify a hypothesis. Rather, the purposes of the case are illustrative, intending to demonstrate the shifting and unpredictable nature of a hostile environment and the ways in which a family enterprise draws upon its social capital to navigate in it. We base the case narrative on a context threat response structure that spans four regime changes. Table 10.1 provides a time line of changing institutional conditions and Liem's adaptations.

Essential Outsiders

Chinese migration in Indonesia started before the colonial times. The numbers peaked in the late nineteenth and early twentieth centuries, as a result of problems in China and diminishing opportunities abroad (Wang, 2000). Despite their long presence in Indonesia, ethnic Chinese have always constituted a separate and comparatively well-organized group. Under colonial rule the Chinese engaged in trading, money lending and middleman functions and, relative to the indigenous population, were economically successful (Wu and Wu, 1980). Both physically and legally, many ethnic Chinese were separated from the indigenous population, they were 'essential outsiders' (Chirot and Reid, 1998).

Ethnic Chinese in Indonesia were not a homogenous group. Scholars typically differentiate between *peranakan*, an assimilated upper class, and

Table 10.1 Institutional change and Salim Group adaptations, 1940–2005

Time period	Macro-institutional context	Salim Group adaptations
1938–64 Genesis	Colonial decline Japanese occupation War of Independence Sukarno's socialist regime Anti-Chinese, Anti-capitalism Nationalization	Zero visibility, small scale trading *Embedded in clan and family networks,* Cultivate relations with military 'All business is good' *First embedded in family, clan and later increasingly in military networks*
1965–96 Growth & Cronyism	Violent regime change Suharto 'New order' Economic development Crony capitalism Sustained economic growth Industrialization Crony capitalism Rapid economic growth Rise of Suharto family business	Industrialization with foreign partners & finance Diversification Emergence as largest business group Internationalization/capital flight Family business succession Establishment of divisions, professional managers, less reliance on connections with Suharto *Embedded in political networks, also ethnic networks and emerging global business networks*
1997–99 Crisis & Regime Change	Asian Financial crisis IMF austerity Regime change Ethnic violence	Negotiations with new government Re-shuffling portfolio *Embedded in global business networks while political networks are declining due to the crisis*
2000–5 Resurgence	Slow growth in Indonesia Business opportunities in India and China Reconstruction of crony environment with new players	More focus on non-Indonesian business Buying back former companies lost in the crisis New investments in China, Australia *Embedded mainly in global business networks, while ethnic networks also play a role*

totok Chinese, recently arrived. The *totok* were further differentiated along dialect groups such as *Hokkien, Hainan, Hakka* and *Hokchia* and also along clan and family lines. Each *totok* group typically specialized in certain trades and occupations. For example, Liem Sioe Liong was a *Hokchia*, a group that initially specialized in the low status occupation of money lending and which, when Liem arrived in the 1930s in Java, was the poorest group among the Chinese migrants. The *Hokchia* were known to be a close-knit group with a strong tendency for developing bonding capital (Post, 1997).

Leaders of independent Indonesia, Sukarno (1949–65) and Suharto (1966–98), sought to diminish the economic power of the Chinese minority and promote the rise of indigenous capital, but without much success. Economic dominance by a visible minority resulted in friction and latent hostility against the Chinese that occasionally erupted into violence, for example in 1942 when the Japanese took over, in 1965 with the regime change between Sukarno and Suharto, and in 1998 during the Asian crisis.

The Years of Living Dangerously

Phase 1: genesis

Fleeing a war in his native China, Liem Sioe Liong migrated to a small town in colonial Indonesia in 1938. His early start in small-scale trading was facilitated by family members and Chinese clansmen (Soetriyono, 1989). The colonial era economy was abruptly disrupted in 1941 by the war and the establishment of Japanese military administration. The occupation created black markets and offered plentiful opportunities for smuggling, which, if detected, carried severe penalties. In this period *Hokchia* became known as 'most daring, most willing to take risks and extremely venturesome' (Twang, 1998, p. 130). To avoid detection *Hokchia* sought to minimize their visibility, operating in a clandestine manner and relying on one another for protection. Liem joined this trade by transporting goods on a bicycle (Twang, 1998). During the occupation the *Hokchia* began to extend their networks beyond Indonesia, and rose in power within Indonesian society. Post (1997) speculates that Liem's fortunes paralleled the general rise of the *Hokchia*.

In the aftermath of World War II Indonesia was plunged into a war of independence, which was a time of great threat for Chinese entrepreneurs who were constantly confronted by nationalist militia with demands for assistance. It was also a time of great opportunity. The disappearance of colonial commercial expertise created a vacuum that Chinese entrepreneurs were willing and able to fill. Nationalists lacked foreign currency, so they permitted Chinese entrepreneurs to engage in foreign trade, importing through their growing regional clan and dialect networks fuel, vehicles, aircraft and munitions. To raise revenues the military also sanctioned the opium trade, which Chinese entrepreneurs also dominated (Twang, 1998).

The period after independence saw a parallel rise of two important institutions: the army and the communist party (Crouch, 1978). Sukarno,

Indonesia's first strong leader, adopted a nationalist agenda that sought to balance the tension between the two. Nevertheless, private enterprise faced difficulties in this environment. Indeed, in the late 1950s foreign business was nationalized; and the largest Indonesian-Chinese conglomerate of that time, the Oei Tiong Ham Concern, followed in 1961 among rising anti-Chinese and anti-capitalist sentiments. The socialist government also made efforts to curb the career possibilities of the Chinese with official discriminatory policies, restricting them to entrepreneurial roles.

In this period Liem engaged in minor industrial activities, such as a soap factory. His motto that 'all business is good' led to diversification into a range of other activities, including trading and banking. Like other successful Chinese entrepreneurs Liem courted local military officers that could offer protection, but who demanded shares, percentages of the profit and/or seats on the board in return. The military, on the other hand, was keen on partnering with businessmen as it had insufficient funds that needed to be complemented in innovative ways. A biography of Liem (Soetriyono, 1989) tells a remarkable story that sheds light on Liem's unique relationship as a trusted trading partner of the army in 1940–50. According to this source, Liem was part of a Chinese organization that provided shelter to the leaders of Indonesian nationalism who were forced into refuge during the last stage of Dutch colonial occupation. Liem was known to be a silent man, and he was asked to accept in his household a person without knowing his identity. According to this biography, this man later turned out to be a major political figure and the father-in-law of Sukarno. It was he who was able to introduce Liem as a trustworthy supplier to the Indonesian army. Liem built on this opportunity and developed good relationships with a local army unit, through which he achieved stability and income as a supplier. From this position, he started to expand his trading activities and, most importantly, established the Bank of Central Asia, which would later become the largest privately owned bank in Indonesia.

Summarizing, Liem migrated into a tightly bonded *Hokchia* community whose members were noted for risk taking and mutual protection. Liem inherited bonding capital advantages due to his ethnic origins and cultivated these ties by linking up with traders within the Chinese community and within his family. Some entrepreneurs in this community also built commercial bridges into a society burdened with conflict. As a young entrepreneur, Liem could not fail to observe that the most successful businessmen depended on both their clan connections and bridging capital with military and political actors to defray the many threats and grasp equally numerous opportunities. Liem too began to cultivate bridging capital with military officials.

Phase 2: growth and cronyism

Sukarno's socialist policies put Indonesia on course for macroeconomic disaster with a huge government debt, and eventually a disintegration of the regime. By 1965 inflation reached 600 per cent per year while poverty remained stubbornly high. Suharto grabbed power and maintained it due to his ability to stabilize and improve the economy. He encouraged private capital and adopted an import substitution policy to develop an industrial base. He also rationed scarce credit and foreign exchange. Both policies presented politicians and state bureaucrats with considerable discretionary power over licenses and resource allocation.

Liem enlarged substantially the range of his business activities. While initially he relied for capital on family and clan members, he broadened his group of close partners, called 'the Liem investors', which now included fellow *Hokchia* businessmen as well as two Indonesian investors, one of them a Suharto family member. But Liem continued to rely upon his clan for senior management; for instance, he entrusted the management of his Bank of Central Asia to a *Hokchia*, Mochtar Riady.

Suharto enlisted an elite group of the most successful Chinese entrepreneurs to build an industrial base. Liem took full advantage of this opportunity. Business volume expanded quickly, often in partnership with army foundations linked to Suharto (Robison, 1986). Liem diversified into an ever wider range of basic industrial sectors, such as food, cement and automotive components. Under the import substitution policy Salim Group businesses produced mostly for the domestic market. To acquire required technology and expertise for the diverse ventures Salim entered into joint ventures with Western and Japanese firms. Without government credit and exclusive licenses this rapid growth could not have been achieved (Robison, 1986; Sato, 1993).

While benefiting from Suharto's profuse patronage, Liem's position nonetheless remained precarious. Suharto used the unpopular Chinese entrepreneurs to achieve economic goals but also as convenient scapegoats. Because they were politically marginal the leading Chinese entrepreneurs could pose no threat to Suharto's power. Indeed, he continued to subject the Chinese minority population to restrictive and discriminatory practices while exempting his elite. In this way Suharto ensured that the Chinese as a group were vulnerable while chosen cronies were used to provide his regime with legitimacy.

In the late 1980s and early 1990s, the Suharto regime became openly corrupt with nepotism and cronyism on the rise. Each of Suharto's six children, their in-laws and grandchildren sought to establish their own business empires. These enterprises increasingly competed with Liem for Suharto's patronage. Moreover reciprocity demanded that the Salim Group assist in the development of the Suharto family enterprises by bringing them into joint ventures. Liem was sometimes expected to bail out Suharto related foundations in financial difficulties. For example, when Bank Duta, the majority of which was owned by several Suharto related foundations, experienced financial difficulties, Liem injected US$ 200 million in 1990. Left unchecked the demands for assistance by Suharto's growing entourage threatened to undermine Salim's position. The need to limit commitments while preserving the relationship with Suharto represented a pressing dilemma for the group.

Liem's solution was to strengthen and expand relations with overseas Chinese networks in Southeast Asia in order to mobilize capital and identify business opportunities. Salim secured major loans from the (overseas Chinese owned) Bangkok Bank and Liem also invested in China. In 1972 Liem's son, Anthony Salim, joined the business and accelerated attempts to steer the firm away from dependency on Suharto. At the same time investing abroad roused criticism of 'capital flight' at home. To cope with this the Salim Group maintained a high level of secrecy when it came to their foreign expansion. Anthony Salim was successful in accessing Western capital, which he mentions as a turning point in the group's development:

> Around 1975 we were able to get access to international contacts. This was our big breakthrough. We were able to get Spanish support for the cement factory, and Italian support for the flour factory. (Anthony Salim)

> From 1979 we started to sort of elevate ourselves from government to market based enterprise. We do understand a lot of political implications, because we try to choose that it is much more on business directions rather than government related business – which is still good. Another characteristic is that of course we start to balance our portfolio. We have no pretension to hide that we have started to invest outside Indonesia since 1975, when we created our Singapore and Hong Kong companies. (Anthony Salim)

> My father is more opportunity driven, we are better organized now, and plus now more complex, larger factories and plus if you go to other countries, that's the next challenge. (Anthony Salim)

During the early 1990s full control of the Salim Group was finally transferred to Anthony Salim. He introduced a formal divisional structure and attempted to professionalize Salim Group management. By 1996 about a third of the revenues came from activities outside Indonesia. Meanwhile, Salim attempted to reduce their exposure to Indonesian political networks. Observers suggest that Anthony Salim appeared less interested than his father in the traditional Chinese networks, and preferred do choose his partners based on business considerations only. Answering a question on the importance of ethnic Chinese networks, a Salim Group executive answered: 'Anthony Salim is not interested in that. The second generation usually does not regard this as important'.

Summarizing, in its rapid growth phase the Salim Group undoubtedly benefited from Liem's bridging capital with Suharto. However this relationship became increasingly costly to maintain as the number of Suharto family dependants increased and made demands upon the group. Liem continued to foster new bridging links with Japanese and Western technology suppliers and with the ethnic Chinese network. Anthony Salim pushed to reduce dependence upon Suharto. In order to do this he did not deemphasize the existing links with the crony regime, but he allocated new investments to ventures with Western partners, thus slowly transforming the composition of the group's business network. Managing his social capital in this manner, Salim strengthened its domestic market position while extending its social capital into an international sphere.

Phase 3: crisis and regime change

When the Asian financial crisis engulfed Indonesia in 1997, the IMF encouraged Suharto to curb protectionist policies favouring friends and family members. Anthony Salim's high status was such that he was invited into the Indonesian government delegation that negotiated with the IMF. However, as the austerity package agreed with the IMF began to take effect, the political crisis deepened and Suharto was forced to resign. This had severe consequences for the Salim family.

In 1998 violent anti-Chinese riots broke out across the country. Closely identified with the fallen regime, the Salim Group became a target for public anger; the family house and several branches of the Bank of Central Asia were destroyed by fire. The Salim family and many senior management fled the country (Liem senior never returned from exile in Singapore), leaving Anthony Salim to salvage the company. The Bank of Central Asia suffered a bank-run that depleted its resources and led to its nationalization. The primary reason for this seemed to be the fact that two of Suharto's children had a sizeable stake in the bank and a position on its supervisory board. The cost of the bank-run and demolition of some of its branches is generally estimated to be over US$ 1 billion. The new government, espousing anti-corruption policies, demanded from the Salim family payment for losses and fines amounting to some US$ 5 billion.

Ethnic Chinese were seen by the public as responsible for the crisis, and the Salim Group was a symbol of a corrupt crony capitalism. The new government came under enormous pressure to bring Suharto's cronies to justice. Several leading entrepreneurs were prosecuted but Anthony Salim escaped arrest. Nevertheless:

> The pressure on the Salim Group became enormous, at the time they were afraid for their lives. When they had some idea that a Chinese hunt would become reality, they tried moving everything out of Indonesia. The government was not fond of Salim, and forced them out of our joint venture. It wasn't even legal; we were not supposed to own 100% of the venture. For some time we were in a very odd situation. We did not want the Salim Group to exit from our cooperation. (Western partner)

The economy continued to deteriorate and many Indonesian businesses became insolvent. The new government set up institutions to restore the financial sector and the economy. Anthony Salim elected to cooperate and he turned over shares in 107 Salim companies as payment of his government debt. The group was structured as a web of separate legal entities, including partnerships with domestic and foreign investors. The government was reluctant to expropriate some of these assets for fear of alienating these investors, and Salim managed to retain ownership of some larger ventures. To repurchase and restructure domestic businesses Salim also sold foreign assets to overseas Chinese groups. Despite physical attacks on Salim and continuing negative media coverage of his activities, the government came to realize that the Salim Group was too large to fail, and that it was needed for Indonesia to rebuild its battered economy. As a result, parts of the group were left intact and Anthony Salim continued to reshuffle assets and reorganize the portfolio. Moreover Salim did not ever consider altering the centralized, flexible organization structure under family control.

> I don't believe in the short term that the family holding will be changed to a listed holding in the future, because there is no necessity. And number two, non-listed holding is more flexible to embark on what we want to do. (Anthony Salim)

Summarizing, the loss of the Suharto relationship produced the most severe crisis in Salim's history. The new government was hostile to the conglomerate and extracted significant resources from it, yet the Salim Group was deeply embedded in the economy and many domestic sectors relied upon it for their own survival. At the height of the crisis Salim could mobilize sufficient resources through overseas Chinese business networks to regain control of some of its assets. Consequently, Salim weathered the economic crisis to re-emerge with a more focused market centred business portfolio (Dieleman and Sachs, 2006).

Phase 4: resurgence

After the crisis subsequent political regimes were all short-lived and unstable. Notwithstanding the financial support of the IMF and its guidance in establishing improved standards of political and corporate governance, a stable, efficient free market institutional infrastructure has not yet materialized (Robison and Hadiz, 2004). In 2004 Transparency International's World Corruption Index ranked Indonesia as 133rd of 145 countries. Relationships with politicians and officials remain essential for every large business. However democratization and frequent regime change make relationship building a precarious and more costly endeavour as more officials need to be considered and their comparatively short tenure has made some impatient and rapacious.

Salim continues to cultivate bridging capital with officials. The group never abandoned Suharto, and walked a tight rope to keep contacts with former power holders while developing relationships with an ever changing political regime. As one official says:

> Anthony Salim spends a lot of time with politicians, ministers, bureaucrats. He has several people from the government on the payroll in order to get secret information. But given a choice he probably prefers not to deal with it. He's not afraid to compete on a level playing field. (Official of IBRA, the government entity in charge of bank restructuring)

In this new political landscape the value of political patronage has decreased but the cost of seeking it has increased. Consequently, the Salim Group now sought to accelerate the balance of the portfolio away from Indonesia by concentrating on investments in Australia, China, Singapore and other Asian countries.

> We have to transform ourselves to manage our resources; to transform our assets. It does not mean money, you see, it is whether we have the contacts. We believe we operate in different markets. (Anthony Salim)

Ironically, the value of Salim's Chinese heritage has recently increased. Mainland China is becoming an attractive market for Indonesia's natural resources but years of official discrimination against the Chinese minority is an obstacle to trade in the eyes of the Chinese government. Salim's extensive experience at bridging and brokering is more valuable than ever to Indonesia's trade officials who have little experience, and even less social capital, in China. Meanwhile, Indonesia continues to require inward investment. Many foreign firms are reticent about acquiring Indonesian assets because they cannot credibly assess the extent of their associated liabilities and few multinational firms have deep connections with officials in the frequently changing government. Yet many multinationals have long established relationships with Salim and are likely to place greater faith in them. In this regard Salim Group's extensive linkages with domestic and international interests effectively serve as a poison pill that deters official predation and leaves Indonesia and Salim hostages to each others fortune.

DISCUSSION

The conduct of routine business transactions can be a dangerous activity in some emerging markets because an abundance of commercial opportunity is often combined with unpredictable and serious threats. In these markets the well intentioned intervention of the international community does not always result in an improved institutional environment. We should perhaps be careful about the optimistic assumptions implied in stage theories of the firm in developing and transitional markets. If hostility is overt, or even if it is latent under a veneer of liberal market institutions, then we should not be surprised if domestic entrepreneurs retreat behind a thick social wall of inherited and self-created relationships to conduct transactions and protect their wealth.

Stage models of institutional development urge firms to adapt their strategies to realities that will occur in the next developmental stage by developing proprietary economic-techno capabilities in place of their contact capabilities (Kock and Guillen, 2001) or inter-organizational links (Peng, 2003). However, if hostile conditions persist then such advocacy may encourage firms to prepare for a future that will not happen. We contend that developments and transitions may not materialize in the manner predicted by stage theory. As illustrated by the case of Indonesia, significant economic development may occur without a fundamental transition in the underlying institutional environment. If so, then firms may be better encouraged to enrich and even deepen the current capabilities that are adapted to institutional underdevelopment.

In addition, stage models of economic development often assume that players in the market cannot exercise agency to maintain (or change) weak institutional settings. However in many emerging markets, in particular those that score high on corruption indexes, economic and political power is centred in the hands of a small elite that benefits from the current set-up. Although this small elite may be in a position to stimulate institutional modernization, they may have little incentive to change the status quo. Rather than being passive actors in emerging markets, large family firms may also be considered to be factors shaping institutional change.

We interpret Salim's social capital strategies in the light of institutional persistence. We consider paramount Salim's long-term, unwavering use and cultivation of both bonding and bridging capital. Figure 10.1 summarizes some of the main linkages highlighted in our case narrative. Links with family, clan domestic and overseas dialect groups (solid lines) represent inherited sources of bonding capital, and links with government, administration, the military, foreign and domestic Indonesian businesses (dotted lines) represent self-created bridging capital.

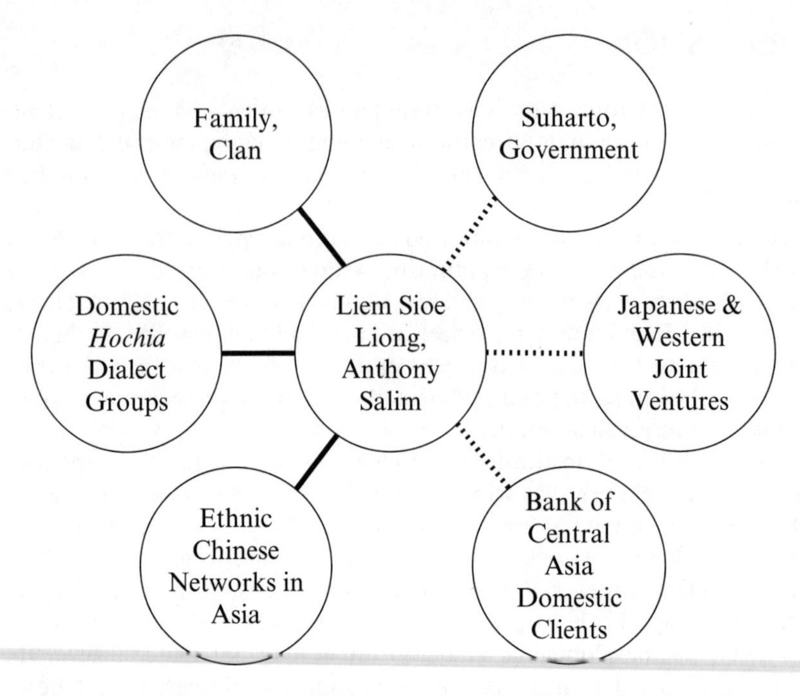

Figure 10.1 Bonding and bridging forms of social capital in the Salim Group

Throughout the period covered in this case study Liem was frequently called upon to draw on the resources contained within kinship, ancestral dialect and overseas Chinese groups. These social groups provided seed capital during Liem's early growth, channels for capital flight in times of acute threat and sources of capital in times of opportunity. These groups were also a source of senior executives and a means for identifying overseas business opportunities. Relationships with Suharto and the international financial community may have eclipsed these linkages in later years, but the bonding capital with overseas Chinese in Southeast Asia remains important in identifying international opportunities, especially in a resurgent mainland China. Reflecting on five decades of corporate history of the Salim Group, we postulate that ethnicity has not always played the same role. While important in the early days, the advantages from bonding capital with other ethnic groups now appear less sizeable in Indonesia, especially in the perspective of the second generation leader. Aside from varying in intensity, bonding capital also changes shape over time: the rise of the Chinese economy offers new opportunities in which the migrant Chinese identity plays an entirely different role than before.

We interpret Salim's links with the military, the administration and politicians as bridging capital that brought numerous opportunities. Moreover the extensive bridging links to indigenous and foreign firms, through the establishment of Indonesia's largest bank and the numerous foreign and domestic joint venture partners, helped establish Salim as Indonesia's largest firm and secure Liem's position when the president was deposed. Bridging relationships beyond the administration served as hostage capital that did much to deter more aggressive movements against Liem's domestic assets in the post-crisis environment. By transcending the closed group, Liem benefited from information, influence and opportunities afforded by bridging capital while attenuating the costs of over-embeddedness within the clan. This dual reliance on both bonding and bridging social capital powered Salim's ascent to become the largest business group in Southeast Asia. Because of this success, the Salim Group was able to exercise considerable strategic choice, and in some instances could directly influence Indonesian institutions.

It is quite possible to interpret Salim's performance as mostly due to a chance meeting with Suharto and his remaining in power for some 32 years. Others attribute Chinese entrepreneurial success to their closed 'maddeningly impenetrable' networks (Weidenbaum and Hughes, 1996). Both explanations contain an element of truth but they underestimate the cosmopolitan quality of Salim's social capital and the skill and deliberation required to maintain the necessary trade-offs between conflicting and potentially unbounded demands for reciprocity.

Despite the promise of institutional reform under post-Suharto regimes, social capital remains the basis of Salim's sustained success. The relative emphasis has shifted over time from bonding towards bridging forms of capital. But relationship management remains critical even as the fortunes of partners rose and fell over time. Hence while the partners change the dance remains the same and Salim's skill in cultivating social capital has maintained its value. One may ask whether these social capital skills are simply the result of adaptation to weak institutional contexts or whether these simultaneously contribute to undeveloped institutional settings. After all, firms like Salim have considerable power and eventually set the standards of local capitalism. Well-adapted players in corrupt countries may have little incentive to change their own strategies nor to change their environment. We feel that it is beyond the scope of this study to reflect upon whether or not the Salim Group contributed positively or negatively to improving Indonesia's institutions.

Anthony Salim often speaks of the need to professionalize Salim's management team. However the persistence of hostile conditions provides little incentive to transform the private and closely held family firm into a dispersed ownership model because family firms continue to maintain a comparative advantage at cultivating social capital. We argued above that in hostile circumstances family firm incentives, authority structure and leadership longevity may offer a 'governance advantage' over professionally managed firms. Reflections on the Salim case suggest some elaboration of those arguments.

Family firm owner-managers have the authority and a greater freedom from accountability to engage in extra-contractual quid pro quo transactions than professional managers. While few professional executives possess the authority to engage in handshake deals this does not mean professionally managed firms cannot cultivate political resources. Chandler (1991) suggests that as the international scope of professionally managed firms grows, the government relations function increases in importance and responsibility and resides at the top of the organization. However in professionally managed firms responsibility for political management is subject to checks and balances due to the demands for accountability and control. The concentration of authority in the family firm permits more flexible and decisive deal making. Second, it is often more costly for managerial governed firms to operate in these conditions: senior managers and their families may demand special security and accommodations, reliable intelligence must be procured, access to channels arranged through local agents; indeed managerial governed firms may need to partner with local firms (such as Salim) precisely because they lack the capacity to efficiently produce and internalize such assets on their own account.

Third, the capacity to engage in personal, reciprocity based transactions depends upon both partners remaining in their positions long enough to reciprocate a favour. The average tenure of a professional CEO in a Fortune 500 company is only some three or four years and there is a probability that they will not be in a position long enough to reciprocate a personal favour. In contrast, family firm CEOs may possess the capacity to reciprocate for much longer periods. Finally, long lived family firms are a generally recognized means of transferring tacit knowledge from one generation to another. The Liem case demonstrates the possibility of the intergenerational transfer of social capital assets through processes of natural immersion as successors gradually assimilate the nuances of network relationships. Indeed the fundamental propositions suggested by this study could be stated as:

Proposition 1: The greater is the threat from the non-market institutional environment the greater is the value of the social capital generated and used by the firm.

Proposition 2: The private family firm is better positioned than a managerial firm in generating and using social capital.

Family firms, such as Salim, possess social capital advantages that allow them to respond to the threats of hostile environments. This, however, may impede the cultivation of other forms of capital. Chandler (1990) suggests that familism impedes the development of organizational and technological capabilities because family firms fail to develop more complex organization structures. Many analysts share this view, suggesting that incentives and accountability processes in family firms can strongly hinder the development of complex routines and systems (Daily and Dalton, 1992; Zahra and Filatotchev, 2004). This may occur because family firms often limit participation in both ownership and decision making to a small cadre of insiders selected on the basis of owner-managers' personal preferences (Chandler, 1990). Such practises can seriously inhibit the accumulation of intellectual capital in the middle levels of an organization (Nahapiet and Ghoshal, 1998). Similarly, the simple organizational structures necessary for the exercise of personal authority rarely possess sufficient capacity to process complex and high volume information and can inhibit a firm's capacity to assimilate outside sources of information, to acquire, integrate and recombine resources.

The use of an extreme case, such as Salim, not only led to the formulation of propositions emerging from the case narrative, but also enabled a teasing out of the particulars of how this family group managed multiple forms of social capital in a hostile institutional setting over a period of various decades.

CONCLUSION

Our longitudinal case study answers calls for research that goes beyond the uncontroversial idea that history matters and explores how it matters (Jones and Khanna, 2006). We suggest that our threat response account of the Salim Group over some 60 years sheds light on the origins of rare and inimitable resources and permits a deeper analysis of the path dependent qualities of competitive advantages. This analysis does not necessarily undercut existing analyses of family firms in emerging markets that emphasize contacts, cronyism or corruption but it treats such phenomena in their historical context. In particular, our social capital analysis permits a more sympathetic treatment of strategic behaviour that is sometimes dismissed as outmoded or unethical.

If institutional underdevelopment and hostility towards certain types of firm persist for long periods, then firms may be advised to continue cultivating their social capital. The continuing importance of these assets suggests that the gradual shift from contact capabilities towards technological and organizational capabilities predicted in stage models of organizational-environment co-evolution may be too optimistic or may represent a special case. Indeed, rather than seeking to develop new forms of competitive advantage, that their governance structures are unsuited to produce, powerful family firms may utilize their existing competitive advantage to maintain the environment in its present state. It may be a hostile state but it is one that the firm has adapted to and one that foreign firms find difficult to navigate, especially foreign professionally managed firms. As such, family firms like the Salim Group, operating in hostile environments, may have found and maintained their own niche in which they possess valued competitive advantages.

REFERENCES

Adler, P.S. and S. Kwon (2002), 'Social capital: Prospects for a new concept', *Academy of Management Review*, **27**, 17–40.
Amit, R. and B. Villalonga (2006), 'How do family ownership, control, and management affect firm value?', *Journal of Financial Economics*, **80** (2), 385–417.
Anderson, R.C. and D.M. Reeb (2003), 'Founding family ownership and firm performance: evidence from the S&P 500', *Journal of Finance*, **58** (3), 1301–28.
Blyler, M. and R.W. Coff (2003), 'Dynamic capabilities, social capital, and rent appropriation: ties that split pies', *Strategic Management Journal*, **24**, 677–86.
Bonacich, E. (1973), 'A theory of middleman minorities', *American Sociological Review*, **38** (5), 583–94.
Burt, Ronald S. (1992), *Structural Holes. The Social Structure of Competition*, Cambridge, MA: Harvard University Press.
Carney, M. (2005), 'Corporate governance and competitive advantage in family-controlled firms', *Entrepreneurship, Theory and Practice*, **29** (3), 249–65.
Chandler, A.D. (1990), *Scale and Scope: The Dynamics of Industrial Competition*, Cambridge, MA: Harvard University Press Boston, Harvard Business School Press.
Chandler, A.D. (1991), 'The functions of the HQ in a multibusiness firm', *Strategic Management Journal*, **12**, 31–51.
Chirot, D. and A. Reid (eds) (1998), *Essential Outsiders: Chinese and Jews in the Modern Transformation of Southeast Asia and Central Europe*, Seattle, WA and London: University of Washington Press.
Coff, R.W. (1999), 'When competitive advantage doesn't lead to performance: the resource based view and stakeholder bargaining power', *Organization Science*, **10** (2), 119–33.
Colli, A., P. Fernandez-Perez and M. Rose (2003), 'National determinants of family firm development: family firms in Britain, Spain and Italy in the19th and 20th centuries', *Enterprise and Society*, **4** (1), 28–65.
Covin, J.G. and D.P. Slevin (1989), 'Strategic management of small firms in hostile and benign environments', *Strategic Management Journal*, **10** (1), 75–87.

Crouch, H. (1978), *The Army and Politics in Indonesia*, Ithaca, NY: Cornell University Press.
Daily, C.M. and D.R. Dalton (1992), 'The relationship between governance structure and corporate performance in entrepreneurial firms', *Journal of Business Venturing*, **7** (5), 375–86.
Davis, K., M.J. Trebilcock and B. Heys (2001), 'Ethnically homogeneous commercial elites in developing countries', *Law and Politics in International Business*, **32** (2), 331–61.
Dieleman, M.H. and W.M. Sachs (2006), 'Oscillating between a relationship-based and a market-based model: the Salim Group', *Asia Pacific Journal of Management*, **23** (4), 521–36.
Dyer, W.G. and S.P. Mortensen (2005), 'Entrepreneurship and family business in a hostile environment: the case of Lithuania', *Family Business Review*, **18** (3), 247–58.
Eisenhardt, K.M. (1989), 'Building theories from case study research', *Academy of Management Review*, **14** (4), 532–50.
Fligstein, N. (1997), 'Social skill and institutional theory', *American Behavioural Scientist*, **40** (4), 397–405.
Flyvbjerg, B. (2001), *Making Social Sciences Matter: Why Social Inquiry Fails and How it Can Succeed Again*, Cambridge, UK: Cambridge University Press.
Friedman, E., S. Johnson and T. Mitton (2003), 'Propping and tunnelling', *Journal of Comparative Economics*, **31** (4), 732–51.
Frynas, J.G., K. Mellahi and G.A. Pigman (2006), 'First mover advantages in international business and firm-specific political resources', *Strategic Management Journal*, **27**, 321–45.
Fukuyama, F. (1992), *The End of History and the Last Man*, New York: Free Press.
Fukuyama, F. (1995), *Trust: The Social Virtues and the Creation of Prosperity*, New York: Free Press.
Granovetter, M. (1994), 'Business groups', in Neil J. Smelser and Richard Swedburg (eds), *The Handbook of Economic Sociology*, Princeton, NJ: Princeton University Press, pp. 453–75.
Granovetter, M. (1995), 'The economic sociology of firms and entrepreneurs', in Alexandro Portes (ed.), *The Economic Sociology of Immigration. Essays on Networks: Ethnicity and Entrepreneurship*, New York: Russel Sage Foundation, pp. 93–130.
Guillen, M.F. (2000), 'Business groups in emerging economies: a resource based view', *Academy of Management Journal*, **43** (3), 362–80.
Gulati, R. (1998), 'Alliances and networks', *Strategic Management Journal*, **19** (4), 293–319.
Harrigan, Kathryn R. (1980), *Strategies for Declining Businesses*, Lexington, MA: Lexington Books.
Jones, G. and T. Khanna (2006), 'Bringing history (back) into international business', *Journal of International Business Studies*, **37** (4), 453–68.
Khanna, T. and K. Palepu (1997), 'Why focused strategies may be wrong for emerging markets', *Harvard Business Review*, **75** (4), 41–51.
Kock, C. and M. Guillen (2001), 'Strategy and structure in developing countries: business groups as an evolutionary response to opportunities for unrelated diversification', *Industrial and Corporate Change*, **10** (1), 77–113.
Kotkin, J. (1993), *Tribes: How Race, Religion and Identity Determine Success in the New Global Economy*, New York: Random House.

La Porta, R., F. Lopez-de-Silanes and A. Shleifer (1999), 'Corporate ownership around the world', *Journal of Finance*, **54**, 471–518.
Locke, E.A. (1999), 'Some reservations about social capital', *Academy of Management Review*, **24** (1), 8–9.
Miller, D. and I. Breton-Miller (2005), *Managing for the Long Run: Lessons in Competitive Advantage from Great Family Business*, Boston, MA: Harvard Business School Press.
Mintzberg, H. (1979), *The Structuring of Organizations*, Englewood Cliffs, NJ: Prentice Hall.
Morck, R. and B. Yeung (2003), 'Agency problems in large family business groups', *Entrepreneurship Theory and Practice*, **27** (4), 367–83.
Nahapiet, J. and S. Ghoshal (1998), 'Social capital, intellectual capital, and the organizational advantage', *Academy of Management Review*, **23** (2), 242–66.
North, D.C. (1990), *Institutions, Institutional Change and Economic Performance*, Cambridge, UK: Cambridge University Press.
Nozick, R.K. (1974), *Anarchy, State and Utopia*, New York: Basic Books.
Peng, M.W. (2003), 'Institutional transitions and strategic choices', *Academy of Management Review*, **28** (2), 275–85.
Perrow, C.A. (1984), *Normal Accidents*, New York: Basic Books.
Pettigrew, A. (1990), 'Longitudinal field research on change: theory and practice', *Organization Science*, **1** (3), 267–92.
Poppo, L. and T. Zenger (2002), 'Do formal contracts and relational governance function as substitutes or complements?', *Strategic Management Journal*, **23**, 707–25.
Portes, A. and J. Sensenbrenner (1993), 'Embeddedness and immigration: notes on the social determinants of economic action', *American Journal of Sociology*, **98** (6), 1320–51.
Post, P. (1997), 'On bicycles and textiles, Japan, South-China and the Hokchia-Henghua entrepreneurs', in L. Douw and P. Post (eds), *South China: State, Culture and Social Change During the 20th Century*, Amsterdam: North Holland, pp. 141–50.
Putnam, R.D., R. Leonardi and R.Y. Nanetti (1994), *Making Democracy Work: Civic Traditions in Modern Italy*, Princeton, NJ: Princeton University Press.
Rajan, R.G. and L. Zingales (2001), 'The great reversals: the politics of financial development in the 20th century', working paper, University of Chicago.
Redding, G. (1990), *The Spirit of Chinese Capitalism*, New York: De Gruyter.
Redding, G. (2005), 'The thick description and comparison of societal systems of capitalism', *Journal of International Business Studies*, **36**, 123–55.
Rindova, V.P. and C.J. Fombrun (1999), 'Constructing competitive advantage: the Role of firm-constituent interactions', *Strategic Management Journal*, **20** (8), 691–710.
Robison, R. (1986), *Indonesia, The Rise of Capital*, North Sydney, Australia: Alan & Unwin.
Robison, R. and V. Hadiz (2004), *Reorganising Power in Indonesia: The Politics of Oligarchy in an Age of Markets*, London: RoutledgeCurzon.
Sato, Y. (1993), 'The Salim Group in Indonesia: the development and behavior of the largest conglomerate in Southeast Asia', *The Developing Economies*, **31** (4), 408–41.
Soetriyono, E. (1989), *Liem Sioe Liong, Kisah Sukses*, Jakarta: Indomedia.
Staw, B. and E. Szwajkowski (1975), 'The scarcity-munificence component of organizational environments and the commission of illegal acts', *Administrative Science Quarterly*, **20**, 345–54.

Steier, L. and R. Greenwood (2000), 'Entrepreneurship and the evolution of angel financial networks', *Organization Studies*, **21** (1), 163–92.

Tan, J. (2005), 'Venturing in turbulent water: a historical perspective of economic reform and entrepreneurial transformation', *Journal of Business Venturing*, **20**, 689–704.

Tan, J. and D. Tan (2005), 'Environment-strategy co-evolution and co-alignment: a staged model of Chinese SOEs under transition', *Strategic Management Journal*, **26**, 141–57.

Telser, L.G. (1980), 'A theory of self-enforcing agreements', *Journal of Business*, **53**, 27–44.

Twang, P.-Y. (1998), *The Chinese Business Elite in Indonesia and the Transition to Independence 1940–1950*, Kuala Lumpur: Oxford University Press.

Uzzi, B. (1997), 'Social structure and competition in interfirm networks: the paradox of embeddedness', *Administrative Science Quarterly*, **42** (1), 35–67.

Wang, G. (2000), *The Chinese Overseas, From Earthbound China to the Quest for Autonomy*, Cambridge, MA: Harvard University Press.

Weidenbaum, M. and S. Hughes (1996), *The Bamboo Network: How Expatriate Chinese Entrepreneurs are Creating a New Economic Superpower in Asia*, New York: Free Press.

Williamson, O.E. (1985), *Transaction Cost Economics: The Governance of Contractual Arrangements*, New York: Free Press.

Woolcock, M. (1998), 'Social capital and economic development: toward a theoretical synthesis and policy framework', *Theory and Society*, **27**, 151–208.

Wu, Y.-I. and C.-H. Wu (1980), *Economic Development in Southeast Asia*, Stanford, CA: Hoover Institute Press.

Xin, K. and J. Pearce (1996), 'Guanxi: good connections as substitutes for institutional support', *Academy of Management Journal*, **39**, 1641–58.

Zahra, S.A. and I. Filatotchev (2004), 'Governance of the entrepreneurial threshold firm: a knowledge-based perspective', *Journal of Management Studies*, **41** (5), 883–95.

[illegible] (2009) [illegible]
[illegible]
[illegible]
[illegible]
[illegible]
[illegible]
[illegible]
[illegible]
[illegible] York: Free Press.
[illegible] Free Press.
[illegible]
[illegible]
[illegible]
[illegible]

Conclusion

Phillip H. Phan, Sankaran Venkataraman and S. Ramakrishna Velamuri

The findings of the research chapters contained in this volume raise a number of issues that merit further research. First, studies of entrepreneurial regions all over the world – Silicon Valley and Route 128 in the USA, Baden-Württemberg in Germany, Scotland, Ireland, Bangalore in India, Shanghai in China, Singapore, among many others – have underscored the critical role of governments at different levels in the emergence of these regions. Such studies, using an evolutionary approach, have shown that the magnitude of governmental influence, which is significant in the early stages of development, seems to decline in later stages relative to other tangible and intangible factors (see Venkataraman, 2004). The explanations for this vary from the traditional 'factor substitution', wherein government 'kick starts' the development of a sector, which then becomes attractive for private capital to accumulate, to the post-modern 'institutionalization', in which the development of such institutions as intellectual property regimes engender capital accumulation. The research in this volume broadly support this approach, with the exceptions of (1) Arıkan's study (Chapter 3) on the rise and decline of the media industry in Silicon Alley (New York tri-state area) during the 1990s, largely as a result of the pioneering actions of a number of private individuals and organizations, and (2) the Castanhar et al. study (Chapter 2) on the development of a furniture cluster in a remote area of Brazil, largely as a result of the efforts of one 'anchor' entrepreneur over a 40 year period, with the support of the mayor and other organizations such as the Furniture Association.

Two questions arise from these two studies. First, it is interesting to note that Silicon Alley's rise was fueled by firms requiring low levels of capital expenditure and very high levels of human capital. Therefore one may conjecture that the role of governmental expenditure and policy in such knowledge based industries might not be as critical, especially if the initial conditions in the form of strong educational institutions, good physical infrastructure, a sophisticated financial services industry and favorable cultural attitudes to entrepreneurship already exist? Second, in recent years there has been a concerted move in the entrepreneurship literature away from the notion of the single entrepreneurial 'hero' toward the concept of entrepreneurial teams and networks as the basis for explaining growth. However, especially in thinly populated areas (such as the Arapongas district in Brazil that is the subject of the chapter by Castanhar et al.), the impact of a single entrepreneur can be significant enough to transform the socio-economic profile of that area. Based on the insights from this Brazilian case, should we reassess the relative role of entrepreneurial agency in promoting regional dynamism?

Supapol et al. in Chapter 8 examine the contextual conditions during three distinct policy eras to examine their impact on the performance of firms that entered in each of these eras. Their study raises the importance of initial conditions but more fundamentally, what remains to be explored are the interactions between these conditions and the human agencies (entrepreneurs, financiers and consumers) that form the entrepreneurial economy. One question we can ask is whether it is possible that the characteristics of the founders who select into entrepreneurial activities are different under different policy regimes? After all, there has long been a theoretical formulation to this effect (see Baumol, 1993). Thus the lingering effects on performance due to the timing of entry that the authors observe might have as much to do with entrepreneurs' characteristics and their interactions with the policy regimes and the resultant contextual conditions.

The case study on the Salim Group by Carney et al. in Chapter 10 traces the evolution of a group of companies that are owned by an entrepreneurial family belonging to an ethnic minority (Chinese) in Indonesia. It offers the reader a nuanced perspective of how the family tries to strike a balance between approximation to and autonomy from the political regime in an environment that the authors characterize as hostile to businesses. The study is commendable for its coverage of a 67 year period from 1938 to 2005. It goes without saying that regimes, such as those of Sukarno and Suharto, that use strategies of patronage to keep economic agents on a tight leash are not unique in emerging markets. Based on a Western notion of a Weberian state, we might be tempted to be critical of the economic players in such regimes for succumbing to political patronage. However what this study shows is that in contrast to entrepreneurs in industrially advanced and institutionally stable economies, who have to deal with high competitive intensity, entrepreneurs in many emerging markets have to deal not so much with competition but with the vagaries of policy making. More fundamentally, the study calls for future research to include an evolutionary perspective when dealing with large sample probabilistic empirical models that attempt to understand how entrepreneurs, especially those belonging to ethnic minorities, are constrained or emancipated by their institutional environments.

The role of international donor agencies in legitimating entrepreneurship and overcoming local institutional rigidities is highlighted by Fletcher et al. in Chapter 4 in their study of the emergence of an organizational field in the Western Balkans. These agencies can prove to be external catalysts or change agents that can provide the initial spark for the launch of new initiatives. They can also serve as linkages to overcome institutional inertia that often promotes a silo mindset in local institutions and impedes the inter-organizational linkages that are so necessary for entrepreneurial regions to flower.

This research volume on entrepreneurship in emerging markets serves as a starting point for a debate on the role of entrepreneurship in promoting 'equitable' or 'inclusive' economic growth. In spite of consistently high economic growth in countries like China and India, policy makers are concerned that large sections of the population are not benefiting from this economic prosperity. This is leading to social unrest and insurgencies in some areas, and to a questioning of the appropriateness of free markets as mechanisms for the uplifting of the economically disadvantaged in others. Although the studies do not directly address the social development issues in the economies they investigated, whether by example or implication, they all point to the importance of the interaction between economic development, social welfare and entrepreneurial action in such economies. These issues are usually hidden or taken for granted in the extant literature on entrepreneurship in developed economies. They cannot be so when dealing with emerging economies or we end up with incomplete models, at best, misleading ones, at worse.

Hence, to understand these issues and explore viable policy options, scholars will have to move away from the predominantly urban settings that have characterized entrepreneurship research to date. We believe that such research, as in all good research, begins with documenting the phenomenon. Are there any instructive case examples of entrepreneurship in rural areas? What are the entrepreneurial capacity building initiatives taking place to endow farmers and rural artisans with the skills they need to improve the quality of life of their families in a sustained way? More specifically, one can ask about the impact of microfinance on the development of capital accumulation capacity and whether this has altered the choices of entrepreneurs in those regions. What is the role of technology in such areas, such as the impact on rural entrepreneurship with the dramatic increase in wireless telecommunication penetration in countries like China, India and the former Soviet satellite states?

In conclusion, the research on entrepreneurship in emerging markets promises to be theoretically interesting, empirically challenging, and may lead to important managerial and policy implications that can have real economic and social impact. Our approach in this book is to show the possibilities and encourage the development of theoretically grounded models that can withstand the test of empirical verification. In doing so, we should probably move away from deriving policy prescriptions using macro-level data, which has heretofore characterized much of the research in this area. Instead, by examining entrepreneurial actors, their agents and the interactions between actors and institutional fields, we are more likely to advance the research and solve real problems.

REFERENCES

Baumol, W. (1993), *Entrepreneurship, Management, and the Structure of Payoff*, Cambridge, MA: MIT Press.

Venkataraman, S. (2004), 'Regional transformation through technological entrepreneurship', *Journal of Business Venturing*, **19**, 153–67.

Index

0 1341 1117089 7